RADIANT FLOOR HEATING

Other Books in McGraw-Hill's Complete Construction Series

Bianchina ▪ *Room Additions*
Carrow ▪ *Energy Systems for Residential Buildings*
Gerhart ▪ *Home Automation and Wiring*
Powers ▪ *Heating Handbook*
Vizi ▪ *Forced Hot Air Furnaces: Troubleshooting and Repair*

Dodge Cost Guides ++ Series

All from McGraw-Hill and Marshall & Swift

Unit Cost Book
Repair and Remodel Cost Book
Electrical Cost Book

RADIANT FLOOR HEATING

R. DODGE WOODSON

McGraw-Hill

New York San Francisco Washington, D.C. Auckland Bogotá
Caracas Lisbon London Madrid Mexico City Milan
Montreal New Delhi San Juan Singapore
Sydney Tokyo Toronto

Library of Congress Cataloging-in-Publication Data

Woodson, R. Dodge (Roger Dodge).
Radiant floor heating / R. Dodge Woodson.
p. cm.
ISBN 0-07-134786-0
1. Radiant floor heating. 2. Floors—Thermal properties.
I. Title.
TH7753.5.W66 1999
697'.72—dc21 99-26465
CIP

McGraw-Hill

A Division of The ***McGraw-Hill*** *Companies*

1 2 3 4 5 6 7 8 9 0 DOC/DOC 9 0 4 3 2 1 0 9

ISBN 0-07-134786-0

The sponsoring editor for this book was Zoe G. Foundotos, the editing supervisor was Caroline Levine, and the production supervisor was Sherri Souffrance. It was set in Melior by Constance M. Tucker of Lone Wolf Enterprises, Ltd.

Printed and bound by R. R. Donnelley & Sons Company.

This book was printed on recycled, acid-free paper containing a minimum of 50% recycled de-inked fiber.

This book is dedicated to Afton and Adam for all they mean to me, and to Tori, Nate, and Jon for giving me a full family experience.

CONTENTS

	Preface	**xv**
Chapter 1	**Why Aren't More Contractors Using In-Floor Heating Systems?**	**1**
	Putting Heat in a Concrete Slab	4
	Thin-Slab Installations	14
Chapter 2	**Comparing In-Floor Systems To Other Systems**	**19**
	Comfort	20
	Efficiency	21
	Dust	22
	Design Issues	22
	Physical Appearance	24
	General Maintenance	24
	Basement floors	25
	Heat Sources	26
	A Perfect System	26
Chapter 3	**Designing Functional, Cost-Effective Systems**	**29**
	Ferrous Components	30
	Tubing Size	31
	The Controls	32
	Zones	32
	No Installation	35
	Heat a Ceiling?	35
	Between the Floor Joists	36
	Worth Looking Into	37
	Software	37
	Tubing Routes	38

Chapter 4 **Combining In-Floor Systems and Baseboard Systems** 43

Why Add Heat? 44
The Temperature Problem 45
Mixing Tanks 49
Heating Exchangers 50
Motorized Mixing Valves 51

Chapter 5 **Establishing Heating Zones** 53

Primary Living Space 55
Bedrooms 56
Bathrooms 57
Specialty Rooms 59
Garages 59
Kitchens 59
General Living Space 60
Outdoor Heating 61
Common Sense 61

Chapter 6 **The Mechanics of Manifolds** 63

Choosing a Location 67

Chapter 7 **Boilers** 71

Cast Iron Boilers 74
Steel Boilers 76
Fuel Oil 77
Operating Fundamentals 79
Gas-Fired Boilers 80
Boiler Selection 81

Chapter 8 **Material Selection** 85

What is PEX? 86
Stress 87
Oxygen Diffusion 89
Stopping Oxygen Diffusion 90
Temperature and Pressure Ratings 92
Tubing Size and Capacity 92

Manifold Systems 93
Fittings 94
Controls 95
Other Materials 95

Chapter 9 Circulating Pumps 97

Pump Design 98
Pump Position 101
Mounting Circulators 103
Circulator Location 104
Understanding Head Factors 105
Performance Curves 106
The Difference 107
Stacking Them Up 108
Lining Them Up 109
Cavitation 109
Picking a Pump 111

Chapter 10 Controls And Control Systems 113

On-Off Controls 114
Staged Controls 116
Modulating Controls 117
Outdoor Reset Controls 117
Electrical Functions 117
Thermostats 118
Aquastats 120
High-Limit Controls 121
Relay Centers 121
Zone Valves 121

Chapter 11 Components For Heating Systems 123

Modern Piping Materials 123
Fittings 124
Heating Valves 126
Gate Valves 126
Glove Valves 127
Ball Valves 127
Check Valves 128

	Pressure-Reducing Valves	129
	Backflow Prevention	129
	Mixing Valves	130
	Zone Valves	131
	Other Types of Valves	132
	Circulating Pumps	132
	Pump Placement	133
	Expansion Tanks	134
	Selection and Installation	136
	Controls for Heating Systems	137
	Thermostats	137
	Controlling in Stages	137
	Modulating Control	138
	Outdoor Temperatures	138
Chapter 12	**Expansion Tanks**	**139**
	Standard Expansion Tanks	141
	Diaphragm Tanks	142
	Matching a Tank	144
	Check the Ratings	145
	A Wide Selection	146
	Where Should the Tank be Placed?	146
Chapter 13	**Domestic Water Heating**	**149**
	Combining a Coil with a Tank	150
	Independent Tanks	151
	Auxiliary Loads	153
	It Makes Sense	154
Chapter 14	**Slab-On-Grade Piping Systems**	**157**
	Fairly Simple	158
	A Sample Installation	163
	Things to Be Mindful Of	165
Chapter 15	**Thin-Slab Piping Systems**	**169**
	Floor Loads	170
	Additional Thickness	171
	Lightweight Concrete	174

	Gypsum-Based Systems	175
	A Little or a Lot	177
Chapter 16	**Dry Piping Systems**	**179**
	Transfer Plates	180
	On Top	183
	Under Subflooring	185
	Obstacles	187
	No Plates	188
Chapter 17	**Radiant Systems for Ice Removal**	**191**
	A Typical Installation	193
	Precautions	194
	Coil Design	195
	Tubing Selection	196
	Not the Same	196
Chapter 18	**Purging Air from Systems**	**199**
	Problems with Air	200
	Air Pockets	201
	Air Vents	205
	The Purging Process	208
Chapter 19	**Solar Heating Systems**	**211**
	Flat Plate Solar Collectors	212
	Installation Methods	213
	Planning a Solar Heating System	216
	The Cost Factor	217
Chapter 20	**Coal-Fired Heating Systems**	**219**
	Types of Coal	220
	Stokers	222
	Start-Up Procedures	222
	Troubleshooting	223
Chapter 21	**Wood-Fired Heating Systems**	**227**
	Many Choices	228
	Efficiency	231

Transfer Storage Tanks 231
Another Option 232

Chapter 22 Troubleshooting Gas-Fired Boilers 235

Standard Safety Measures 235
The Troubleshooting Process 237
No Pilot Light 240
When a Pilot Light Goes Off During a Standby Period 240
If a Pilot Light Goes Off When the Motor Starts 241
Motor Won't Run 242
It Runs but Doesn't Heat 243
A Short Flame 243
Long Flames 244
A Gas Leak at the Regulator Vent 244
Won't Close 244
Be Careful 244

Chapter 23 Troubleshooting Oil-Fired Boilers 247

Noise 248
A Smoky Joe 249
It Stinks 250
Noisy Operation 251
No Oil 252
Leaks 253
Bad Oil Pressure 253
Pressure Pulsation 253
Unwanted Cutoffs 254

Chapter 24 Selling Radiant Heating Systems 255

No Concrete Needed 257
Can't Combine 258
Won't Last 259
Lower Operating Cost 259
Space-Saving Features 260
Maintenance 260
Comfort 261

It's New 262
Information 262

Glossary 265

Appendix A–Pressure Losses and Flow Charts 269

Appendix B–Temperature Charts 289

Index 297

PREFACE

Radiant floor heating systems have been around since the days of ancient Romans. People throughout the ages have recognized the comfort associated with these systems. However, even with their long and successful history, radiant floor systems have only recently begun to gain major attention in modern construction, and there is still market resistance to what is an excellent, multipurpose heating system. Understanding the features and benefits of the system will enable one to appreciate this quality form of heating.

Old-school heating contractors can be stubborn when it comes to touting the advantages of little plastic tubes installed under a plywood subfloor. It's common for tradespeople to assume that radiant floor heating systems must be installed in concrete to achieve a suitable heat delivery. This is because in the "old days" that's the way "it was always done." Not so, today. It's very common to find radiant floor systems installed without the use of any concrete. Heat tubing can be stapled to the bottom of subflooring, it can be installed on top of subflooring, and it can be installed in thin-slab applications. And, of course, conventional slab installations are still quite popular. The fact is, almost any job is a potential job for radiant floor heating.

Did you know that radiant floor heating systems can be combined with hot water baseboard heating systems? Is it possible for one boiler to serve both types of heating system? Certainly. Minor modifications allow the temperature of supply fluids to be controlled for both types of systems. Radiant floor heating can be used in additions, remodeling, and new construction. Both commercial and residential buildings can be heated with radiant heat tubing in floors. And, special heating needs, such as snow and ice melting systems, can be combined with the general heating system.

Radiant floor heating systems require only a low temperature fluid to do their job. This translates into lower operating costs. The PEX tubing used to create most radiant floor systems is very inexpensive when compared to copper tubing. Since radiant floor heating rises from a floor, it actually heats more space than perimeter heat emitters do. There is little dust movement with a radiant floor heating system. Radiant heating systems deliver a clean, comfortable, even heat in all applications.

The last few years have seen surges in the use of radiant floor heating. This trend is likely to not only continue, but to expand. The systems available today offer a wide range of opportunities for contractors and installers. Radiant floor systems can be tailored to meet any number of heating needs. But, installers and contractors must be aware of the wide array of options open to them. This is where the concept for this book originated. Here you will find the guidance needed to launch your career or heating business into the future with the power of radiant heating systems at your disposal.

Did you know that gypsum-based materials can be used to create a thin-slab, wet-system radiant floor heating system? Well, they can. Lightweight concrete is another option for thin-slab systems. But, no slab is required. Modern radiant floor heating systems can be installed in direct contact with plywood subfloors. This makes the cost of an installation less, and the weight that can be a problem with a slab system is not a consideration. Heat transfer plates boost the effectiveness of dry systems. Modern materials and methods, which are discussed in detail in the following chapters, allow for all sorts of creative heating systems.

Why do you need this book? Maybe you don't need it, but you won't know what you're missing without it. Can you afford to move into the next decade without having a full grasp of radiant floor heating systems? Progress calls for innovation and improvement, which is exactly what modern radiant floor heating systems deliver. Consider this book your survival guide in the everchanging world of heating systems. With it, you have plenty of opportunity to grow and prosper. Without it, you are taking an unnecessary risk. It's time for you to make the call. Look over the table of contents. Thumb through the pages. Look at the numerous illustrations. Scan the reader-friendly text. It will only take a few moments for you to see that this book may be the most valuable tool in your arsenal of money-making tools. You are holding the power of the future in your fingertips. Don't let it get away.

CHAPTER 1

Why Aren't More Contractors Using In-Floor Heating Systems?

Radiant floor heating systems have been around for quite a while. If you want to go back to the origins of heating buildings by having heated floors, you can trace the process all the way back to the Roman Empire. The Romans heated their floors by directing exhaust gases from wood fires to open space under raised floors. Today's radiant floor systems depend on hot water, rather than exhaust gases, to heat floors. For many years, the concept of heating homes and buildings by putting heat in the floors of the structures was largely ignored. This fact is changing. With modern technology, the cost of installing radiant floor heat is lower. Additionally, the problems that were often associated with threaded pipe connections and steel pipe, as well as the pinhole leaks sometimes experienced with copper tubing, have been greatly reduced with the introduction of PEX tubing.

The benefits of radiant floor heat are numerous (Figs. 1.1 to 1.6). Such systems give unsurpassed thermal comfort, produce no noise, can operate with low-temperature water, and often use less energy to function. Another advantage of radiant floor heating is the fact that the heating pipes are out of sight and out of mind and don't interfere with furniture placement. Dust can be a problem with fin-tube baseboard heat, and it can be a big problem with some fan-assisted heating units.

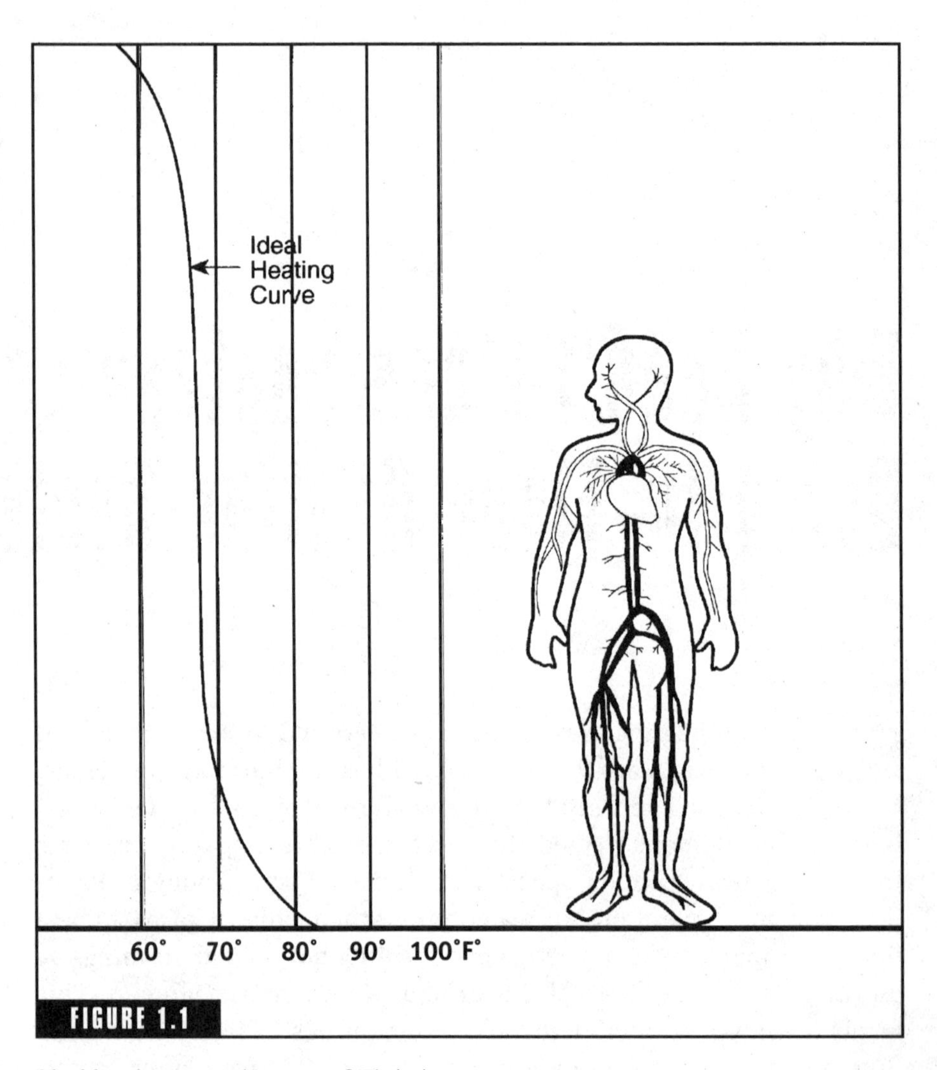

FIGURE 1.1

Ideal heating curve. (*Courtesy of Wirsbo.*)

This is not a problem with radiant floor heating. Stratification is another problem that may occur with baseboard heating systems but is not a problem with radiant floor heating. The key to having an enjoyable experience with radiant floor systems is to have the system designed and installed properly.

Radiant floor heating is not new, but the methods and materials used to install the systems are new. When this type of heating system first gained acceptance in the United States, it was most often installed in concrete slabs. Concrete is a good place to embed radiant heat pipes.

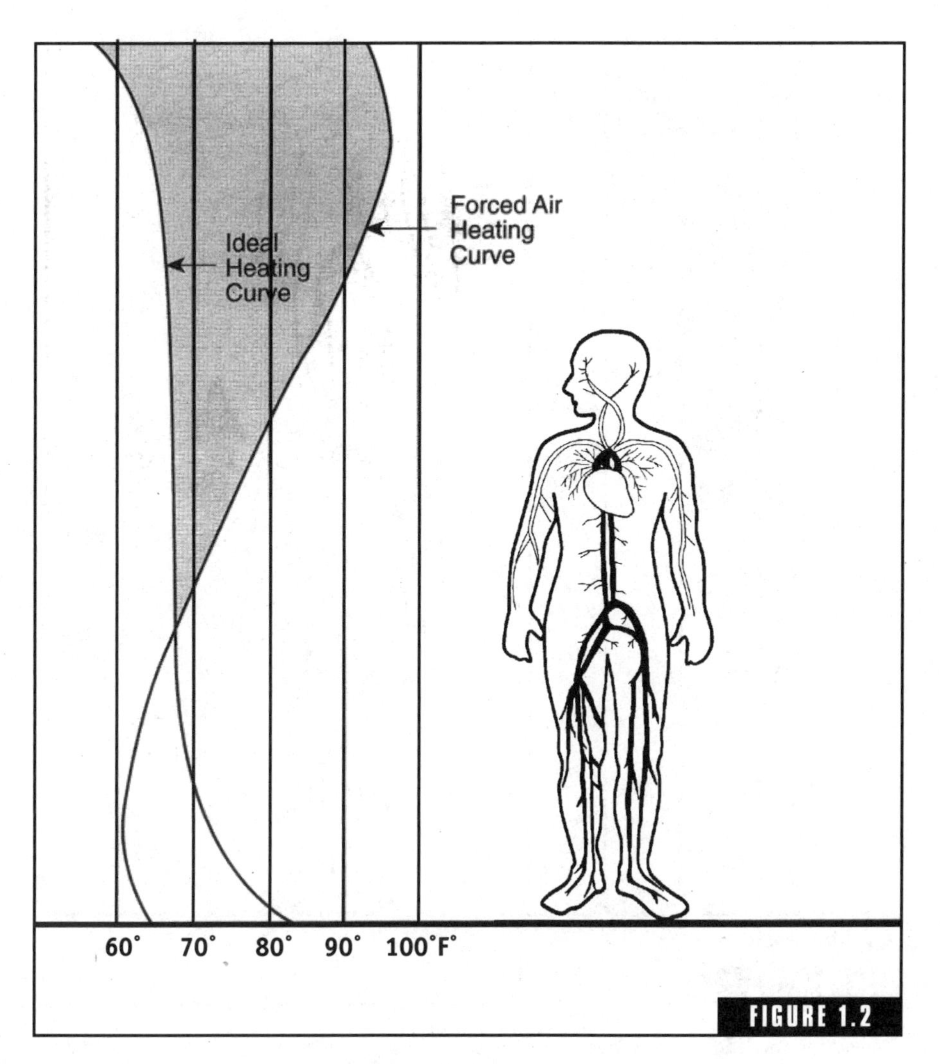

Forced air heating curve. (*Courtesy of Wirsbo.*)

The concrete gathers the heat and stores it. As the desire for in-floor heating increased, so did new installation methods. The three types of installation used today are slab-on-grade systems, thin-slab systems, and dry systems. For a long time, slab-on-grade systems were the most often used type of radiant floor heating. This is still a popular type of heating installation, but other ways are growing in popularity, as people want more of the advantages possible with radiant floor heating (Figs. 1.7 and 1.8).

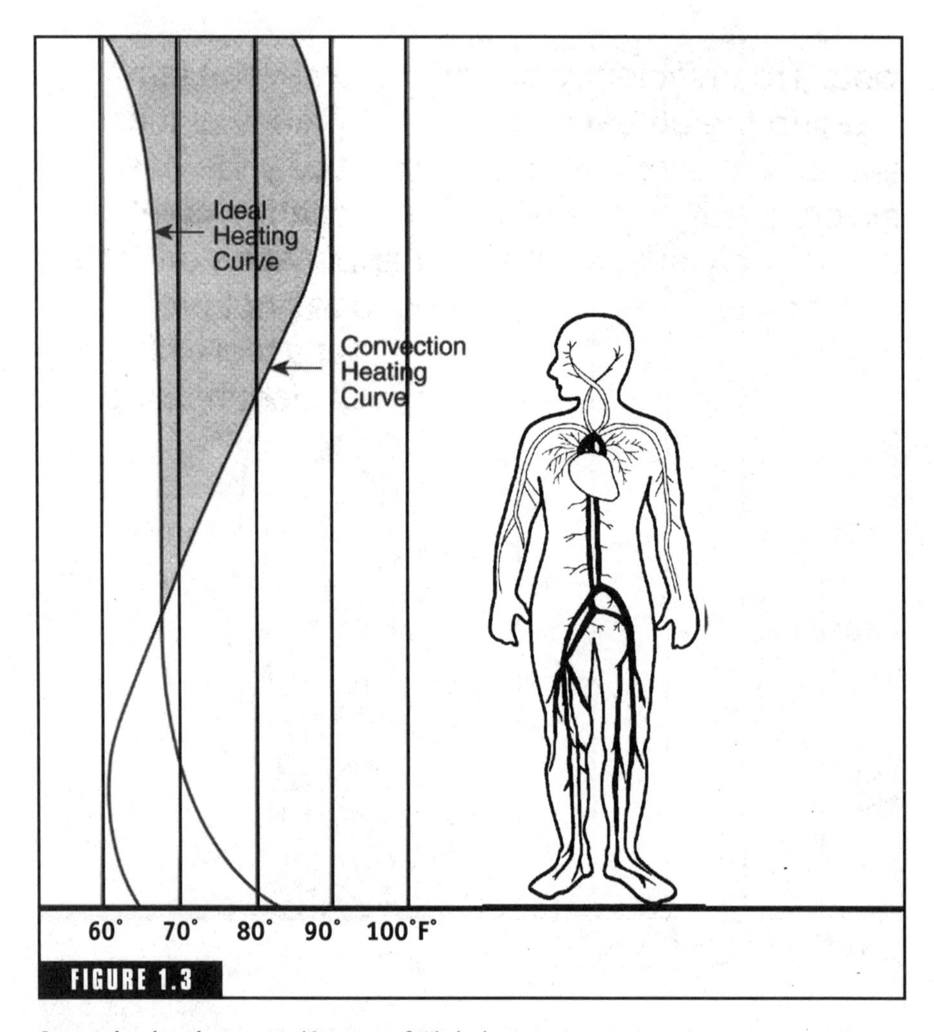

Convection heating curve. (*Courtesy of Wirsbo.*)

Putting Heat in a Concrete Slab

Putting heat in a concrete slab is not a huge undertaking. When you or a customer is paying to have a concrete slab installed, it makes a lot of sense to consider putting radiant heat in the slab. Since there will be a slab regardless of whether there is heat or not, the only real cost of adding the heat is some inexpensive PEX tubing and some labor cost in having it installed. This, of course, is assuming that some form of hot-water heat will be used for the remainder of the heating system. If the slab will hold the entire living space of a home, the situation is

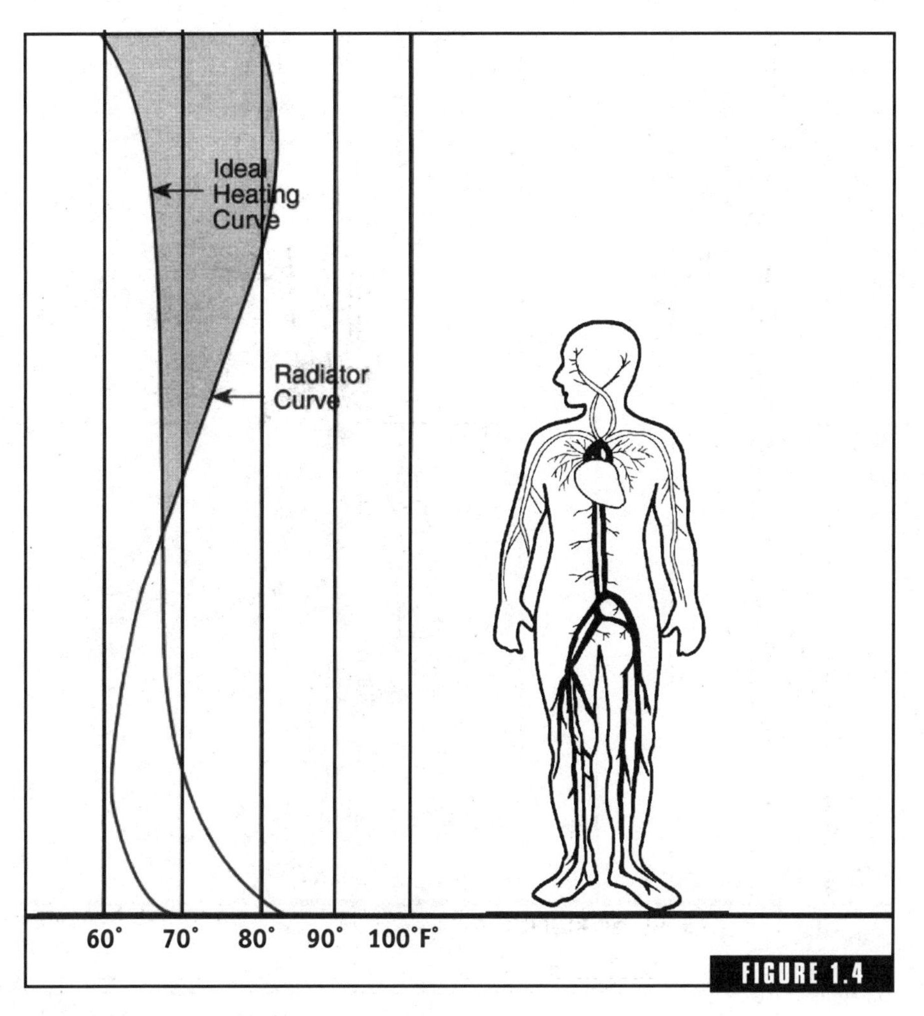

Radiator heating curve. (*Courtesy of Wirsbo.*)

even better. This means that there will be no need for any other type of above-floor heating unit.

A very important part of installing radiant heat in a concrete slab is the protection of the tubing used to transport hot water through the concrete. It is normal to install PEX tubing on the reinforcement wire that is generally placed in the slab area prior to pouring concrete. The tubing is looped throughout the slab area and can be attached to reinforcing wire with wire ties that are specifically designed for the purpose (Fig. 1.9). Another method, and one that many contractors feel is better, is to use special clips to support the PEX tubing. The clips are

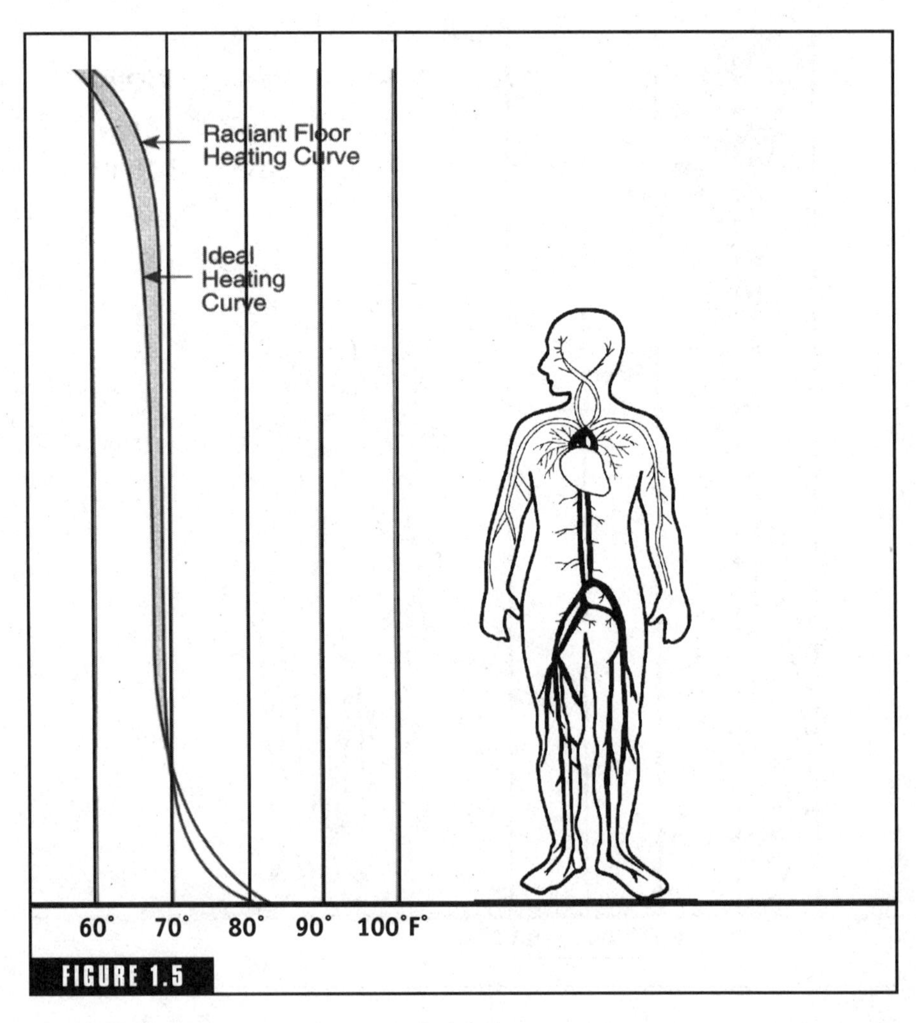

FIGURE 1.5

Radiant floor heating curve. (*Courtesy of Wirsbo.*)

available as individual clips or as bars that are notched to accept the tubing. Regardless of your preference for securing the tubing, refer to the instructions provided by the tubing manufacturer. Failure to secure the pipe in compliance with manufacturer's recommendations can result in a voided warranty.

QUICK»TIP **Failure to secure the pipe in compliance with manufacturer's recommendations can result in a voided warranty.**

The spacing between loops of radiant piping can vary a great deal. Loops that are placed close together will boost heat output. Loops may be placed as far as 2 ft apart in some jobs and much closer in oth-

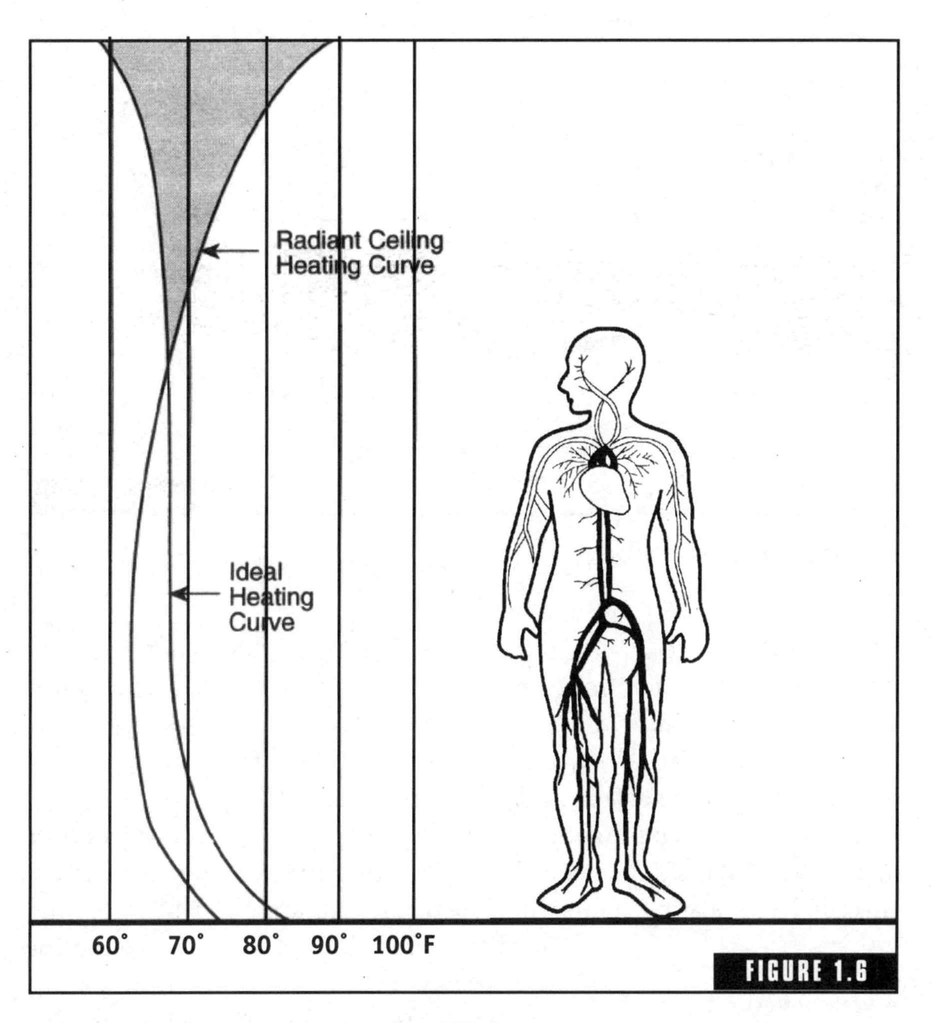

Radiant ceiling heating curve. (*Courtesy of Wirsbo.*)

ers. Most installers attempt to keep the loops about 1 ft apart. This makes bending PB or PEX tubing much easier, without as much fear of kinking the tubing. The distance between loops is determined when a heat load is computed and a system is designed.

Just as the distance between loops varies, so does the depth at which radiant tubing is installed. It's generally considered best to position tubing so that it is concealed approximately midway in a slab. However, tubing is often set lower in concrete. This can be due to many factors, such as reinforcing wire sinking during the pouring of concrete. When tubing is placed deep within a concrete slab, the heat-

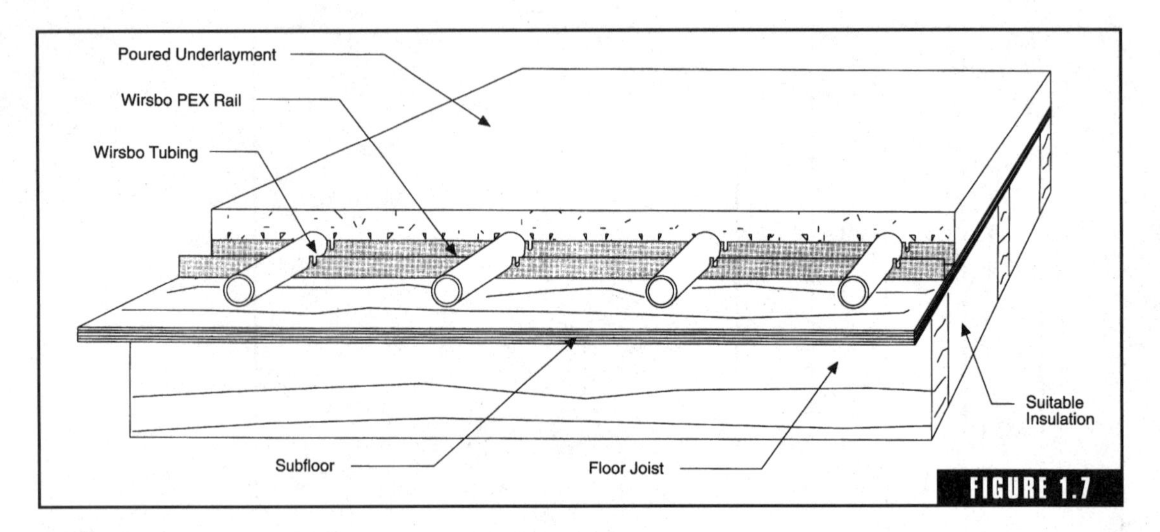

Thin-slab radiant heating system. (*Courtesy of Wirsbo.*)

ing efficiency is reduced. This means that hotter water is needed to produce the same amount of heat that cooler water would produce if the tubing were closer to the surface of the slab. If a system has been shut down for a while, it will take longer for tubing that is deep within a slab to warm the surface of the concrete (Fig. 1.10).

In order to maximize heating efficiency, rigid foam insulation should be installed between the earth under the slab and the tubing installed for heating. Older homes where radiant floor heating was installed often did not have the benefit of such insulation. As a result, the heating systems were forced to work much harder to maintain a comfortable heating temperature. Without the insulation, much of the heat produced by the tubing is lost to the earth. In time, the ground reaches a level temperature that allows the tubing to heat the slab fairly well, but this lost heat can be greatly reduced when insulation is installed below the tubing. A foam insulation board with a thickness of just 1 in. will make a huge difference in how well a radiant floor heating system performs.

TRADE SECRET

A very important part of installing radiant heat in a concrete slab is the protection of the tubing used to transport hot water through the concrete. It is normal to install PEX tubing on the reinforcement wire that is generally placed in the slab area prior to pouring concrete. The tubing is looped throughout the slab area and can be attached to reinforcing wire with wire ties that are specifically designed for the purpose (Fig. 1.9).

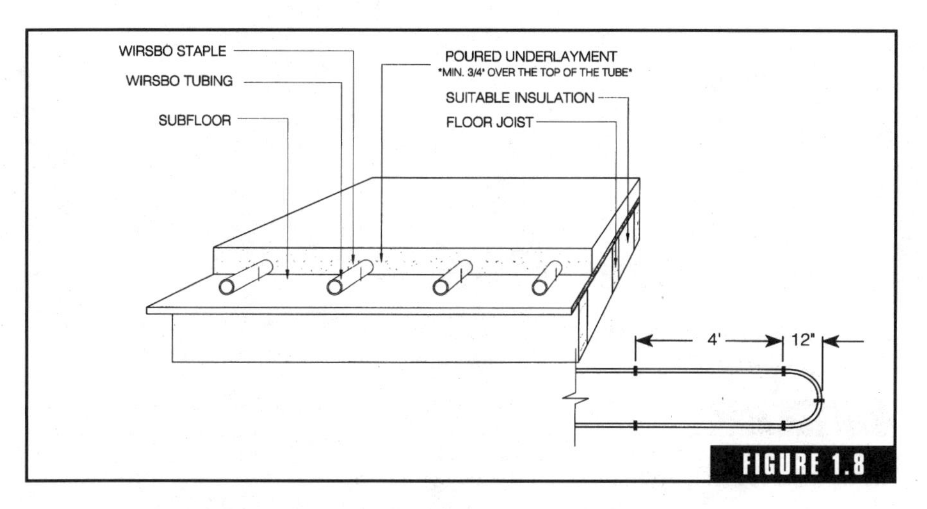

Radiant heating system on a suspended wood floor. (*Courtesy of Wirsbo.*)

Site Preparation

Before heating tubing is installed in anticipation of concrete being poured, there are some site preparations to be attended to. It is common for plumbing and electrical work to be installed in a way to be covered with concrete. When this is the case, make sure that all of the plumbing and electrical work is completed before installing heating tubing. The earth in the slab area should be fully prepared for concrete prior to heating tubing installation. Before tubing is placed for heating, all aspects of the slab preparation should be complete. Check to make sure that foam insulation is in place, that reinforcing wire is installed, and that no rocks or dirt clumps are lying on top of the insulation board. Sweep the insulation, if necessary, to provide a clean surface for your tubing. Rocks or other sharp objects could cut or crimp tubing when concrete is poured.

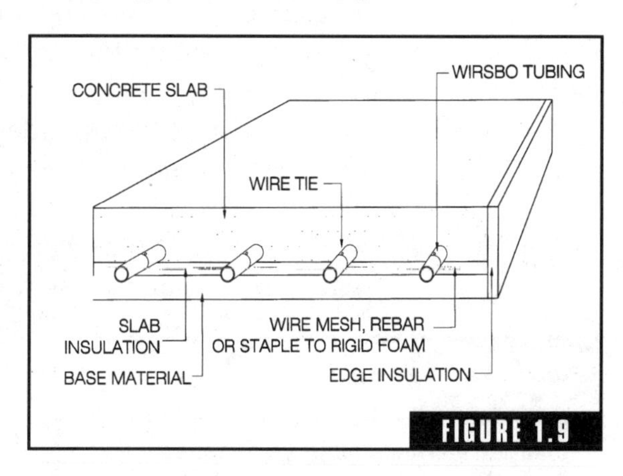

Attachment method for heat tubing when wire ties are used. (*Courtesy of Wirsbo.*)

Manifold Risers

Manifold risers are the sections of tubing that are turned up to be exposed above a

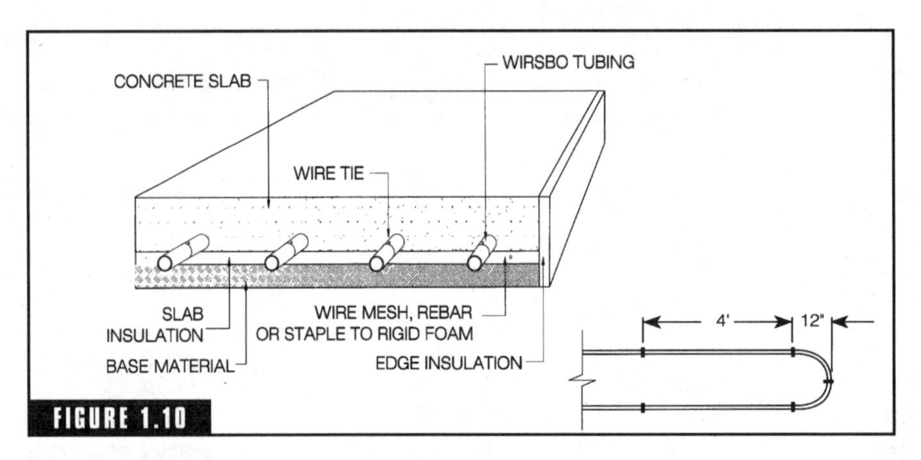

FIGURE 1.10

Tubing installed deep within a slab. (*Courtesy of Wirsbo.*)

slab when concrete is poured. These are the feed and return pipes for the heating system. Assuming that you are working with a detailed heating design, and you should be, the locations for manifold risers will be shown on the layout drawing (Fig. 1.11). Precise placement of the manifold risers is usually critical. These pipes are typically designed to turn up in wall cavities. Since installers are working prior to full construction, the walls that the risers are to turn up in are not yet in place. This requires the installer to read plans perfectly and to create mock walls. The mock wall locations can be created with string and stakes. Some installers use what are called block guides. These guides are usually the same width as the walls to be installed. Not everyone likes the idea of using wood below grade. This is due to the potential risk of termite infestation. For this reason, a lot of installers pull strings to indicate wall locations and then stake the risers with nonwood stakes.

Plastic bend supports are used to turn risers up, through the concrete. The supports act as sleeves to protect tubing that is to penetrate concrete. Another purpose of the supports is to guide the tubing through a tight bend to an upright position without kinking the tubing. If stakes are used to hold risers in place, the supports are generally attached to the stakes with several wraps of duct tape. It is essential that the risers not move

TRADE SECRET

In order to maximize heating efficiency, rigid foam insulation should be installed between the earth under the slab and the tubing installed for heating.

during the pouring and finishing of the concrete. If the tubing is moved accidentally, it may wind up sticking up through the finished floor in a place that is not acceptable. The result of this could be using a jackhammer and losing a good deal of money to correct the problem. To avoid this, make sure that all risers are secured in a way that is sure to produce desired results.

MATERIALS

Manifold risers are the sections of tubing that are turned up to be exposed above a slab when concrete is poured. These are the feed and return pipes for the heating system. Assuming that you are working with a detailed heating design, and you should be, the locations for manifold risers will be shown on the layout drawing (Fig. 1.11). Precise placement of the manifold risers is usually critical.

Follow the Design

When you begin to lay out tubing in a groundworks (Fig. 1.12) (the prepour slab area), you should always follow the heating diagram provided by your design engineer. After finding locations for manifold risers, it's time to run the tubing system in the groundworks. All tubing should be installed without joints or couplings. When it is absolutely necessary to install a joint below a slab, there are approved fittings for the job, but they should be used only in extreme circumstances. Use full lengths of tubing whenever you can. When joints don't exist, they can't leak, and this is a big advantage, especially in underfloor systems (Fig. 1.13).

A typical slab will contain a lot of heating tubing installed in it. PEX is the tubing most often used for heating, but PB tubing is also used, and was used more than PEX for a long time in the United States. Both types of tubing are easy to work with, and neither of them crimps quickly. But crimping or kinking is a possibility that must be avoided. Since a large volume of tubing is needed for most jobs, you need some way to manage the tubing. A good way of doing this is using an uncoiler. The uncoiler is basically a rotating spool that allows tubing to come off a large roll with little likelihood of kinking. It's similar to the uncoilers so often used by electricians for their large rolls of electrical wire.

All tubing installed should be installed in compliance with the manufacturer's recommendations. In typical straight runs, the tubing should be secured in increments of 24 to 30 in. It is common practice to install three fasteners on return bends. But it is important to read and follow the fastening instructions provided by manufacturers. If you are securing tubing to reinforcing wire, make sure that the wire

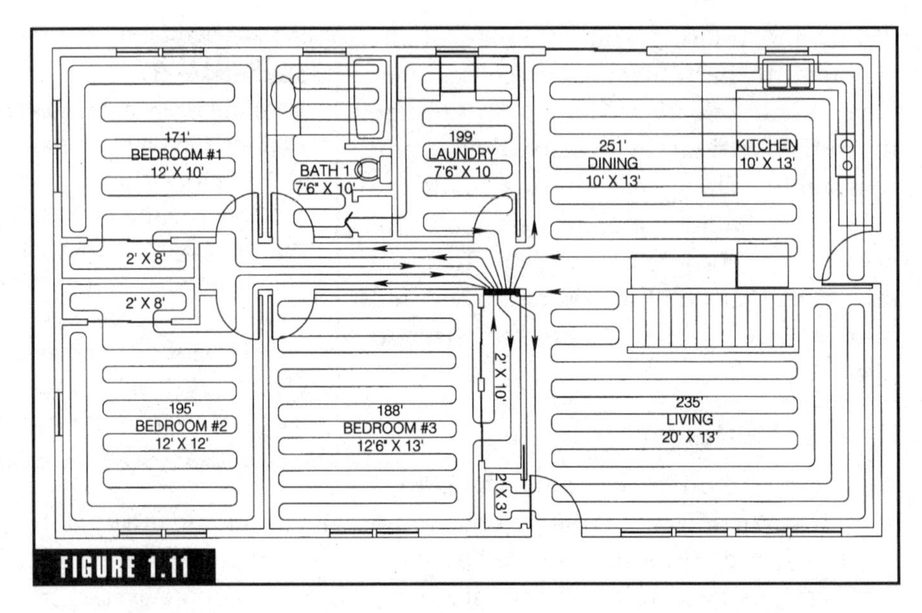

FIGURE 1.11

Heating design. (*Courtesy of Wirsbo.*)

where it comes into contact with tubing is not sharp.

All tubing should be protected from any anticipated damage. If you have tubing in an area where damage may occur, such as if it is passing through an area that will become a foundation wall or is in an area where expansion joints or sawn control joints might exist, sleeve the tubing. Any sleeve used should be at least two pipe sizes larger than the tubing it is protecting. The pipe used by plumbers for drains and vents works well for sleeving PEX or PB tubing. This could be polyvinyl chloride (PVC) or acrylonitrile butadiene styrene (ABS) plastic pipe. If tubing is already installed before you realize the need for a sleeve, you can cut a section out of the sleeve material and place it over the tubing. The main thing is to make sure that all tubing is protected from potential damage.

Testing

Testing a new installation before concrete is poured is important. Even when there are no joints in a system, the tubing should still be tested with air pressure to make sure that there are no defects in the system. Procedures for testing are easy, don't take long to perform, and assure a much better job. Once you create a test rig, or rigs, the testing process

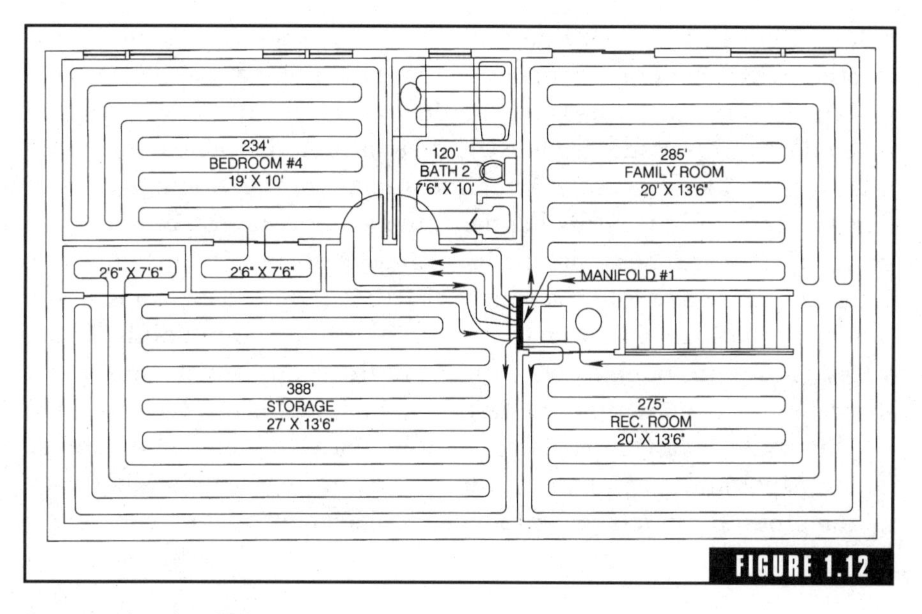

FIGURE 1.12

Basement heating design. (*Courtesy of Wirsbo.*)

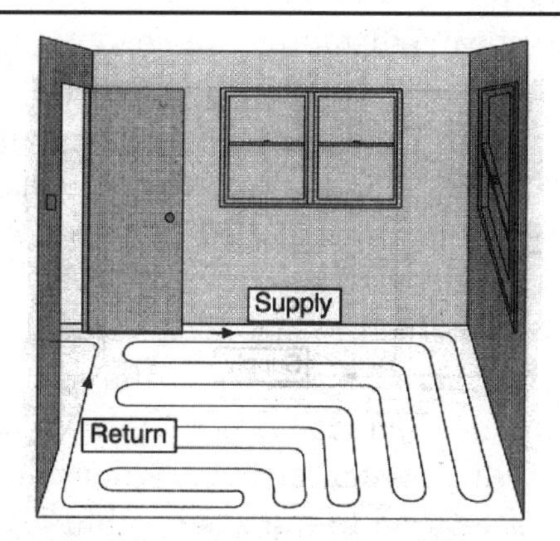

Double Wall Serpentine—Used when there are two adjacent walls representing the major heat loss of the room. The supply is fed directly to either of the heat loss walls and then serpentined toward the lower heat loss area in an alternating pattern against the two heat loss walls. Start tubing runs 6" from walls or nailing surfaces. A 6" on center tubing run is often installed along outside walls to improve response time.

FIGURE 1.13

Supply and return heating layout. (*Courtesy of Wirsbo.*)

is fast. You can test each line individually, or you can tie the tubing together with temporary connections to test all of it at once.

What amount of air pressure should you use when testing a system? A standard test pressure is 50 psi. A system should be able to maintain this pressure for 24 hours. It's not uncommon for the tubing to expand during a test period. If this happens, the air pressure on an air gauge will drop. It is also possible that leaks may be present in the temporary connections used to join a system for testing. If you suspect a leak at a test joint, use some soapy water to paint the connection points. When a leak is present, bubbles will appear in the soapy water. Don't omit the test stage. Even without underfloor joints, tubing can have holes in it. This could be a factory defect, a hole that was incurred in storage and transportation, or a hole that was made during installation. Check all the pipes. It's much easier to make replacements before concrete is poured.

Thin-Slab Installations

Thin-slab installations are pretty much what they sound like. Radiant heating systems installed above a large slab can be done with a thin-slab approach. This means installing the heating tubing on a wooden floor and then pouring a thin slab of concrete over the heating tubing. This procedure is common and effective. A thin-slab system is not the only way to use radiant floor heating above a slab, but it is a fine choice. The thickness of concrete used in thin-slab systems doesn't usually exceed 1 1/2 in.

Piping procedures for a thin-slab system are about the same as those used for larger concrete slabs. One difference is how the tubing is attached. In a slab-on-grade system, the tubing is usually attached to reinforcing wire or with special clamps. When installing a thin-slab system, the tubing is secured to the wood subflooring. The floor joist cavities below the subflooring should be filled with insulation. Batt, glass fiber insulation is the typical choice for insulation.

Manifold risers are needed with a thin-slab system, just as they are with a large slab. The risers usually turn up either in walls or inside closets. When the risers are in a wall, an access door should be provided for service and repair. All tubing must be attached securely to the subflooring. Since thin slabs don't have much concrete cover for tubing to be protected and covered, the tubing must be held tightly to the floor.

Any loose-fitting tubing can rise up and protrude above the finished concrete covering. General principles call for tubing to be secured at intervals of no more than 30 in. on straight runs. It is a good idea to keep the fasteners closer together.

Construction Considerations

Special construction considerations have to be addressed when a thin-slab system is used. Weight is one major concern. The weight added when concrete is poured over subflooring can be quite substantial. This has to be considered during the planning stage of construction. Larger floor joists or other means of additional support are needed to accept the added weight of concrete. While weight is a major consideration with thin-slab systems, it is not the only thing to think about.

Adding about 1 1/2 in. of concrete to a subfloor raises the level of a finished floor. This is not a big problem when it is planned for in advance, but it can mean trouble if allowances are not factored in for the increased height. For example, door openings and thresholds will be off if the added height is not allowed for. Plumbing fixtures and base cabinets will be affected by the increased height. As long as adjustments are made during rough construction, the finished product should turn out fine.

Concrete or Gypsum?

There are two types of cover to place over heating tubing in a thin-slab system. In my region, lightweight concrete is the leading cover for thin-slab systems. But gypsum-based underlayment is another popular material for covering heating tubing. Lightweight concrete in a typical mix will add up to about 14 psf to the dead-load weight rating of a floor. Believe it or not, this is actually a little less weight than what gypsum-based material will add. Either type of cover material will work fine; it's a matter of regional preference as to which material is most likely to be used.

Special construction considerations have to be addressed when a thin-slab system is used. Weight is one major concern. The weight added when concrete is poured over subflooring can be quite substantial. This has to be considered during the planning stage of construction.

When gypsum-based materials are used, a sealant is generally sprayed on the subflooring. By spraying a sealant and bonding agent on the subflooring, the floor surface is strengthened and made more resistant to moisture problems. The spraying process is done after all tubing is installed. Once the floor is prepared, the gypsum material is mixed, in a concrete type of mixer, and pumped to the location of the heating system through a hose. The material is loose enough in consistency to flow under and around the tubing. Material is pumped in until the level of it is equal to the tubing's diameter. This is called the first coat or the first lift. More material will be needed after the first coat is dry.

The first layer of cover material should dry, depending upon site conditions, within a few hours. After the material can be walked on, (it could be ready in as little as 2 hours), you can apply the second lift or coat. When the second level is distributed, it should cover the tubing by at least 3/4 in. After a few hours, the second coat should be dry enough to walk on, but this does not mean that it is dry enough for a finished floor covering to be installed. Depending upon site conditions, meaning air temperature, humidity, and so forth, it can take up to a full work week for the material to dry adequately for finish flooring.

There are pros and cons to gypsum-based cover material. A big advantage over concrete is the ease of application offered with gypsum-based material. Another advantage of the gypsum material is that it doesn't shrink or crack as badly as concrete might. The strength of a gypsum floor material is strong enough to support foot traffic and light equipment, but the material is subject to cuts and gouges, which must be avoided during construction. Any type of consistent water leak can ruin the gypsum material. Even a minor leak that goes undetected for a period of time can turn the material to mush. It is common for a finished gypsum-based cover material to be sealed with a sealant. Another disadvantage of the gypsum material is that it does not have the thermal conductivity that concrete does. This means that the water temperature in heating tubing covered with gypsum material is likely to be higher than it would need to be in a concrete floor.

MATERIALS

There are pros and cons to gypsum-based cover material. A big advantage over concrete is the ease of application offered with gypsum-based material. Another advantage of the gypsum material is that it doesn't shrink or crack as badly as concrete might.

Lightweight Concrete

Lightweight concrete is a common cover material for thin-slab systems. The concrete can be delivered to the point of installation in buckets or wheelbarrows. It can also be pumped to the distribution point with a hose and grout pump. Since concrete is what it is, the material is not affected greatly by moisture, as a gypsum-based product would be. But concrete does have a tendency to crack, and this can be a problem for finished floor coverings. Most contractors install control joints in the material to reduce, or to at least control, the effects of cracking. Control joints are often placed under door openings.

The finished floor covering on a floor containing radiant heat must be taken into consideration. Some types of floor covering allow heat to rise better than others. For example, a tile floor will allow much more heat to escape from a radiant system than a padded, carpeted floor covering. To control this, insulation can be placed under the subflooring of a radiant system. When the R-value of the insulation under the floor is greater than the R-value of the finished floor covering, heat will rise through the floor, rather than escaping below it.

Dry Systems

When radiant heat is installed beneath a floor and is not covered with concrete or gypsum-based material, the installation is known as a dry system. The reason for this is simple. The fact that no material is poured over the tubing makes the system dry. These systems are great in that they don't add any appreciable weight to a flooring system. Since weight is not added in substantial amounts, a dry system can be used in remodeling without the need for additional floor support. However, a dry system does need some help in producing heat in desirable quantities.

Since dry systems do not have concrete or gypsum to conduct heat, it's common to install heat-transfer plates for lateral heat conduction. The plates may be installed above or below subflooring, depending upon the design. Heat plates are made of aluminum and work very well. Installation of below-floor systems is less expensive than that of above-floor systems. Much less labor is involved in below-floor systems, and this is one reason why the cost is less.

Above-Floor Systems

Above-floor systems are installed above subflooring and below finished flooring. Since space is needed for tubing, a sleeper system is required for above-floor systems. The sleeper system is a series of wood strips that provide cavities for the tubing to run through and give flooring installers a nailing surface for an additional layer of subfloor-

QUICK»TIP **Below-floor systems require less time and labor to install. They also require less material, since sleepers and a second subflooring are not needed.**

ing. Heat-transfer plates are installed for straight runs of tubing. The plates are placed between the sleeper members and stapled down on one side. Once the plates are in place, the tubing can be installed in them. Since sleepers and a second subflooring are installed in an above-floor system, the floor height is raised by an inch or more. This can be a problem for doors, plumbing fixtures, and base cabinets. Just as with a thin-slab system, these problems can be avoided with proper planning and adjustments.

Below-Floor Systems

Below-floor systems require less time and labor to install. They also require less material, since sleepers and a second subflooring are not needed. Whether remodeling or building, below-floor systems make a lot of sense. Heat tubing is usually placed in heat-transfer plates that are stapled to the bottom of subflooring. When tubing penetrates floor joists, the holes should be drilled near the middle of the joist, rather than the top or bottom edge. This helps to maintain structural integrity. Also, the holes should be somewhat larger than the tubing diameter, to avoid squeaking as the tubing expands.

When heat tubing is attached to the underside of subflooring, it is at risk of damage from the work of others. A plumber or electrician who is not aware of the heat tubing may drill through it accidentally. It's not practical to protect all tubing with nail plates, but it is wise to make all workers aware of the tubing under the subfloor. If a section of tubing is damaged, repair couplings can be used to correct the problem. However, whenever possible, avoid joints and connections in heating systems. It's best to maintain full lengths of tubing, rather than coupled sections.

Radiant floor heating systems are gaining popularity quickly. They can be quite cost-effective to install and operate. The comfort offered from a radiant system is, in many ways, unmatched. If you are not familiar with this type of heating system, you owe it to yourself and your customers to learn more about it. Radiant heating systems in floors are well worth consideration.

CHAPTER

2

Comparing In-Floor Systems to Other Systems

Many people think that radiant floor heating systems are new. Well, they are just now gaining popularity with modern builders and homeowners, but the fact is, radiant floor heating has been around for a very long time. Radiant floor systems were in use with the Romans as early as 60 A.D. As a society, we have learned much from the past, and radiant floor systems are no exception. The Romans knew how to create a comfortable indoor environment, and modern builders are taking a page from history. There is a boom in radiant floor heating systems. It's no accident that they are gaining in popularity. Word-of-mouth advertising is some of the best advertising in the world, and it has done wonders for radiant floor heating systems.

Contractors have been selling and installing radiant floor heating systems for many years. The systems have generally been well accepted. There have been some problems in the past with the materials used for these systems. Copper tubing can have a bad reaction to contact with concrete. This can be a problem with radiant floor systems, since concrete is a common medium for heat transfer. Polybutylene (PB) tubing was used for many years, but it too had its problems. The joints made when using PB tubing sometimes didn't hold up. This created leaks and problems. Nowadays, a new type of plastic tubing is used. It's called cross-linked polyethylene, and its trade name is PEX. This tubing is very similar in appearance and in installation procedures to PB tubing. But the cross linking of the tubing makes it better.

PEX tubing came onto the market in 1971. Until recently, PB tubing was very popular, but now PEX is leading the pack. The Wirsbo company introduced PEX to the heating industry and continues to be a major provider of quality heating components. Some reports indicate that PEX tubing is used in nearly one-half of all radiant floor heating installations. It's very uncommon to find copper tubing in a new installation. This is due to both cost and effectiveness. Frankly, PEX tubing is hard to beat when putting together a radiant floor heating system.

A lot of people question using plastic tubing for a heating system. Some people feel that the plastic can't take the heat, so to speak. Fact is, radiant floor systems don't require the high water temperature that a baseboard system does. The lower water temperature makes using plastic tubing both easy and effective. Radiant floor heating is an effective source of heat for both residential and commercial uses. Fuel economy is another desirable feature of radiant floor systems. In fact, there are many benefits to radiant floor heating systems and very few disadvantages. With this in mind, let's explore some of the primary benefits.

Comfort

The comfort associated with a radiant floor heating system is extremely nice. Some would say that it is unique. Radiant floor heating warms a room entirely, from the floor up. Baseboard heating systems warm the perimeter of a room, but they do not offer the overall heating comfort that a radiant heating system does. Radiant floor heating systems essentially turn an entire floor into one large radiator. The temperature of the heat is low, but quite effective. Since heat rises from all of the floor surface, everything in the room is warmed in a fairly equal manner.

If you ask an average person what comfort from a heating system means, you will probably be told that it's the act of keeping a person warm. The thought of keeping a person from feeling cold is common among average people who are thinking about what an effective heating system is. While neither of these concepts is really wrong, they are also not exactly right. Comfort from a heating system is largely a matter of controlling the rate at which a body loses heat.

People generate a lot of personal heat. In fact, the amount of heat generated is often more than is needed. This creates a need for a way to give off excess heat. Under normal conditions, an average human loses about 400 Btu per hour. That's a lot of heat, and it's lost in three basic ways. A human body loses heat through convection. This is done as air currents pass over the body. Heat is also lost through evaporation when breathing and sweating occurs. Radiation is the third means of heat loss. This occurs when a warm body is subjected to colder temperatures. Based on these factors, radiant floor heating systems seem to be the most effective form of heating system to maintain a comfort level for human beings.

Efficiency

Efficiency is a word that has gained a lot of attention in recent years. With energy cost and desires to protect natural resources, energy use is a big issue. When it comes to efficiency, radiant floor heating systems have an edge. It is widely believed that they are the most efficient form of heating. This is due in part to the fact that radiant floor systems use low-temperature water and that the heat output can be adjusted on a room-by-room basis. Radiant heating systems in floors heat people and objects rather than heating the air in a room. This is a more efficient means of heating. Some reports indicate energy savings for radiant floor systems ranging from 20 to 40 percent over forced hot-air systems.

Some types of buildings are more difficult to heat than others. For example, a building with a tall ceiling can require more heat at the occupant level to maintain an even heat, since much of the heat rises into the ceiling. If large windows are installed in a building, the amount of heat output required to maintain a comfort level is likely to be higher. But radiant floor heating systems overcome these problems better than other types of heating systems. Since the source of the heat is in the floor, all the heat rises around the people in the room.

When heat comes from a source higher than a floor, some of the room space is not heated. Baseboard heat is low and close to a floor, but it is perimeter heating. Having heat come from a perimeter system makes it difficult to keep the center of a room comfortable. Radiant floor systems heat entire rooms and radiate heat in all areas of the room. Energy efficiency varies from building to building, but it is true that radiant floor heating systems are believed to be extremely effective in heating buildings of all sizes and types.

Dust

Dust is a big problem with forced-air heating systems. The problem is not large with radiant baseboard heat, but it's almost nonexistent with radiant floor heating. People often have problems with dust in their homes and businesses. Many new filters have been created for forced-air systems, and the filters do a good job. Still, the dust issue can be reduced by avoiding forced-air systems. Since there is no need for circulating the heat from a radiant floor system, there is no reason to have increased airborne dust.

Some people assume that no real airborne dust is associated with convection baseboard heat. This is a misconception. Since the baseboard systems are in the living space, they collect dust. As heat rises from the units, so does dust. True, it's not as bad as fan-forced systems, but the dust still gets into the air of the living space. With radiant floor systems, the heating system is concealed below the living space, thus eliminating the rising dust problem. When you eliminate dust, you greatly reduce the spread of pollen indoors. This makes radiant floor systems a healthy choice for homes and offices.

Design Issues

Design issues come up during the planning, construction, and decorating of buildings. Some types of heating systems hamper design preferences. Forced-air systems require ducts and return grilles. Baseboard heat eats up wall space. Radiant floor heat allows total freedom in furniture placement and other design issues (Fig. 2.1). This is yet another major benefit to radiant floor heating systems. Being able to arrange furniture, office equipment, and other elements of a building without impairing a heating system is without a doubt a big advantage.

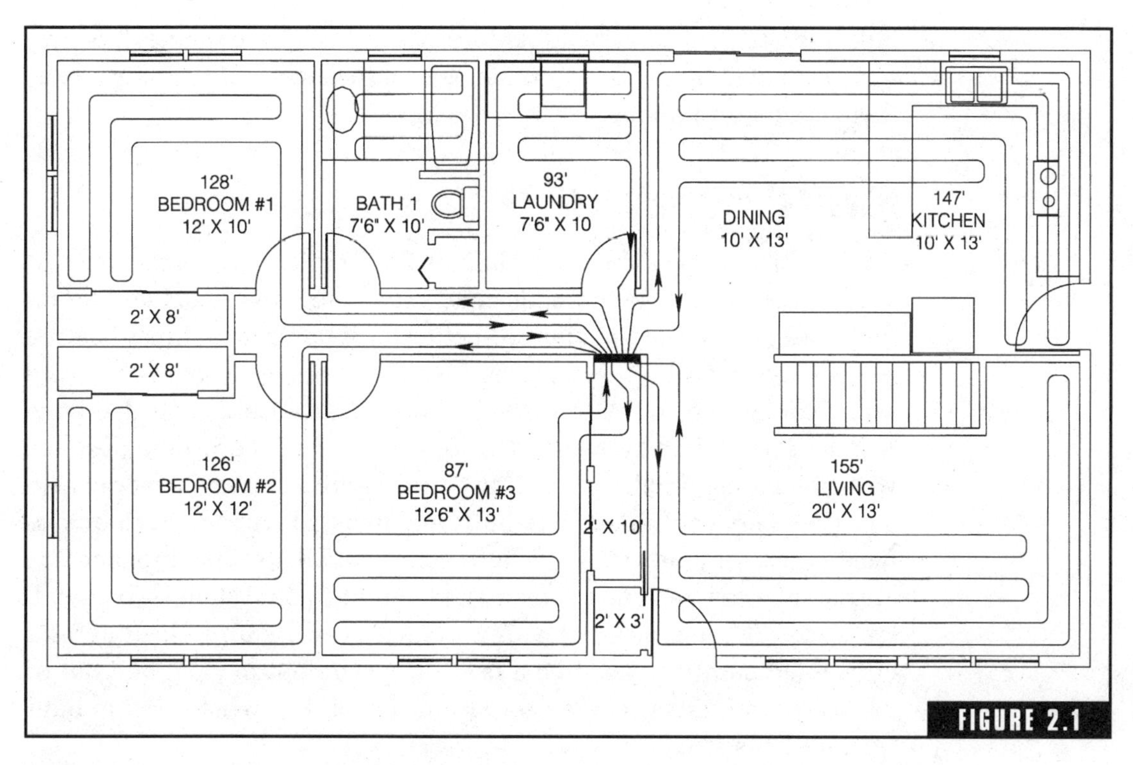

Underfloor heating doesn't compromise furniture placement. (*Courtesy of Wirsbo.*)

Most people don't think much of safety risks with convection heating units, but they do exist. If a child manages to touch a copper tube in a heating element, a severe burn can occur. This doesn't happen with heating systems that are embedded in a floor. The sharp edges of ducts can be a safety hazard for kids who remove registers and grilles. This is something that doesn't come into play with radiant systems in a floor. While safety problems are not usually a big problem with other types of heating systems, they simply don't exist with in-floor systems.

> **QUICK TIP** **More living space is available when radiant floor heating systems are used. Radiators, baseboard heating systems, and forced-air heating systems all restrict the amount of space that can be used in a building. Radiant floor heating doesn't do this.**

More living space is available when radiant floor heating systems are used. Radiators, baseboard heating systems, and forced-air heating systems all restrict the amount of space that can be used in a building. Radiant floor heating doesn't do this. Since the heating system is situated in a floor and radiates heat throughout the living space, there are no limitations.

Physical Appearance

The physical appearance of a radiant floor system is great. In fact, it doesn't have one. The tubing is concealed below the floors and therefore doesn't offer any visual indication of its presence. Most heating systems are seen by residents and guests. Floor heating systems are not. If you have ever seen rusted radiators or baseboard units that have lost their paint, you know that the look can be bad. Forced-air systems with major dust problems hanging around grilles and registers can also be ugly. Many contractors and building owners love radiant floor systems since they are felt and not seen. Overall, the maintenance of appearance can become a factor. When an in-floor system is installed, it's done. Other types of heating systems can require cleaning and occasional painting. Keeping a heating system out of sight and out of mind may not always be an important factor, but when it is, radiant floor systems are the answer.

General Maintenance

General maintenance for a radiant floor system is minimal. Some maintenance is required for the boiler supplying heat to the system, but the outreaching elements of the system are nearly maintenance-free. If the tubing doesn't rupture and if joints don't break down, the system takes care of itself. When compared to a forced-air system, the maintenance requirements are much lower. Ducts don't have to be vacuumed. Belts, blowers, fans, and other elements don't have to be oiled or replaced. There are elements of boilers that must be maintained, and we will talk about them in later chapters.

When a radiant floor system is used, it doesn't require cleaning. No one has to go around dusting heating units, because there are no visible heating elements. There are no grates, grilles, or registers to clean. The savings on cleaning may not be heavy in monetary terms, but they count. Plus, floors tend to dry cleaner after cleaning, owing to the heat, and this reduces the risk of slips and falls, not to mention the pure comfort of a warm floor.

Basement Floors

QUICK»TIP **Basement floors are an ideal place to install radiant floor heating. Concrete is an excellent mass for radiant floor heating.**

Basement floors are an ideal place to install radiant floor heating. Concrete is an excellent mass for radiant floor heating. Radiant heating systems can be used in both thick and thin slabs. When concrete is used, it allows heat to disperse laterally and vertically (Fig. 2.2). This is a definite advantage. Without under-floor heating, a concrete slab creates a comfort problem. Even an insulated slab gets cold. If a heating system is used in a perimeter system above a floor, the center area of the slab will give off plenty of cold. The only way to effectively control cold in the center of a room is to put the heating system within the slab.

Some people are concerned about installing heating pipes and tubing in a concrete floor. Assuming that the heating materials are properly selected, this should not represent a problem of any real concern. There is some risk that a pipe will burst or that a connection will break. If connections below the floor are minimized, and they can be, the risk is greatly reduced. Basically, if floor systems are installed properly with approved materials, the risk of problems is minimal. Using a quality material like PEX tubing is the best way to avoid problems. Cutting corners with inferior materials is a major mistake.

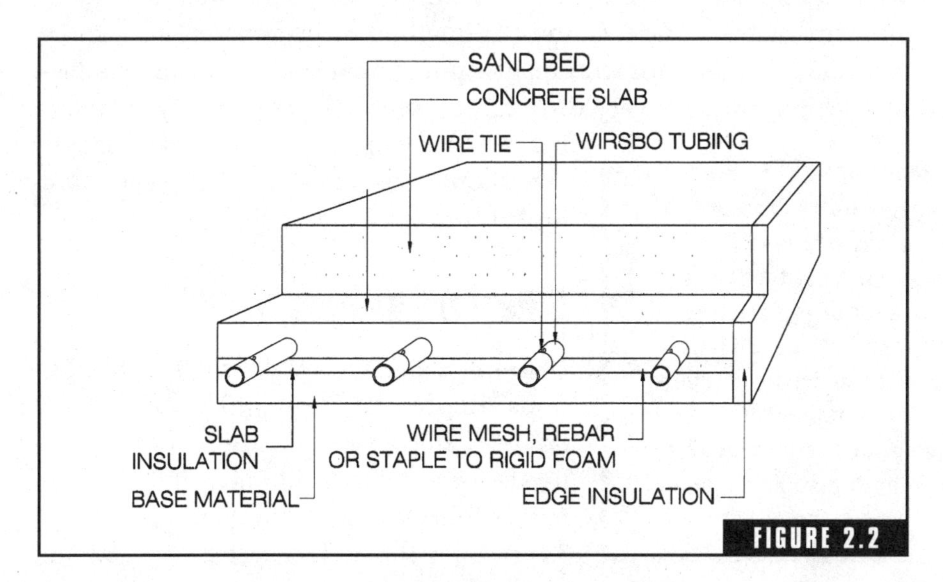

Typical heating installation in a concrete pad. (*Courtesy of Wirsbo.*)

Heat Sources

Heat sources for radiant floor heating systems are not limited. Gas- and oil-fired boilers are probably the most common types of heat sources for radiant heating systems, but they are not the only ones. Geothermal heat sources can be used with radiant floor heating systems, as can heat sources fueled by electricity. Wood is another possible fuel for creating the warm water needed for a radiant heat source. It is also possible to use combination boilers, where heat can be generated by a combination of wood and oil or wood and gas. Essentially, any cost-effective heating source may be practical for a radiant heating system.

In the far north, oil-fired boilers are the most common means of heating water for radiant heating systems. Gas boilers are also quite common, especially where natural gas is available, but bottled gas can also be used as a fuel source. Wood is less common, but it is used, sometimes exclusively but normally with a boiler that is of a combination design. Geothermal systems are less common, generally owing to the expense of these systems and their limited ability to heat water in some harsh climates.

How long will a system last for a radiant floor heating system? This is a question that is, at the least, difficult to answer. However, PEX tubing is guaranteed to last 25 years. This is as long or longer than the warranty on many boilers. The actual lifetime of a system is affected by many factors. The location and method of installation have to be considered. Quality of workmanship during installation can be a factor. Realistically, it's very difficult to put a life span on a system, but it's safe to say that a good radiant floor system should last for at least 20 years, and probably much longer.

Flexibility is a cornerstone of the radiant floor heating system. A contractor can install a little or a lot of radiation tubing to account for climate conditions. Since contractors can make adaptations on a job-by-job basis, radiant floor heating systems are serious contenders for all types of jobs, both residential and commercial.

A Perfect System

Defining a perfect system for all applications simply isn't feasible. The needs for a home in Florida are very different from the needs for a home in Maine. Virginia's climate differs greatly from that of Montana. There is no one perfect system, but radiant floor heating is a solid place to begin the search for an ideal heating system. Flexibility is a cornerstone of the radiant floor heating system. A contractor can install a little or a lot of radiation tubing to account for climate conditions. Since contractors can make adaptations on a job-by-job basis, radiant floor heating systems are serious contenders for all types of jobs, both residential and commercial.

It would not be fair to say that radiant floor heating systems are perfect, but they can come very close to perfection. Contractors need to be aware of this. It's a strong sales tool and can do a lot to build a contractor's reputation. Any contractor who is not familiar with in-floor systems is at a definite disadvantage. The growth pattern being established by radiant heating systems indicates that it may become the premier choice of heating systems in the near future. Now, let's move to the next chapter and talk about design issues for functional, cost-effective systems.

CHAPTER 3

Designing Functional, Cost-Effective Systems

Designing functional, cost-effective systems can be easy when you are working with radiant floor heating systems. The flexibility available for this type of system allows for plenty of creative input. Installing other types of heat can be much more difficult. For example, you can't install baseboard heating units where there are no walls. Air ducts can't be installed in some areas. Kick-space heaters can be needed in kitchens if you are not using radiant floor heat. There are almost no limits to what can be done with radiant floor heating. If a room has a floor in it, you should be able to get heat in the room. However, routing the tubing to the floor can pose some problems. This is usually not a problem, but it can be.

How important is a heating design? It's very important. Installing heat in a building without a known heating plan is risky. There are limits to what is too much heat and what is too little heat. Contractors should learn how to design their own heating systems, although the systems may rarely be sized or drawn by the contractors. In commercial buildings a heating diagram is usually provided with the working blueprints. This is sometimes the case with residential jobs, but not as often. Even if there is no heating plan on file with blueprints, a contractor can usually have a heating plan drawn by a supplier of heating materials. Most material suppliers have people on hand who can create feasible heating plans for all sorts of buildings.

A lot of contractors don't enjoy doing the math associated with figuring a heat installation. It's easy and more cost-effective to put the burden on someone else. In the case of small jobs, the other person is often someone associated with a material supplier. Big jobs generally have detailed heating plans drawn long before construction, so they are not a bother to bidding contractors. It is the smaller job that can require a personal touch. Remodeling jobs are another example of where a detailed heating design probably will not be readily available.

Contractors who wish to do their own heating designs must consider eight points during the process. These elements are as follows:

- The number of Btu per square foot of heat loss for each room in a building
- The floor surface temperature
- The method of installation
- The finished floor material and its R-value
- Tubing spacing and water supply temperature
- The loop length
- The rate of fluid flow, based on gallons per minute
- The pressure loss of the system

Computer programs exist to make sizing a heating system as simple as filling in the blanks. A design can be created using pen and paper, but the time and effort required is more, and in this modern world, computers are kings. In small jobs, some rules of thumb should be considered in a designing process. Let's talk about them for a few moments.

Ferrous Components

If you will be installing a radiant floor heating system that will involve ferrous components, you must make adjustments in your tubing material. What is a ferrous component? It could be a cast-iron boiler or circulator. If a system is to contain ferrous components, you should use a tubing that has an oxygen diffusion barrier. This is no big deal as long as you know it before you bid a job or install materials, but it can become a big deal later in life if you don't know that tubing with an oxygen diffusion barrier should be used whenever there are ferrous components in a heating system. As you might imagine, tubing with the barrier costs more than tubing without it, so don't bid the higher-costing tubing unless it is needed.

> **MATERIALS**
>
> **If you will be installing a radiant floor heating system that will involve ferrous components, you must make adjustments in your tubing material.**

If you are installing a system that is free of ferrous components, you can use a less expensive tubing. The tubing is generally of the same quality as the tubing with a barrier, but there is no barrier in the cheaper version. This may not seem like much to worry about, but the cost on a big job can be considerable, certainly enough to make a bid too high if the right tubing is not used in the bidding process. And not using a tubing with a barrier when you should could cause problems long after an installation is completed. No contractor wants headaches that can be avoided.

Tubing Size

Tubing size for residential jobs can basically be figured on a rule-of-thumb basis. In most cases, tubing with a diameter of 1/2 in. is all that is needed. There are times when larger or smaller tubing is more effective, but 1/2-in. tubing is the standby size. Contrary to the belief of some contractors, tubing with a larger diameter does not give off more heat per square foot of a radiant panel. Some contractors prefer to use tubing with a diameter of 5/8 in. This may be due to the old days of using copper tubing of this same size. But, the 5/8-in. tubing costs more than 1/2-in. tubing, is not as flexible for bending and installing, and does not produce more heat in a usable manner. There is one advantage to the larger tubing. The bigger tubing doesn't suffer from as much pressure loss when the same loop length is used.

Trying to work 1/2-in. or 5/8-in. tubing into floor joists can be tricky. The tubing is large enough to make bending a chore. You can remedy this problem by moving down to a tubing with a diameter of 3/8 in. Keep in mind, however, that the small diameter translates into more pressure drop. The drop is considerably more than it would be with a larger tubing size. Keeping loop lengths to 200 ft or so is the way to compensate for the pressure drop. Remember, too, that this is general information and that each job presents its own set of circumstances to be considered.

The Controls

The controls used with a radiant floor heating system can be simple or complex. A standard on-off switch can be used as the only control for the system, but this is rarely desirable. At the other end of the spectrum, you could find a weather-responsive reset control. When a system is being designed, the controls play a role in the overall cost. The cost can be looked at in two ways. First there is the start-up cost of installation. The second way to consider cost is the expense of operating the system over time. Additional money spent at the time of installation may well be recovered over coming years, owing to a more efficient system. As a contractor, you should give your customers options to consider along these lines. In a bidding war, the lowest price might be all that matters. But, when the opportunity exists, explain all aspects of the control costs to your customers so that they can make informed decisions.

Common sense should be used during the process of choosing suitable controls. For example, a customer who wants a heated garage might be happy with only an on-off switch, and perhaps a tempering valve (Fig. 3.1) if one is needed. But the owner of an office complex will most likely want a more sophisticated control system. Projections on cost-saving controls can be computed to give customers an idea of what their payback term will be if they opt for more expensive controls. If you don't have your own computer software for design purposes and don't wish to purchase any, talk to your material supplier. The chances are good that the supplier will run numbers for you and your customers with a variety of factors to produce different outcomes.

Zones

Heating zones are excellent ideas when designing a heating system. With the right zoning, a building can provide great comfort and economy from a heating system. Buildings can be broken into zones in

TEMPERING VALVES:

Taco Tempering Valves are available in ½-inch and ¾-inch sweat connections. The basic construction is brass and stainless steel. The Tempering Valve saves energy by reducing the outgoing water temperature at the hot water source.

SIZE & DIMENSIONS

508, 526

PRODUCT NUMBER	SIZE CONNS.	TYPE CONNS.	TYPE ADJUST.	RATINGS		TEMP. RANGE	MAXIMUM TEMP.	MAXIMUM PRESSURE	LENGTH	SHIP. WT. LB.
				BATH	GPM					
508	½"	Sweat	External	1-2	6	120-160	200°F	125 psig.	3¾"	.5
526	¾"	Sweat	External	1-3	12	120-160	200°F	125 psig.	3¾"	1.0

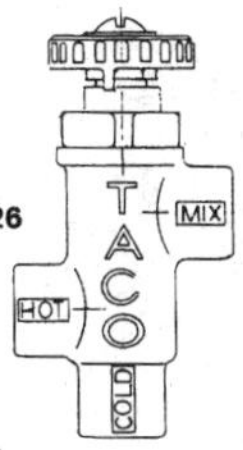

FIGURE 3.1

Tempering valve. (*Courtesy of Taco.*)

many ways. It's certainly possible to have every room in a building on a separate zone, but this is usually not practical, and it gets expensive. If you are working with customers, you can consult them on the number of zones desired. If you are working with a spec house, the most common way to zone it is with individual zones for types of rooms, rather than every room. For example, the bedrooms would all be on one zone, the living room and halls might be on a zone, while the kitchen and dining area is on another zone. This type of three-zone installation is common.

The biggest benefits of zones are comfort and economical use. Why heat a bedroom to full temperature when it will not be used for hours to come? If you are going to be spending 4 hours watching television in the living room, you can cut the zones for the kitchen and bedrooms down. It might be desirable to have bedrooms on independent zones. Some people like their bedrooms cooler than others. How far you go in breaking down a building into zones is up to your customers.

When zones are created they can be equipped with zone valves, circulators (Fig. 3.2), or telestats on a single manifold (Fig. 3.3). Zone valves are the least expensive option, and they work well. Circulators are more than zone valves and less than telestats. The establishment of zones can involve extensive planning and a fair amount of expense.

TRADE SECRET **It's certainly possible to have every room in a building on a separate zone, but this is usually not practical, and it gets expensive.**

Typical homes have two or three zones. But a house could have a great many zones. For example, there might be two

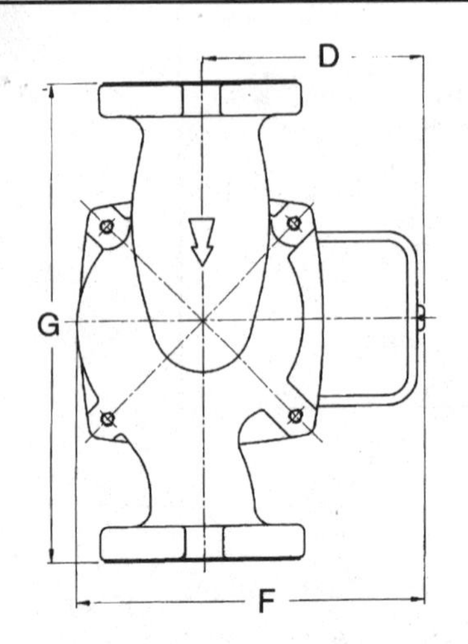

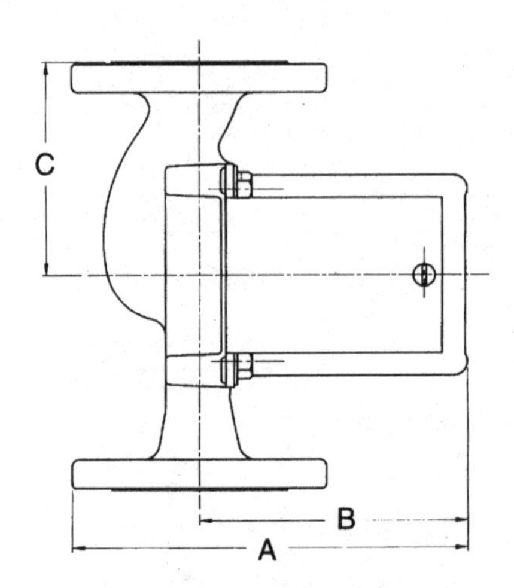

SPECIFICATIONS	UL & CSA LISTED
CASING	Cast Iron or Bronze
IMPELLER	Non-Metallic
SHAFT	Ceramic
BEARINGS	Carbon
CONNECTIONS	Flanged ¾", 1", 1¼", 1½"
PRESSURE RATING	125 psi Maximum
TEMPERATURE RATING*	230°F Maximum
MOTOR TYPE	Permanent Split Capacitor
HP	1/35 HP @ 3250 RPM
ELECTRICAL CHAR:	115/60/1 .53 Amp Rating Impedance Protected

FIGURE 3.2

Circulating pump. *(Courtesy of Taco.)*

zones for bedrooms, a zone for living space, a zone for kitchen and dining areas, a zone for a heated garage, a zone for melting ice from a roof, and a zone to melt ice on a sidewalk. The list of possibilities for independent zones is a long one. The cost of installation for a lot of zones can be substantial, but the zones can increase heating efficiency to a point where the upfront cost is well justified.

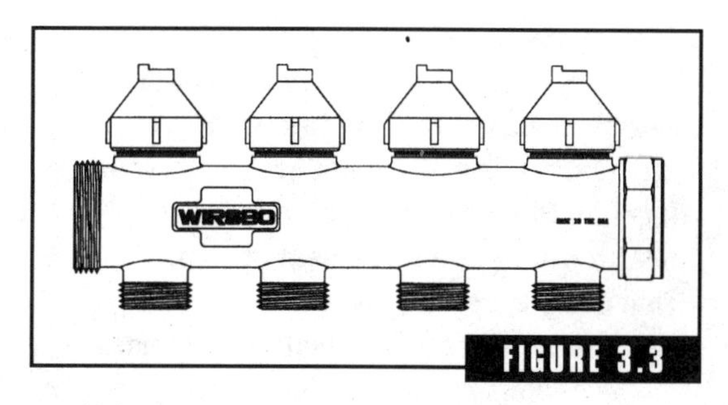

FIGURE 3.3

Manifold device. (*Courtesy of Wirsbo.*)

No Installation

If a room has no heat loss rating, no installation of radiant floor heating is usually needed. This could be the case where a small room, say a photography darkroom, is surrounded by heated space and has no exposure to cause a heat loss. If this were the case, the floor in the darkroom would not need to be heated. There is an exception, however. If the room was built on a slab, where downward heat loss is possible, the floor should have heat tubing in it. Regardless of whether the slab is on grade or below grade, the entire slab should be heated.

It's not unusual for bathrooms to be built with interior locations that will not suffer heat loss. This can mean that the floor doesn't require any heat tubing. But adding heat to the room can make the floor more comfortable for bare feet, and the heat will help to dry up moisture on the bathroom floor. Adding heat to a bathroom floor where the finished floor covering is ceramic tile is very sensible. Tile can be cold to the touch, even when it's in the interior portion of a building. Putting heat tubing in the floor can make a big difference.

TRADE SECRET

Putting heat in a ceiling is not something that a lot of people think about. It is a somewhat uncommon practice, but there are times when it's sensible. Installing heat tubing in a ceiling is one way to deal with difficult conditions during a remodeling job.

Heat a Ceiling?

Putting heat in a ceiling is not something that a lot of people think about. It is a somewhat uncommon practice, but there are times when it's sensible. Installing heat tubing in a ceiling is one way to deal with

Here's a tip that's worth looking into when you are heating a small area. Check with your local building codes to see if a water heater can be used as a heat source. That's right, I said a water heater, just like the type used for plumbing systems. Water heaters cost a fraction of what boilers do, and where acceptable, water heaters can produce enough hot water to service a small heating system.

difficult conditions during a remodeling job. When tubing is installed in a ceiling, there is no need for underlayment. Generally, less tubing is required for a ceiling installation, and the components of the system are frequently fewer in number. This all adds up to cost savings. While ceiling heat is not usually considered the best means of heating a room, the process should not be ruled out completely. Typically, this type of installation is used most often in retrofitting and remodeling jobs.

Between the Floor Joists

Installing 3/8-in. tubing between floor joists is not difficult, and it doesn't have to be very expensive. The tubing can be clipped or stapled to the subflooring. To keep costs down, you can avoid installing aluminum heat-emission plates. Not having the plates does have disadvantages, but they are not too great. For example, a job without plates will require a higher water temperature and the response time when heat is called for will be a bit slower. But the omission of the plates can save a lot of money at the time of installation.

Some builders prefer to have radiant floor heating installed between floor joists to eliminate the need for additional underlayment or thin slabs. The amount of tubing used for a joist system is more than what would be required if underlayment were used, but when the cost of underlayment and installation is compared to the cost of extra tubing, it's easy to see that a joist installation is usually much less expensive. The plastic tubing used for radiant floor heating is not very expensive. Components used with the tubing can be pricy, but the tubing itself is quite affordable. Underlayment, on the other hand, is outrageously expensive, and the labor cost required to install it can be substantial.

Worth Looking Into

Here's a tip that's worth looking into when you are heating a small area. Check with your local building codes to see if a water heater can be used as a heat source. That's right, I said a water heater, just like the type used for plumbing systems. Water heaters cost a fraction of what boilers do, and where acceptable, water heaters can produce enough hot water to service a small heating system. Another advantage offered by a water heater is that a tempering device is not needed to obtain and maintain the proper water temperature. Since water heaters are equipped with thermostats, it's an easy matter to maintain a stable water temperature. However, water heaters are limited in their capacity and are not suitable for large jobs. Make sure you verify that code permits the use of a water heater in your area before you offer it as an option.

If water heaters are acceptable heat sources in your region, you can choose from a variety of fuel types. Electric water heaters are quite common but can be expensive to operate. A gas-fired water heater can be a good option, with either natural gas or bottled gas. If you are heating a space that doesn't have suitable electrical service for a water heater, a gas-fired unit might be ideal. Oil-fired water heaters are also available. Matching a water heater to a heating job should not be done in haste. Choose a water heater that has an adequate capacity and recovery rate as well as a good efficiency rating. The venting of gas-fired water heaters is usually a simple matter, and the vent can often be constructed of plastic pipe rather than a metal or masonry chimney. Check all local codes and comply with them whenever you are performing an installation.

Software

If you are going to figure many heating jobs on your own, you should seriously consider buying computer software to do the work for you. One software package that I can recommend is the Wirsbo Radiant Express package. Computer software has its limitations, but it can do calculations for you and answer a variety of questions that will save you time and trouble. In order to compute heating loads and figures, you need some base points to work from. This is true with or without

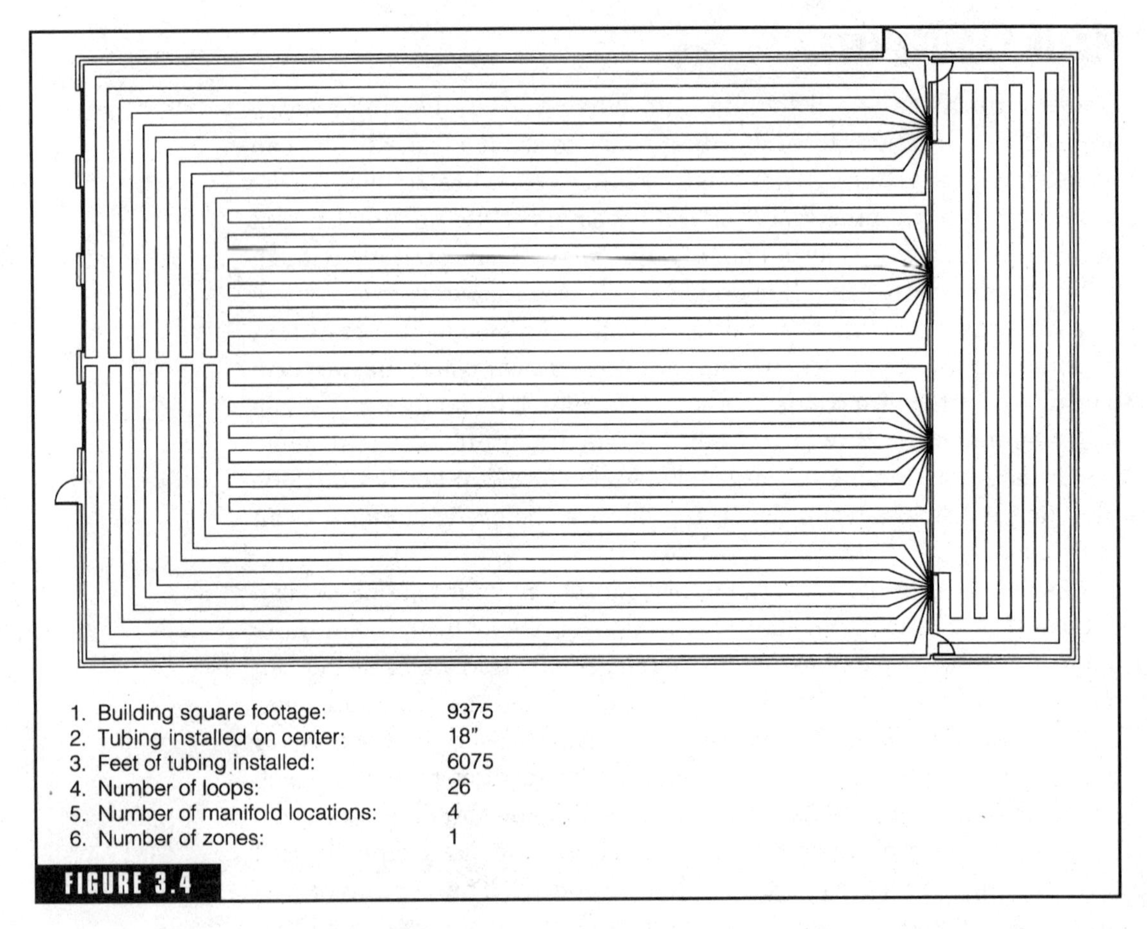

FIGURE 3.4

Installation Diagram. *(Courtesy of Wirsbo.)*

computer software. Therefore, I will provide you with some of the key information needed for such work before we leave this chapter.

Tubing Routes

Tubing routes are usually pretty simple to establish (Fig. 3.4). But there are times when getting tubing from one place to another is a challenge. For example, you might be running your heat tubing and doing just fine, until you come to a steel beam that you were not expecting. Maybe you will be installing your heat tubing for a slab and run into a conflict with the plumber or electrician. Because of the diminutive

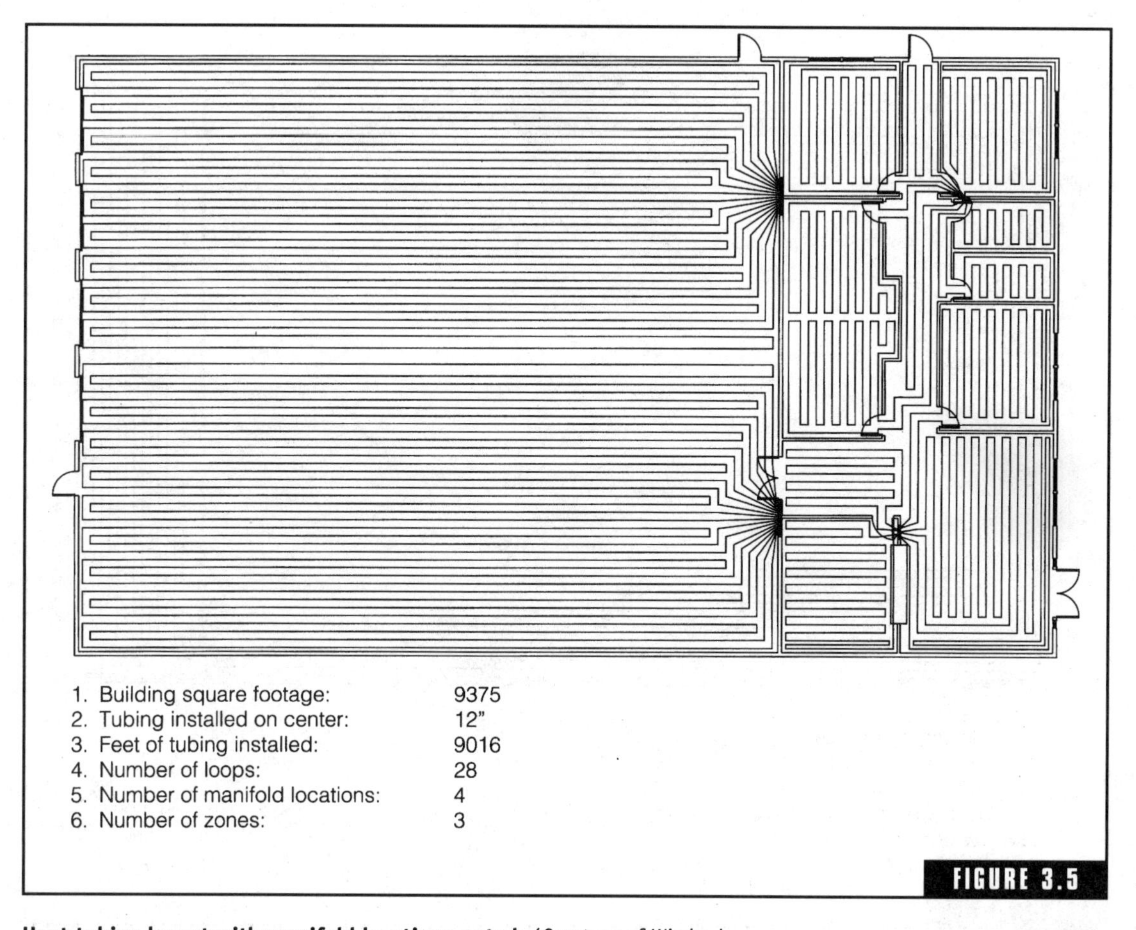

Heat tubing layout with manifold locations noted. *(Courtesy of Wirsbo.)*

size of heat tubing, it can be routed in almost any place and in almost any way. The key is planning your route in advance. Proper planning before an installation not only makes the job faster, easier, and more enjoyable to do, it can make it much more profitable (Fig. 3.5). Sit down with a set of blueprints, if they are available, and draw in your tubing diagrams. Find out what roadblocks you might hit, such as stairs, support posts, beams, and foundation walls.

Not all jobs have blueprints drawn for them. Sometimes contractors have to work with what they can see and hope for the best. This is usually the case in retrofit and remodeling jobs. Experienced remodelers have done enough jobs to have a very good idea of what to expect, but

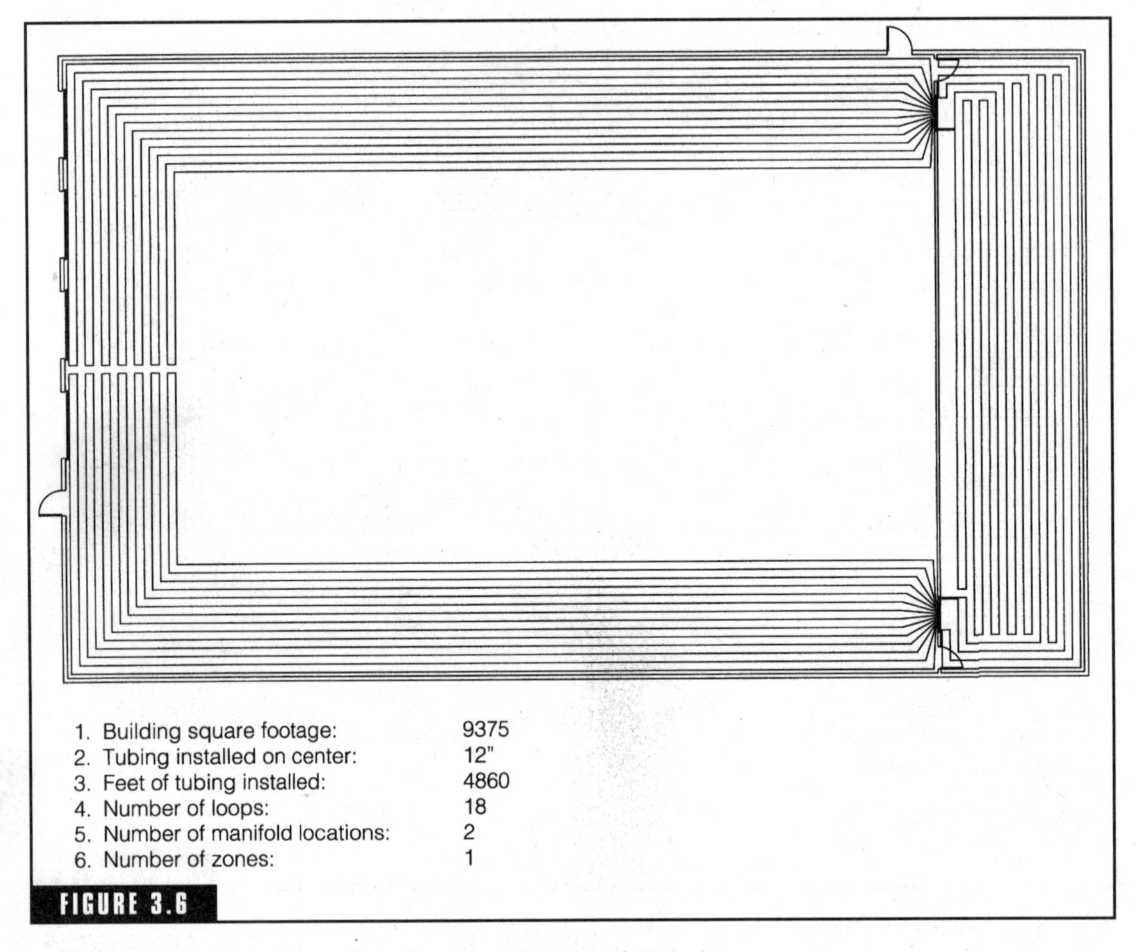

FIGURE 3.6

Perimeter heating system for radiant floor heat. *(Courtesy of Wirsbo.)*

even they can't be sure of what they will find until they run into it. This does not mean, however, that planning a heat design for a remodeling job is not practical. It absolutely is, and you should dedicate time to plotting a course before you begin an installation (Fig. 3.6).

In the case of new construction, blueprints are usually available. The prints for residential construction may not have a heating plan. If this is the case, you or one of your contacts must create one. Once you know what you'd like to do, you have to find out if it can be done. This might require talking to the builder, the carpenters, the plumbers, and so forth. Making a heating system work in new construction can be much easier than making one work in a remodeling job.

If you have to figure out a way to get heat tubing into a remodeling job, you will have to do a site inspection. When you do this inspection, go prepared and be willing to spend an hour or two at the job site. You should take basic inspection tools and either a tape recorder or a notepad and pen. Chances are good that you will need to keep many notes pertaining to the property. Not all remodeling jobs are difficult to fit with radiant floor heating. Sometimes you can gain good access under the flooring. Then again, you can run into old houses where there is virtually no room to work beneath the floors. In a case like this, you might install heat tubing on top of the existing floors and have a sleeper system installed with new underlayment on it. There is almost always some way to get around obstacles when you are working with plastic heat tubing.

Creating an effective heating system with plastic tubing is not usually difficult to do. If you have a good plan, the installation should be a breeze. When you do hit a snag, the flexibility of the tubing should make compensating for the unexpected problem fairly easy. Get a good heat design to work with and then look carefully at the routing of all tubing. If you arrive on a job with a well-researched plan, you shouldn't have any serious setbacks.

CHAPTER

4

Combining In-Floor and Baseboard Systems

Combining radiant floor systems with baseboard heating systems is not as difficult as some people think it is. In fact, it's fairly simple and can be quite effective. Baseboard heating units are very effective in their own right. Areas where extreme climates make winters brutally cold are ideal places to find baseboard heating units. In a state like Maine, baseboard heat is the number one choice among homeowners. Some houses are equipped with forced-air heat and a few have electric baseboard heat, but a majority of homes have hot-water baseboard heat and oil-fired boilers. Recently, radiant floor heating has become increasingly popular in Maine. Houses that are already equipped with hot-water baseboard heat are ideal candidates for radiant floor heating. The same heat source can be used for both types of heat, but the supply water for the radiant system must be cooler than the supply for the baseboard units. This is easy enough to accomplish, and we will explore the process a little later in this chapter. A large number of people don't realize that radiant floor heat can be used in combination with hot-water baseboard heat. It absolutely can.

It's unlikely that a heating contractor would design a combined system for new construction. The most common use for a combined system is in a retrofit or remodeling situation. For example, if a remodeler is adding an addition to a home, it could make a lot of sense to install radiant floor heating in the addition. If the home already has a boiler, the process is fairly simple and quite cost-effective. There could be some type of circumstances that would suggest using a combined system in new construction, but they would be few and far between. Retrofitting and remodeling are the ideal times to consider combining baseboard with radiant floor heating.

Why Add Heat?

Why add heat to a house with an existing heating system? There can be many reasons. The existing heating system may not be very efficient. By adding new radiant floor heating, the existing system can fall into a backup role. Sometimes houses are not fitted with enough heat when they are built. This happens more often than it should. If a heating system is inadequate, adding radiant floor heating to selected areas can make a big difference. Some homeowners want radiant floor heat added to warm floors, specifically in bathrooms. Comfort is always a good reason to add new heat. It's not uncommon for people to decide after building a home to heat parts of the property that were not heated during construction. Garages are a good example of this type of situation. In a retrofit situation with a garage, radiant heat in the floor might not be sensible. A good space heater hung in the garage should cost less and be easier to install. Radiant floor heating in a garage is fine when the garage floor has not yet been poured, but a retrofit is tough. It can be done, but it's rarely sensible. A more likely space could be a porch that has been converted from seasonal use to year-round use. This might be an ideal situation for installing radiant floor heat.

People come up with all types of reasons for adding heat to both homes and commercial spaces. Having the skills to offer your customers a combined system can make your business more profitable or your employer happier with your job performance. Mixing hot-water baseboard heat with radiant floor heat is not difficult. Anyone with general heat installation skills can combine the two types of systems. Some special modifications are needed, however, to use a boiler that is designed to work with baseboard heat when you want to use the same boiler to supply water to radiant heat. The major part of the process is getting cooler water to the radiant heat without hurting the heat source by returning water that is too cold to it.

The Temperature Problem

Factors that influence the temperature of supply water include:

- **The spacing of heat tubing**
- **The method of installation for a system**
- **The type of finished floor material being used**
- **The heat load for the building being heated**

The temperature problem of using a single heat source for both hot-water baseboard heat and radiant floor heat is one that many people don't understand fully. It's true that the temperature of water supply to the two types of systems must be considerably different. This does not mean that a single heat source can't serve both systems. Confusion on this issue sometimes influences people to look for alternative heating options. Minor modification of the existing heat source for hot-water baseboard heat is all that's needed to make it produce water of a suitable temperature for a radiant system.

Water temperature for water being run through hot-water baseboard heat must be much hotter than the water needed for a radiant floor heating system. Copper tubing is the most common type of distribution piping used with baseboard systems. The fin-type heating elements collect heat from the water passing through the copper tubing and then radiate the heat to living space. Radiant floor systems don't use fins. Plastic tubing is the most common material used to transport water in a radiant floor system. These differences require different water temperatures. If a boiler is being used for hot-water baseboard heat, the temperature of the water coming out of the boiler and going to the baseboard units must be high. But this same high-temperature water is too hot for a radiant system. So what can be done? Well, let's see.

Radiant floor heating is considered a low-temperature heat, in terms of the supply water being used. A standard water heater is all that is needed to produce adequate hot water for a small radiant system. The exact temperature of the water supplying a radiant system can depend on many factors, but in all cases, the water temperature should be much lower than the water used to run a hot-water baseboard heating system. Factors that influence the temperature of supply water include:

- The spacing of heat tubing
- The method of installation for a system
- The type of finished floor material being used
- The heat load for the building being heated

There are three standard categories for radiant floor systems. Type 1 systems don't require additional temperature control. Single-temperature tempering is required when a type 2 system is used. For a type 3 system, a weather-responsive reset control is needed. A type 1 system might be one where a standard plumbing water heater is used as the heat source. Since the water heater is thermostatically controlled, that is the only heat control needed. For example, the thermostat for the water heater could be set at 110° and left to function independently. This, of course, doesn't apply when a boiler for a hot-water baseboard system is being used as a heat source.

A condensing boiler (Fig. 4.1) is another type of heat source that is used with type 1 systems. These boilers are designed to operate properly with extremely low return water temperatures. This is not the case with boilers used for baseboard heating systems. Condensing boilers use low-temperature return water to condense flue gases. The heat gained from the flue gases helps to heat supply water. This is quite different from the operation of a noncondensing boiler, such as the type used with baseboard heating systems.

Even if an entire heating system consists of radiant floor heating, a noncondensing (Fig. 4.2) boiler should not be used as a heat source. These boilers are meant to work with higher water temperatures. If the return water to a noncondensing boiler is colder than 135°, the flue gases within the boiler can condense. If this happens, the condensation is highly acidic and can damage the flue or the boiler.

Installation of a type 2 system requires the use of a tempering valve. This valve must be installed between the boiler and the radiant heating system. When a tempering valve is installed, it protects the boiler from low-temperature return water and provides water as a supply for the radiant system at a proper temperature. A three-way tempering valve is the simple and effective way to achieve type 2 control. This type of valve gives a constant, fixed water temperature for radiant floor heating, without affecting boiler operation.

MATERIALS

A condensing boiler (Fig. 4.1) is another type of heat source that is used with type 1 systems. These boilers are designed to operate properly with extremely low return water temperatures.

How does a three-way tempering valve function? Goods ones have valves inside that contain elements that expand and contract to control water temperature. One port of the three-way valve supplies the hot-water baseboard zones. A second port supplies radiant zones. The third port is a bypass that allows hot water to return to the boiler. A good valve can have a dial on top of it that allows water temperature to be set to a desired level. Expansion and contraction of the interior element opens or closes a shuttle valve, as needed, to maintain a steady temperature. A good tempering valve is a reactive valve. This means that the valve will maintain a constant water temperature even if there is a drop in boiler supply water. A

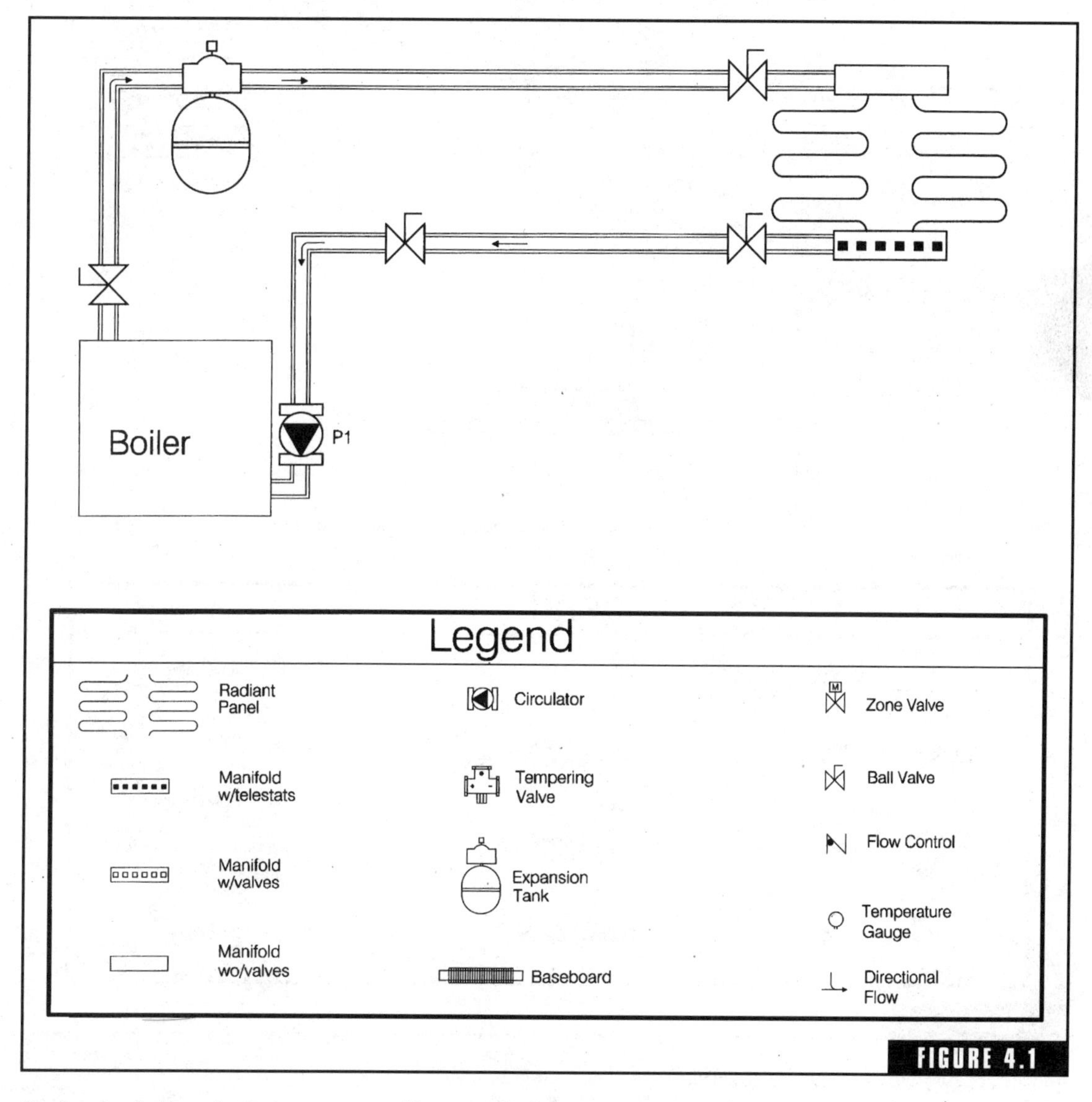

FIGURE 4.1

Condensing boiler, single temperature. (*Courtesy of Wirsbo.*)

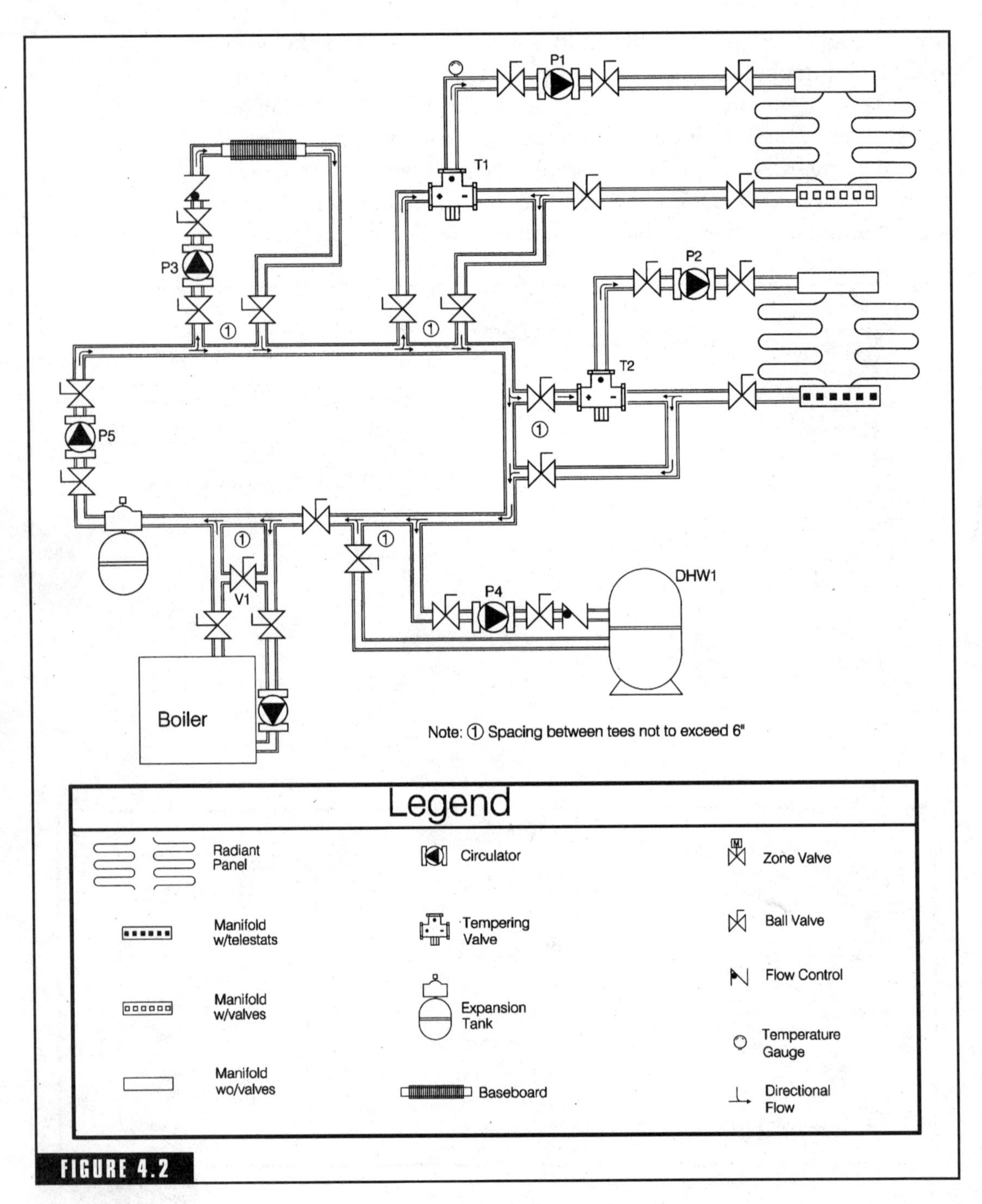

FIGURE 4.2

Noncondensing boiler. (*Courtesy of Wirsbo.*)

reactive valve of this type can be used with intermittent, zone, or on-off controls.

A three-way tempering valve offers many advantages in a type 2 situation. The valve doesn't cost a lot, requires no electrical wiring, offers reactive service, is easy to install, and can provide a wide range of operating temperatures. A three-way valve can be installed in any position, but a circulator must be installed on the radiant side of the valve to ensure proper flow to the radiant system. The circulator should be installed between the MIX port of the tempering valve and the supply manifold for the radiant system. A temperature gauge should be installed downstream of the MIX port so that the supply water temperature can be monitored.

Of all the options available for making an noncondensing boiler work with both a hot-water baseboard system and a radiant system, a three-way tempering valve is the least expensive and the simplest to install. There are, however, other options for a type 2 system. Out of fairness, let's spend a few minutes looking at the more complex methods of making a suitable conversion.

Mixing Tanks

Mixing tanks (Fig. 4.3) can be used to create type 2 heating systems. Some people refer to these tanks as side arms. Water from a boiler that is serving hot-water baseboard heat is sent into the mixing tank. Return water from a radiant heat system is also channeled into the tank. The combination of cool water and hot water creates warm water for the radiant supply water. A boiler loop circulator and a radiant panel loop circulator are both required when a mixing tank is used. An aquastat is also needed. It can be either a strap-on unit or an immersion unit. The aquastat senses supply water temperature for the radiant system. Any insulated water tank can be used as a mixing tank.

The advantages of a mixing tank are that they are not extremely expensive, provide energy storage, can be installed with simple piping, and provide water mass to reduce potential boiler short cycling. Mixing tanks are especially well suited to

QUICK»TIP **The advantages of a mixing tank are that they are not extremely expensive, provide energy storage, can be installed with simple piping, and provide water mass to reduce potential boiler short cycling.**

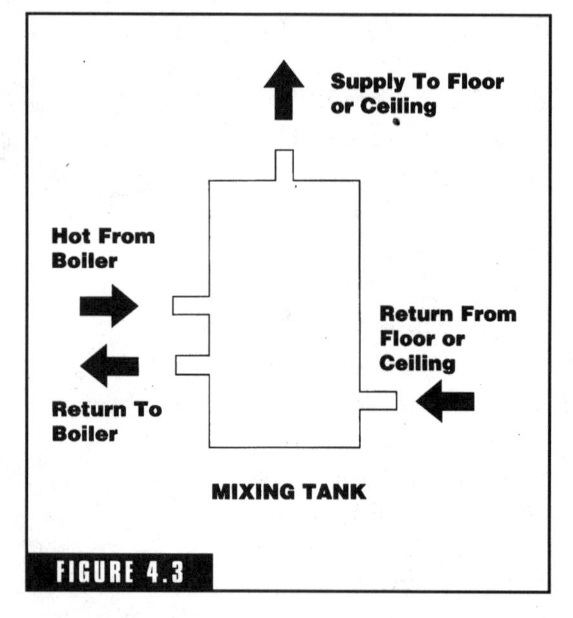

FIGURE 4.3

Mixing tank. (*Courtesy of Wirsbo.*)

wood-fired boilers, but they can be used with a variety of heat plants.

Heat Exchangers

Heat exchangers (Fig. 4.4) can be used to create type 2 systems, but they are more commonly used in connection with systems designed to melt snow. One side of a heat exchanger contains boiler water while the other side of the unit contains water for a radiant heating system. When boiler water is pumped through a heat exchanger, the water warms the walls of the exchanger. Radiant water is pumped through the other side of the heat exchanger and is warmed as it comes into the exchanger. Since both types of water are in separate containers, the water types never mix. A circulator and expansion tank are both required on the radiant side of a heat exchanger.

Heat exchangers are used to deal with the issue of oxygen diffusion corrosion when nonbarrier tubing is used. Nonferrous components must be used on the radiant side of the system. Bronze or stainless steel circulators with nonferrous flanges should be used. Any standard expansion tank for potable water is suitable for use with a heat exchanger. A brass or bronze air separator and all nonferrous hard piping is required on the radiant side of the system. The boiler side of the heat exchanger can be piped with any type of approved piping. Since the heat exchanger doesn't allow boiler water to mix with radiant water, there is no risk of oxygen diffusion corrosion in this type of setup.

Heat exchangers are universally acceptable, keep different types of water from mixing, protect the heating plant from cold return water, and even allow the isolation of systems using high-glycol mixes. The expense of heat exchangers is a disadvantage. But, when used for snow melting or applications where domestic water heaters are part of the system, heat exchangers can be an ideal selection.

Motorized Mixing Valves

Motorized mixing valves perform the same basic service as a standard three-way mixing valve, but they use motors to get their job done. A four-way motorized mixing valve (Fig. 4.5) responds to control input from electronic sensors to maintain fixed water temperature. A sensor checks radiant supply water temperature. If the temperature falls below the desired temperature level, a control fires a circulator on

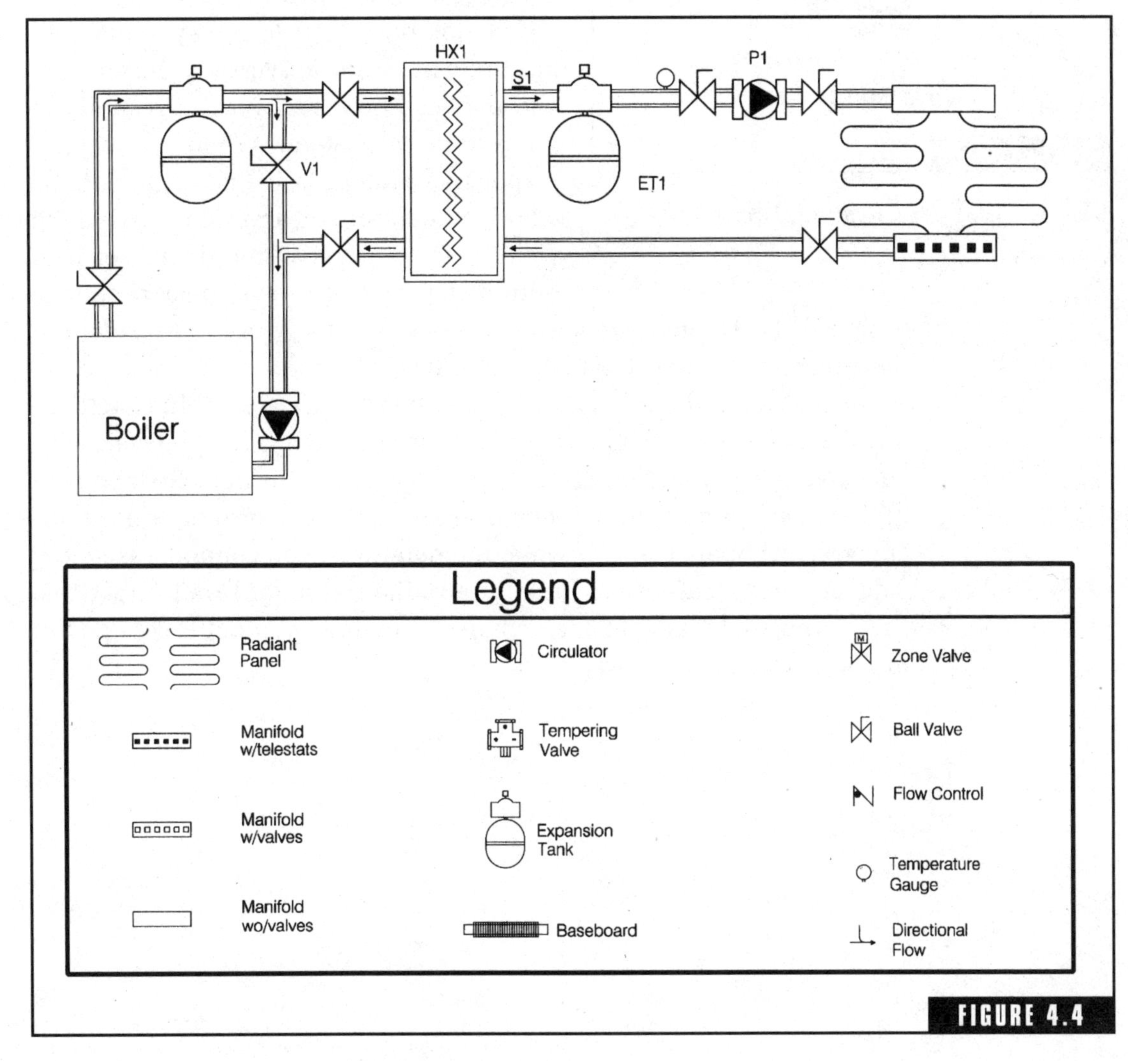

Noncondensing boiler with heat exchanger. (*Courtesy of Wirsbo.*)

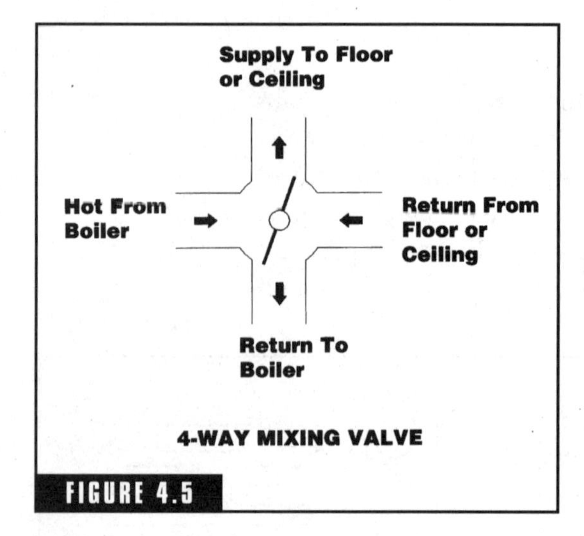

Four-way mixing valve. (*Courtesy of Wirsbo.*)

the boiler side of the valve and triggers the mixing valve to adjust the temperature of the water. This, of course, is done by mixing hot water with colder water to create warm water.

Four-way mixing valves are dependent on their electrical functions to be reactive. The additional cost of a four-way valve is a consideration when a three-way valve can get the job done. But four-way, motorized mixing valves are universally acceptable, and they can be made weather-responsive for a type 3 system. When a weather-responsive system is in use, it maintains water temperature on a routine basis. If the weather is cold, the system increases heat output. On warmer days, heat output is reduced. These high-end systems are not normally used in residential heating applications. Cost is a major factor for this.

As you should now see, it's not really difficult to combine a radiant heating system with a hot-water baseboard system when you start with an existing hot-water baseboard system. The same is not true if you are trying to add hot-water baseboard heat to a radiant system. Since radiant systems operate at low water temperatures, you cannot reasonably add hot-water baseboard units to a radiant system. Don't forget this fact if you are looking at a job where the main heat plant is set up for a radiant system.

CHAPTER

5

Establishing Heating Zones

Establishing heating zones can be very simple. In fact, a single heating zone is all that is needed. There is nothing that says a building must be equipped with multiple zones. But, for comfort and efficiency, multiple zones are desirable. Once it is determined that multiple heating zones are wanted, the planning for the zones becomes important. Each zone adds expense to a heating system during the installation process. But the zones may save enough money during the operation of a heating system to more than offset the cost of the zone.

Typical homes that use zoned heating systems have two or three zones. A two-story home might have one zone for the upstairs living area and another zone for the downstairs area. This same house might use three zones, so that each level has its own zone and then the bedrooms are on a zone of their own. The combination of zones is limited only by one's desire and financial ability. Business and commercial properties may have many zones to accommodate different types of usage or individual office areas.

In terms of installation, setting up different zones is not complicated, but it can get expensive. Supply and return piping must be run for each zone. The piping cannot be tied into piping for other zones. Wires must be run to thermostats for each zone. Multiple circulators or zone valves are needed in a system with more than one zone (Fig. 5.1). The labor and material costs to create a number of independent zones can be intimidating. So, why do it? Because a building that is zoned properly is much more efficient and costs less to heat. The additional money spent during an installation should be recovered in cost savings during operation.

A lot of contractors don't give much thought to creative use of zones. These contractors tend to do things the way they have always been done or in a way that is most consistent with common everyday practices. This is not a good attitude. Every building or home can have individual needs that should be assessed when planning for heating zones. Contractors who want happy customers should take the time to talk to their customers about all available options in regard to heating zones. Most consumers are not that well informed on the use of zones, so they may not know to ask for specific zones. More often than not, consumers will leave design suggestions up to their contractors. This is fine, as long as the contractors are willing to share their knowledge and experience with their customers before installations are made.

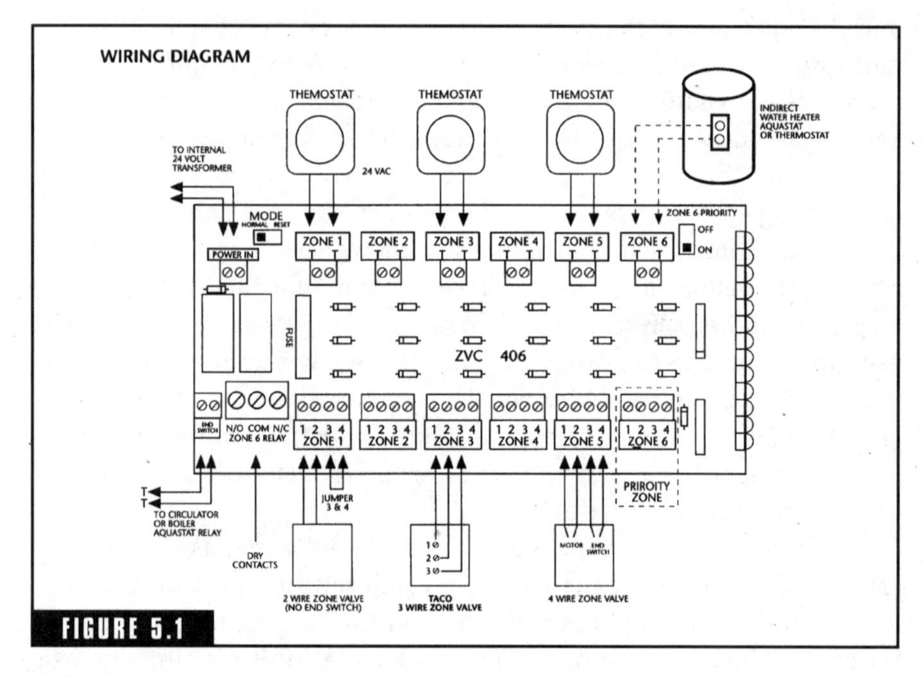

FIGURE 5.1

Wiring diagram for zone valves. (*Courtesy of Taco.*)

Let's say that you are going to install a heating system in a house that contains 3000 sq ft of living area (Fig. 5.2). The house has four bedrooms, three bathrooms, two levels of living space, a sunroom, and other normal rooms. Wouldn't three heating zones be enough for this house? It certainly could be, but it may not be the optimum way to lay out the heating system. Are all of the bedrooms on the same level of living space? If not, this could be reason for a fourth heating zone. Will anyone sleeping in the bedrooms be extremely young or old? People who are at one extreme or the other of age might need warmer sleeping conditions. This could be a reason for yet more zones. The sunroom probably will not need heat all of the time, and it may not need as much heat as other rooms in the house. Here's a situation for still another zone. Are you starting to see how the desire for additional zones can grow quickly? Well, let's look at some specific conditions that might warrant added heating zones in a home. (Fig. 5.3).

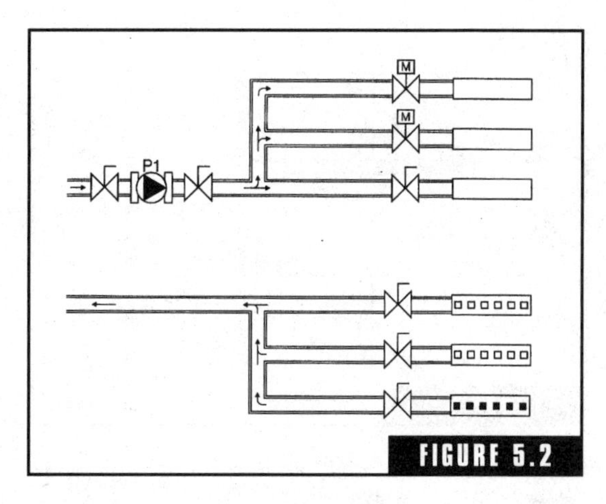

Multiple zones on multiple manifolds with telestats and zone valves. (*Courtesy of Wirsbo.*)

Primary Living Space

Primary living space as it is discussed here refers to various levels of living space in a home. For example, a two-story home with a full basement would have three primary living spaces in the context of our conversation. Each level of living space should have its own heating zone (Fig. 5.4). The reasons for this are pretty obvious. If residents spend a majority of their time on the main level of the home, it's not necessary to heat the upstairs area and the basement as often as it is to heat the main level. Having the heat turned down in the basement is not going to make the majority of the living

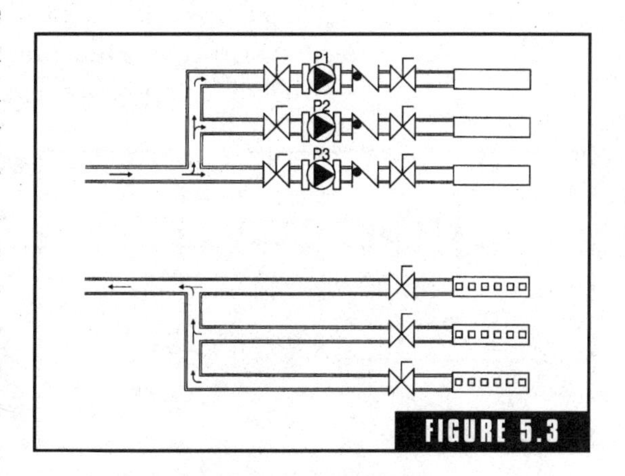

Single zones on multiple manifolds with circulators. (*Courtesy of Wirsbo.*)

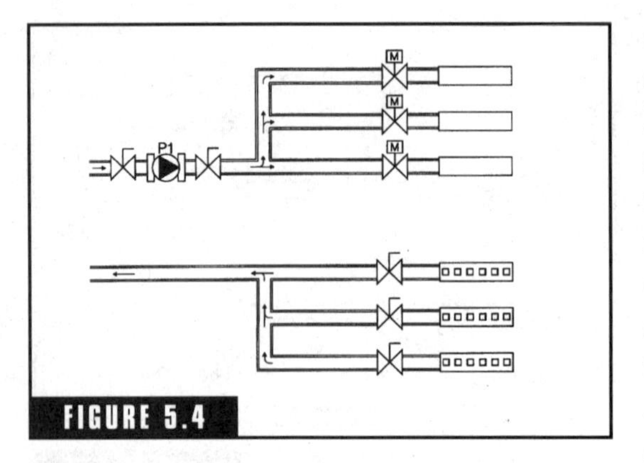

FIGURE 5.4

Multiple zones on multiple manifolds with zone valves. (*Courtesy of Wirsbo.*)

space uncomfortable. Assuming that bedrooms are located on the upper level of the home, the heat upstairs can be turned up shortly before bedtime. This means that most of the heating time is used to heat the main level. If all three levels were heated equally, the cost of operating the heating system would be much higher. There is no reason to waste money or natural resources, so the installation of different heating zones makes a lot of sense (Fig. 5.5).

In my opinion, each level of living space in a home should have its own heating zone. If a basement is used frequently, such as a play or work area, the heating zone can be turned up as needed and left at a cool setting when the basement is not being used. Most contractors do place different levels of living space on separate heating zones. It's basically standard procedure. This is the easy part, the rest of the planning that goes into zoning a house for heat can be more complicated to predict.

Bedrooms

Bedrooms are often placed on one zone of heat that is separate from main living areas. This is because bedrooms are generally used only at night, so there is no need to keep them very warm until they are needed. However, the practice of putting all bedrooms on the same heating zone can be a mistake. Don't get me wrong, bedrooms should be zoned differently from main living areas, but maybe independent zones should be installed for each bedroom.

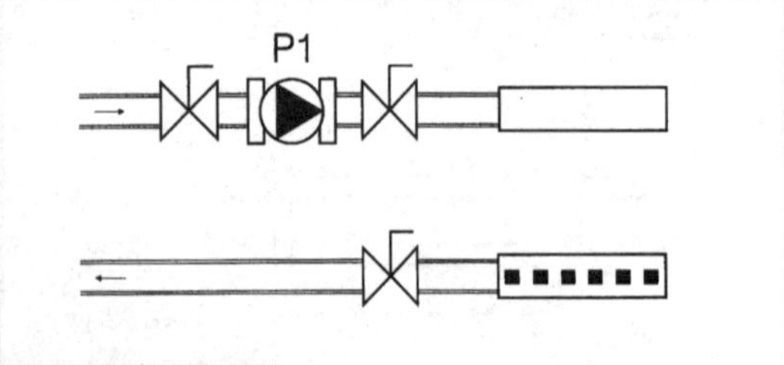

Adding heating zones to a system increases the cost of installation. This is why many builders and heating contractors limit the number of zones in a house. But if a customer is after total comfort and satisfaction for everyone who will be using a home, the decision to cut corners on zones could be a poor one. Having every

FIGURE 5.5

Multiple zones on single manifold with telestats. (*Courtesy of Wirsbo.*)

bedroom on an independent zone could be considered extravagant, and maybe it is. But there are certainly times when circumstances warrant putting at least some of the bedrooms in a home on separate zones (Fig. 5.6).

QUICK»TIP **Bedrooms are often placed on one zone of heat that is separate from main living areas. This is because bedrooms are generally used only at night, so there is no need to keep them very warm until they are needed.**

If a family has two adults, two teenagers, and a young child, that family might want at least one of the bedrooms on an independent zone. The young child is more likely to be uncomfortably cold than the older children and adults if all of the bedrooms are kept at the same temperature. Putting a separate thermostat in the young child's room, with an independent zone, gives the parents a lot of control when it comes to keeping the child comfortable.

Elderly people often get cold more quickly than younger people. If a person is building a home and is providing a bedroom for a parent or in-law who has advanced in years, it could be very sensible to provide a separate heating zone of the older person's bedroom. The issue of elderly residents and very young inhabitants is not often thought of. Common practice puts all bedrooms on one zone, so if the selected temperature is 70°, all of the bedrooms are maintained at that temperature. Personally, I prefer a cooler bedroom temperature, but others like their rooms very warm. This is not just a matter of age; it is also a matter of personal preference.

Having bedrooms on independent zones makes them easy to control. But setting up a zone for every bedroom in a large house can be quite expensive. It might be reasonable to set up one zone for heating rooms occupied by parents and older children while other zones are established for young children and older adults. Of course, to do this a plan has to be in place before the heating system is installed. If you are heating a home for a customer who has an infant, you should mention the possibility of putting the baby's room on a separate zone. The same advice applies if your customer is building a home with in-law quarters in it.

Bathrooms

It's not uncommon to find bathrooms with little or no heat in them. Some bathrooms have wall-mounted heating units, others have radiant floor heating systems, and many have no heat. When a bathroom is

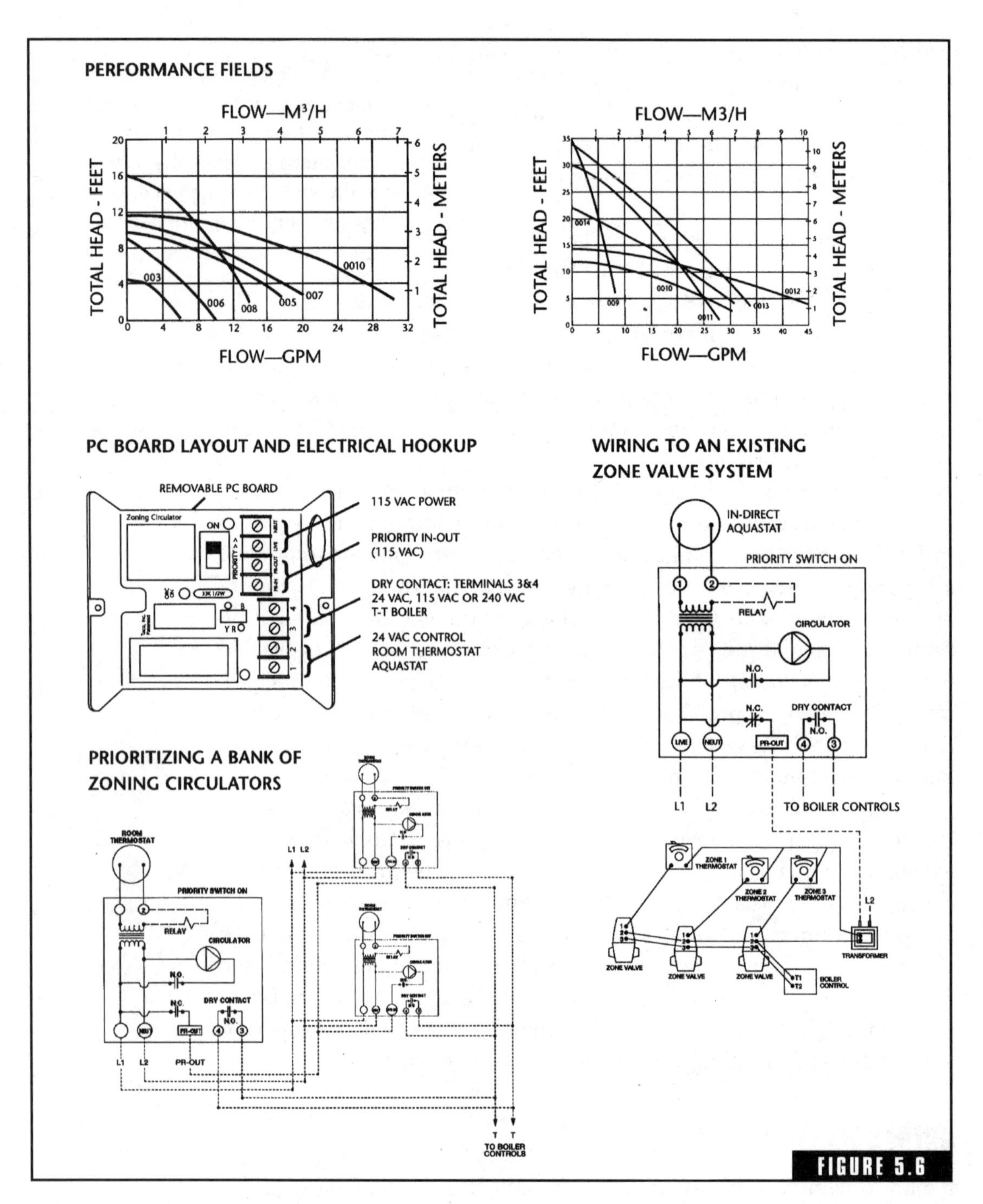

FIGURE 5.6

Priority zoning. (*Courtesy of Wirsbo.*)

located within the heated confines of a home, the bathroom may not need any heat. Yet people take showers, step out to dry off, and get cold. Having heat in a bathroom or at least having it available is a good idea. Radiant floor systems are particularly nice in bathrooms, since people using the bathroom are likely to be doing so without shoes or slippers. Since most bathroom floors are not carpeted, the floors can be cold to the touch. Radiant floor heating eliminates this problem.

Normal procedure has bathrooms on the same heating zone as the rooms around the bathroom. This is usually okay, but it can be nice to have the bathroom on an independent zone. Bathrooms don't usually require much heat unless people will be bathing. If a bathroom has its own zone, the heat can be kept turned down until it is needed. This helps to reduce the operating cost of the heating system.

Specialty Rooms

Some houses have specialty rooms, such as a billiard room, a woodworking room, a sewing room, or whatever. This type of living space can benefit from a separate heating zone, since the room is not used regularly. Any part of a house that is not used routinely is a candidate for a separate heating zone. For example, a kids' playroom will not need much heat during the day if the children are in school. A sunroom that collects a lot of solar heat during the day will not need much heat unless it is being used when there is little to no sunlight to warm the room. An exercise room would normally be kept cooler than standard living space. Any situation like this could be justification for adding another heating zone to a system (Fig. 5.7).

Garages

In cold climates it's not unreasonable to expect some garages to be heated. Since most garages have concrete floors, they are an excellent place to install radiant floor heating. While a house temperature might be kept near 70°, a garage can do very nicely at a much lower temperature. Why heat a garage to 70° if 50° is suitable? Following this line of thought, a separate heating zone should be used for a heated garage.

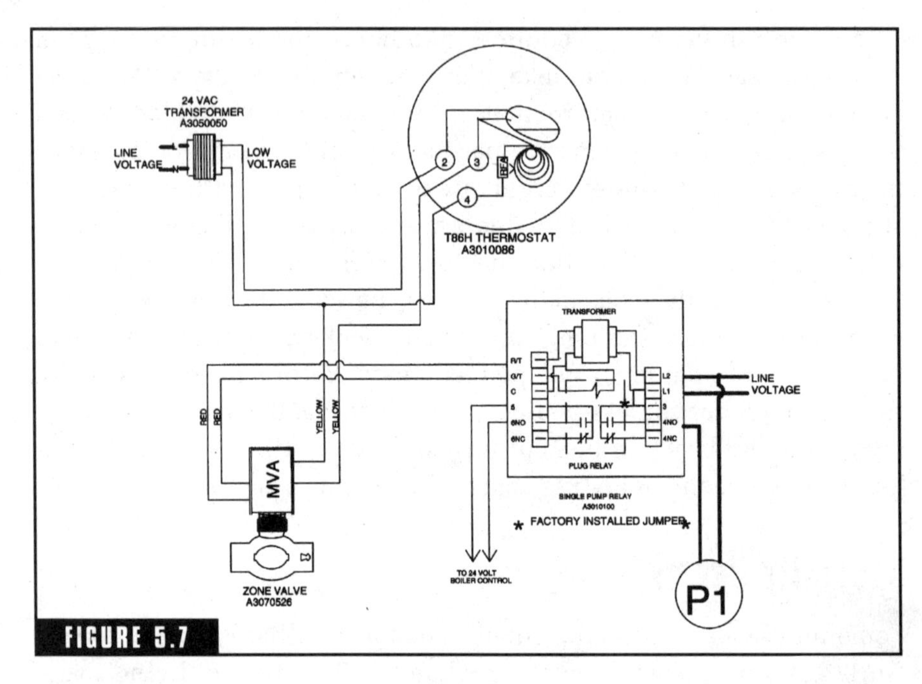

FIGURE 5.7

Wiring detail for a zone valve. (*Courtesy of Wirsbo.*)

Kitchens

Kitchens are usually placed on the same heating zone that is used to heat living space around a kitchen. In most cases, this is perfectly acceptable. However, kitchens can build up a lot of heat if extended cooking is being done. For this reason, some cooks might like a cooler room temperature when they begin to cook. If a kitchen is already at a comfortable temperature and then heats up as a result of cooking, the cook could become overheated and uncomfortable during the cooking process. It's rare to find a kitchen with its own heating zone, but there is no reason why a kitchen couldn't be placed on an isolated heating zone.

General Living Space

General living space and common areas, such as halls, will normally be on the same zone if the space is on the same level of a home. An exception might be made in a large home, where the living space is separated by a great deal of distance. In an average home, a dining room, living room, family room, and hall would all be on the same heating zone. The kitchen and even the bathrooms might also be on this one zone. If a room can't be closed off from other living space, it makes little sense to provide the room with a private zone. For example, an open dining area should not be zoned individually. But a dining room with French doors that can be closed could be a good candidate for a separate zone, since the dining area is used only for short periods of time.

Outdoor Heating

Outdoor heating with radiant heating systems is not rare. Many homes are built with radiant heat installed to melt ice from roofs, walkways, and parking areas. Obviously, this type of heating installation should be separated from the controls used for interior living space. Normally, the outdoor system will be equipped with a weather-responsive control to maximize energy efficiency on the overall heating system. But a simple on-off switch can be used for the outside system, depending on a customer's desire (Fig. 5.8).

Common Sense

Common sense is the best tool to use when laying out heating zones. But a thoughtful planning process should be incorporated into every heating system before it is installed. You've just seen examples of different situations where additional zones could be used. The person who is going to live in a house should have a say in what rooms have independent zones. Contractors should sit down with customers and plan heating zones from top to bottom before any work is started. Once a heating system is installed, it's not cost-effective to break the system down into additional independent zones. The time to decide on zone locations is before any work has begun. Only the people who will be living with a heating system know for sure how they would like the dwelling zoned off.

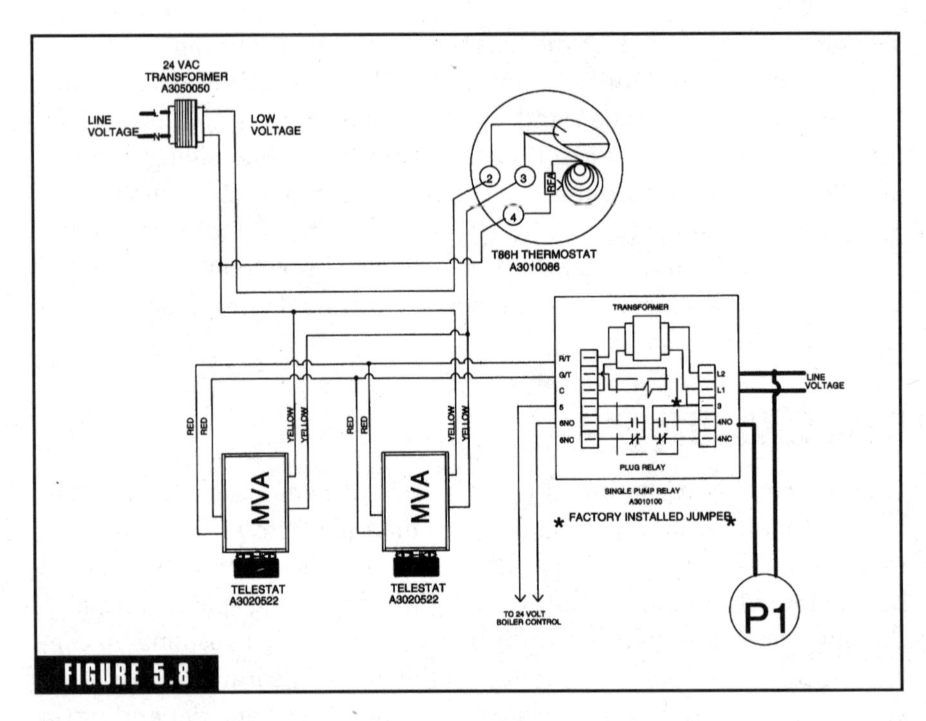

Wiring detail for a zone valve. (*Courtesy of Wirsbo.*)

CHAPTER

6

The Mechanics of Manifolds

Manifolds play an important role in a radiant floor heating system. Generally, a supply manifold and a return manifold are used with every system. In simple terms, a manifold is a device where all supply or return tubing comes together and where a larger pipe as a part of the manifold either supplies the tubing with water or returns water from the tubing to a heat source. Sounds simple enough, and it is pretty simple, but it's also important. The location of a manifold can have a lot to do with the overall cost of installation for a system. Planning the location of a manifold requires the system designer to consider access and general piping locations. Manifolds should be installed in centrally located areas to reduce installation cost for tubing.

The location chosen for a manifold is affected by a few factors. A first consideration should be the piping layout. Place the manifold in a location that will minimize the amount of tubing required to complete a job. Since manifolds have valves on them, the manifolds should be accessible. If it's practical to make the manifolds readily accessible, that's even better. Closing a manifold up in a wall is a bad idea. It's possible that access to a manifold might not be needed for a long time, but reasonable access should always be available.

MATERIALS

A manifold might house only one piece of tubing, but it may have 10 or more tubing lines connected to it (Fig. 6.1).

Some contractors, myself included, build their own manifolds, at least some of the time. Other contractors rely on stock manifolds that are available from material suppliers. There is no great advantage to building your own manifold. It might be a little less expensive to fabricate a manifold than it would be to buy a preassembled unit, but the savings are usually not worth the additional time required to make a manifold. Of course, if you are short on work and long on time, building your own manifold might increase your profit by a little. The main motivation for building a manifold on a job site is control. There can be times when a custom-made manifold simply suits a job better.

Manifolds confuse some contractors. There is no reason why this should be the case. In reality, manifolds are simple in nature and function. You can think of a manifold as an intersection where some water is coming in and some water is going out. Supply water is piped to a manifold so that it can be distributed throughout a heating system. Part of the confusion can be that a single manifold handles both ends of the job. But there are two manifolds, as a minimum, for each heating system. There is a supply manifold and a return manifold. Since some contractors refer to the pair as a manifold station, people sometimes think of a single unit. The two units are often located close together and can even be supported by the same brackets.

A manifold might house only one piece of tubing, but it may have 10 or more tubing lines connected to it (Fig. 6.1). If a contractor requires a manifold that will support more than an average number of connections, the manifold must either be made on the job site or ordered with specifications which detail the exact number of connections. When a contractor has any reason to believe that a heating system may be enlarged later, manifolds with spare connection ports can be installed. The unused connection points can be capped off until they are needed for the expansion of a heating system.

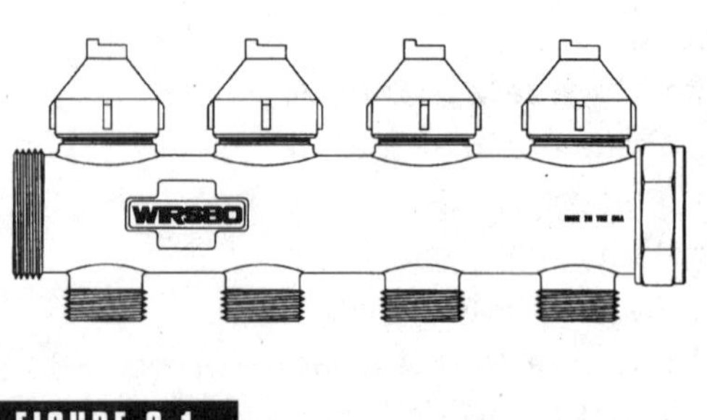

WIRSBO

FIGURE 6.1

Return manifold. (*Courtesy of Wirsbo.*)

Balancing valves are frequently installed with manifold systems. The valves permit adjustment of individual flow amounts in the various sections of heating tubing. When a preassembled manifold is used, the exact settings for each balancing valve might be specified. The purpose for this is to assure the proper flow in terms of design criteria. Additional accessories for a manifold station can include a flowmeter, air-venting devices, thermometers, and purging valves that allow air to be removed from the system, as well as providing a means by which water can be drained from the system.

MATERIALS

Balancing valves are frequently installed with manifold systems. The valves permit adjustment of individual flow amounts in the various sections of heating tubing.

Telestats are installed in conjunction with manifold stations. The telestat makes is possible to control individual zones with a room thermometer. Telestats consist of a small 24-V ac heat motor. The units screws onto a manifold valve. This allows the piping circuit to respond to demands from individual locations that are equipped with independent thermostats. Telestats can be purchased with or without isolated end switches, just as zone valves are. When a room thermostat calls for heat, the telestat slowly opens a valve on the manifold circuit. This process can take up to 5 min to complete. By installing telestats on every connection of a manifold system, you can essentially zone off a building so that every connection creates its own zone. Of course, this procedure can run the cost of a heating system up by a considerable amount. The use of a large number of connections can require the installation of multiple manifold stations.

One reason for using more than one manifold station (Fig. 6.2) can be floor coverings. If a building has carpeting as a floor covering in one part and tile as a floor covering in another part, the water temperature needed to provide comfortable heat will be different. Tubing installed beneath a tile floor will not have to produce as much heat as tubing installed under carpeting. If this type of situation exists, it's desirable to use two manifold stations, one for the carpeted area and one for the tiled area.

The design and installation of manifold systems varies. One variable is the heat source. If the heat source being used can accept low-temperature return water without problems, a direct piping system can be used. A boiler that requires high-temperature return water

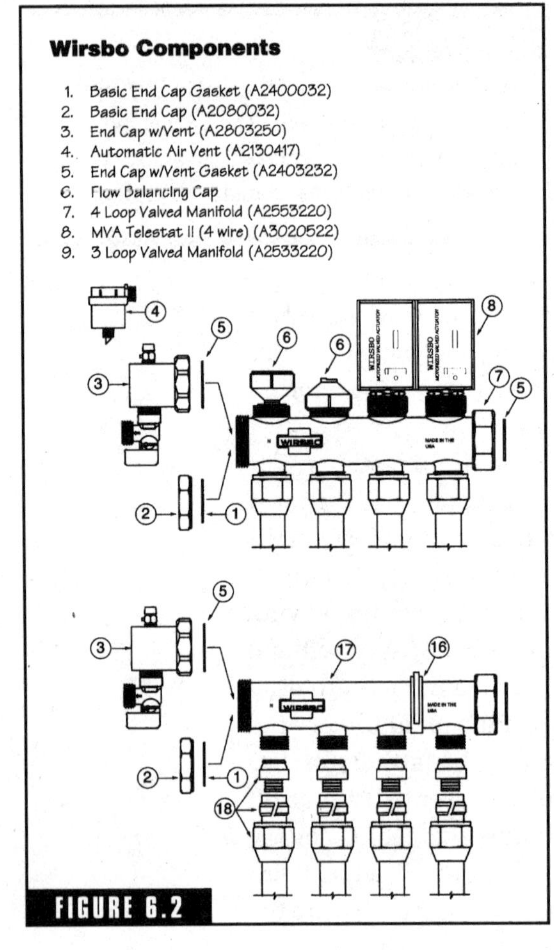

FIGURE 6.2

Manifold components. (*Courtesy of Wirsbo.*)

makes a direct piping system unfeasible. There are several types of heat sources, however, which can be used with direct piping systems. Condensing boilers are the most frequently used heat source for a direct piping system. Hydronic heat pumps and electric boilers can also be used with direct systems. Water heaters and solar collection systems can both be used to heat water in a direct piping system.

A radiant heating system that is using a direct piping installation should have a circulator that runs continuously. The system can be controlled by either an on-off control or a reset control. When an on-off control is used, it is wired in series with a temperature-limiting setpoint control, and this is important. When heat is called for, the heat source is turned on. The heat source will continue to produce heat until the thermostat that called for heat is satisfied. If the temperature of the supply water rises too high, the setpoint control will sense it and shut the heat source down. Once the temperature of the supply water drops to an acceptable level, the setpoint control restarts the heat source.

Setpoint controls protect concrete heat masses from cracking. If a heating system is operating with a heat source that is oversized, the water temperature in the heating tubing can become too hot. A room thermostat cannot sense this heat and therefore will not shut down the heat source. There have been occasions when hot water temperatures in radiant heating systems have cracked concrete. Setpoint controls prevent this from happening.

Electronic reset controls can be used in place of setpoint controls. A reset control maintains a calculated supply water temperature to supply tubing. Heat input to a system is in direct response to prevailing heating loads. When you want to minimize fluctuations in the temperatures of heated space, reset controls work very well. If you are after a direct piping system, use a heat source that will accept low-temperature return water, that has continuous circulation, and that is equipped with either a setpoint control or a reset control.

TRADE SECRET

If you are after a direct piping system, use a heat source that will accept low-temperature return water, that has continuous circulation, and that is equipped with either a setpoint control or a reset control.

Choosing a Location

Choosing a location for a manifold station doesn't have to be a complicated process. It is, however, a decision that should not be taken lightly. One consideration to account for is the accessibility of the manifold station (Fig. 6.3). It's common for the stations to be installed in wall cavities. When this is the case, the location should be accessible. Closets are an ideal place to install a manifold station. It's acceptable to hide a manifold station behind a wall covering, but there should be a removable access panel to make working with the manifolds easy. Since access panels are not always extremely attractive, it's advantageous to place the manifolds in a closet.

Houses that have basements can have the manifolds housed somewhere within the basement. It's common for heat sources to be installed in basements when basements are available. Keeping manifolds in the general proximity of the heat source can be cost-effective. If manifolds are installed in a mechanical room, they should be installed in an exposed manner. This makes the valves easily accessible.

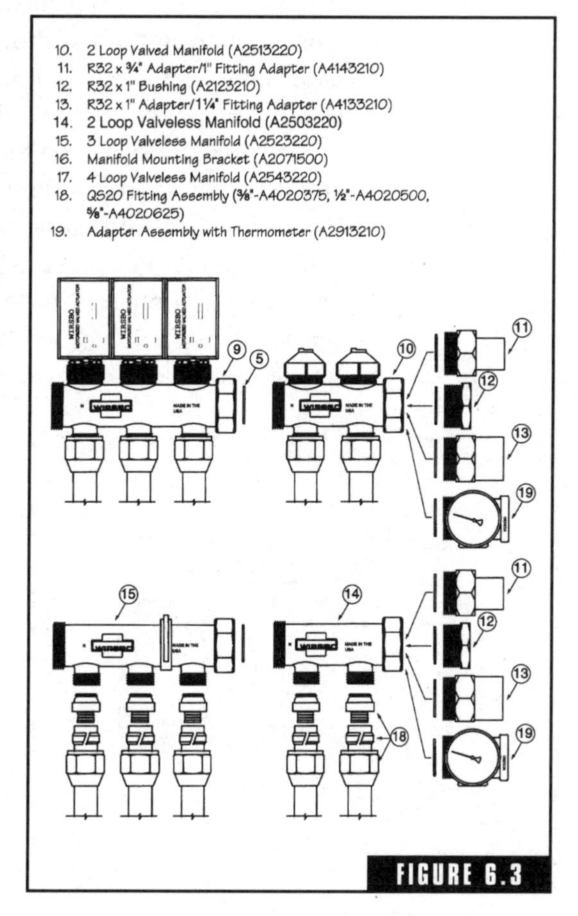

Manifold components. (*Courtesy of Wirsbo.*)

Whenever it's reasonable to make a manifold station readily accessible, you should do so.

Preassembled manifolds (Fig. 6.4) are sometimes sold with custom cabinets to house them. These cabinets are similar to the type used for washing machine hookups. The metal cabinet is installed in a recessed manner in a stud wall. Both the supply and return manifolds are installed within the cabinet. A hinged cover on the front of the cabinet makes access to the manifold station fast and easy. This is a very good way to go when leaving a manifold station completely exposed is not practical.

Closets are not always available in areas where manifold stations should be installed. If a manifold can't be installed in a basement, a mechanical room, or a closet, you have to get a bit more creative. Consider installing the manifold station in a base cabinet in a kitchen or bathroom. Manifold stations require space and access, but a medium-sized base cabinet can accommodate most manifold stations.

Creative construction tactics can go a long way in hiding manifold stations. For example, you might build a window seat in a room to conceal a manifold station. The hinged top on the window seat can provide plenty of access to the manifolds. Plant shelves can be built to spruce up a room and hide a manifold station. A removable panel on the front of the plant shelf or a bookcase can hide a manifold. If a house has a heated garage it may be possible to install a manifold station in the garage. There are plenty of ways to hide a manifold, so don't compromise on a desirable location of manifolds just because there is not an obvious place to hide the heating system components.

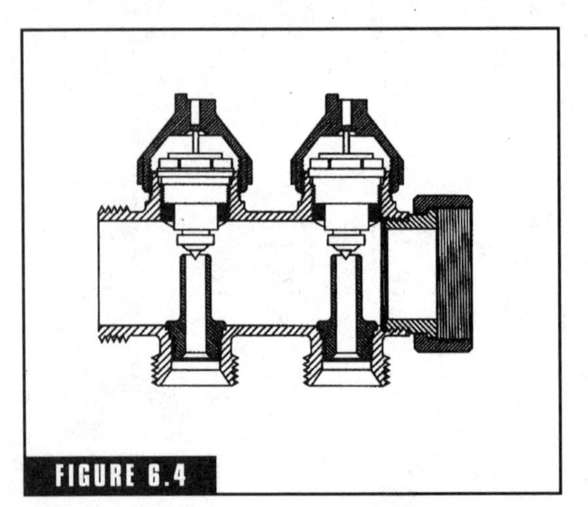

FIGURE 6.4

Manifold cutaway. (*Courtesy of Wirsbo.*)

Tight connections at manifold points are critical. If you buy a preassembled manifold, don't make any assumptions regarding connections. Inspect all existing work yourself and test the manifold connections before you trust them. Some contractors are lax in testing the manifold connections that they make. This may be due to the assumption that the manifold can be accessed whenever necessary. Don't fall into this trap. A very small leak can do substantial damage over time. Many manifolds are not in plain sight. This means that small leaks can go undetected for weeks or months. Even if you are insured, you can't afford to let little leaks ruin your business and reputation. Never conceal a manifold station before testing it completely.

QUICK»TIP **Tight connections at manifold points are critical. If you buy a preassembled manifold, don't make any assumptions regarding connections. Inspect all existing work yourself and test the manifold connections before you trust them.**

The mechanics of manifolds are not complicated. Even when multiple manifolds are used, the basic principles are the same. Sometimes it's more economical to split a heating system off into different manifold stations. This could be due to long distances in pipe runs or varying types of floor coverings. In simplistic terms, a manifold either receives water from a heat source and divides it up among supply tubing or it receives return water from tubing and consolidates it into a single pipe that returns the water to the heat source. There really aren't any mysteries tied to manifold stations.

CHAPTER 7

Boilers

Oil-fired boilers are the most common type used where I live, in the state of Maine. Why is this? Part of the reason is that natural gas is not available in many parts of the state. If it were, I suspect that gas-fired boilers would be much more popular. With the exception of some of the southern cities in Maine, natural gas is mostly unavailable. If a gas-fired unit is wanted, it is likely to have to be a liquefied petroleum (LP) gas unit. Fuel oil is, without a doubt, the prime source of fuel for boilers in Maine. This is not the case in all areas of the country. Gas-fired boilers are very efficient and quite popular where natural gas is available. However, remote areas don't always enjoy gas service. It is often colder climates where the most remote regions exist, such is the case in Maine. Therefore, oil-fired boilers are frequently used in hydronic systems.

Are oil-fired boilers good heat sources? Yes, they are fine heat sources. When I lived in Virginia, I installed heat pumps in my personal homes and in the homes that I built for most of my customers. Over the last decade, I've installed oil-fired boilers almost exclusively. This is because I relocated to Maine about 11 years ago. I used an oil-fired boiler in the home I built for myself and in all the homes I've built for customers. In addition to my work as a builder, my plumbing and heating company has installed numerous oil-fired hydronic heating systems. Simply put, oil-fired hydronic heating systems rule the roost in most of Maine.

What makes the oil-fired system so desirable? Well, in Maine, it is partially due to the fact that natural gas is not readily available in most areas. Plus, fuel oil is a common source of heating combustion, and Maine likes to do things the way they've "always" been done. By this, I mean that Mainers are not fast to accept change. It's sort of like the saying, "If it ain't broke, don't fix it." Certainly, a gas-fired boiler is a great choice when the fuel is available. I probably would opt for natural gas over fuel oil, but this is a personal opinion.

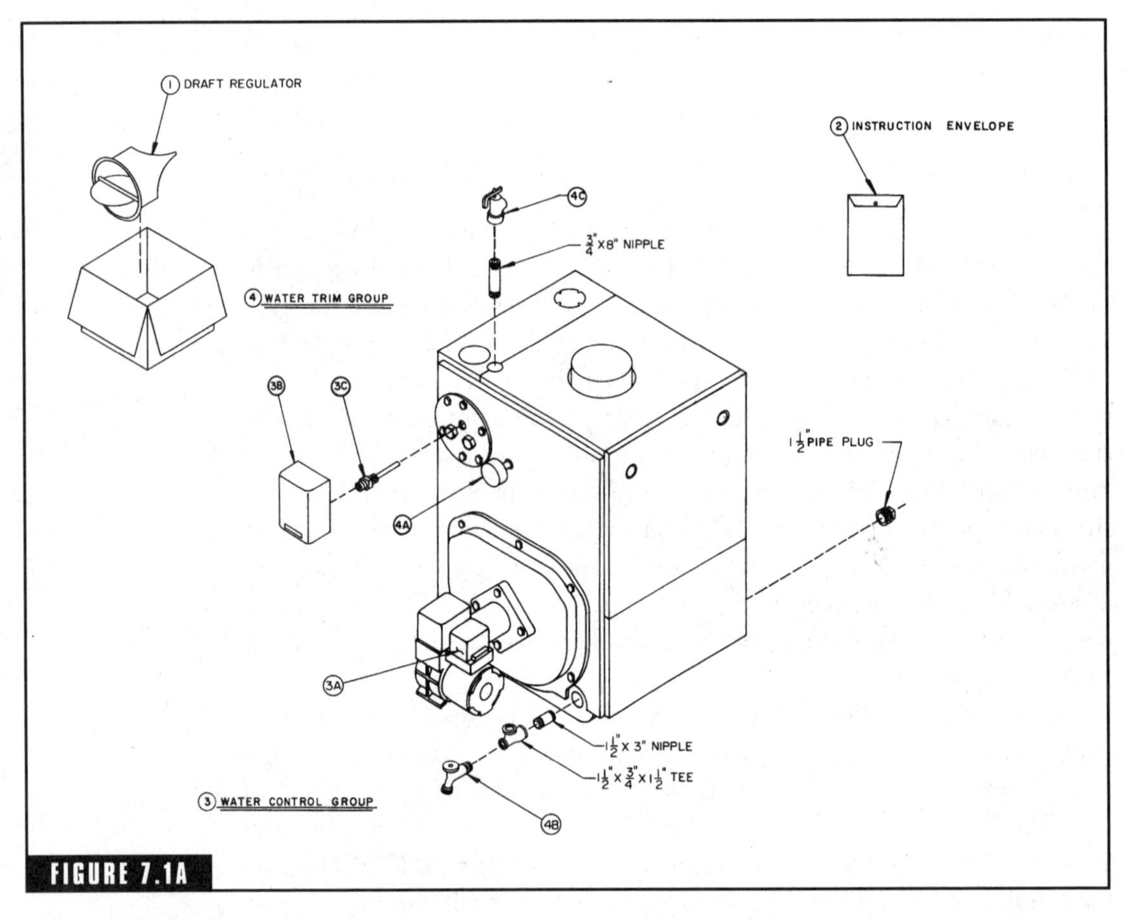

FIGURE 7.1A

Boiler detail. *(Courtesy of Bard.)*

There are some distinct differences between oil-fired boilers and gas-fired boilers (Figs. 7.1*A* and *B*). We are going to talk about gas-fired boilers in the next chapter, but this chapter is dedicated to oil-fired units. When you talk of oil-fired boilers, you must consider both cast-iron boilers and steel boilers. This is the basic difference between the two types of oil-fired boilers, but many, many other components are involved in such heating systems. This chapter is going to explore the major elements of such systems. We will talk about controls, valves, flue requirements (Fig. 7.2), installation requirements, nozzles, operating fundamentals, and much more. So, get ready for a crash course in the installation and use of oil-fired boilers.

ITEM NO.	DESCRIPTION	Part No.	V72	V713	V73	V714	V74	V75	V76	V77	V78	V79
V72 Thru V79, V713 and V714 WATER BOILERS - TRIM AND CONTROLS (See Page 42 for Illustration)												
1. DRAFT REGULATOR												
1A	DR-6 Draft Regulator	8116029	1	1	1	1	1	1	1	----	----	----
1B	DR-7 Draft Regulator	8116001	----	----	----	----	----	----	----	1	1	1
2. INSTRUCTION ENVELOPE CONTAINING:												
2A	Installation and Operating Instructions	8142711	1	1	1	1	1	1	1	1	1	1
2B	Limited Warranty Mailer (Water Boilers)	81460135	1	1	1	1	1	1	1	1	1	1
2C	I=B=R Pamphlet	81460061	1	1	1	1	1	1	1	1	1	1
3. WATER CONTROL GROUP												
3A	Honeywell R4184D (1027/1001) Protectorelay	80160473	1	1	1	1	1	1	1	1	1	1
3B	Honeywell L8148A1090 Hi Limit Circ. Relay (WB)	80160449	1	1	1	1	1	1	1	1	1	1
	--OR -- Honeywell L8124C1102 Hi & Lo Limit, Circ. Relay (WBT)	80160406	----	1	1	1	1	1	1	1	1	1
3C	Honeywell #123870A Immersion Well, ¾NPT x 1½" Insulation (WB)	80160426	1	1	1	1	1	1	1	1	1	1
	-- OR -- Honeywell #123871A Immersion Well, ½NPT x 3" Insulation (WBT)	80160497	----	1	1	1	1	1	1	1	1	1
4. WATER TRIM GROUP												
4A	2½" Dia. Temp/Pressure Gauge, ENFM #41042.5210	8056169	1	1	1	1	1	1	1	1	1	1
4B	¾NPT Drain Cock, Short Shank (WB and WBT), Conbraco #31-606-02	806603011	1	1	1	1	1	1	1	1	1	1
4C	¾NPT, F/F 30 LB. Relief Valve, Conbraco 10-408-05	81660319	1	1	1	1	1	1	1	1	1	1

FIGURE 7.1B

Boiler parts description. (*Courtesy of Bard.*)

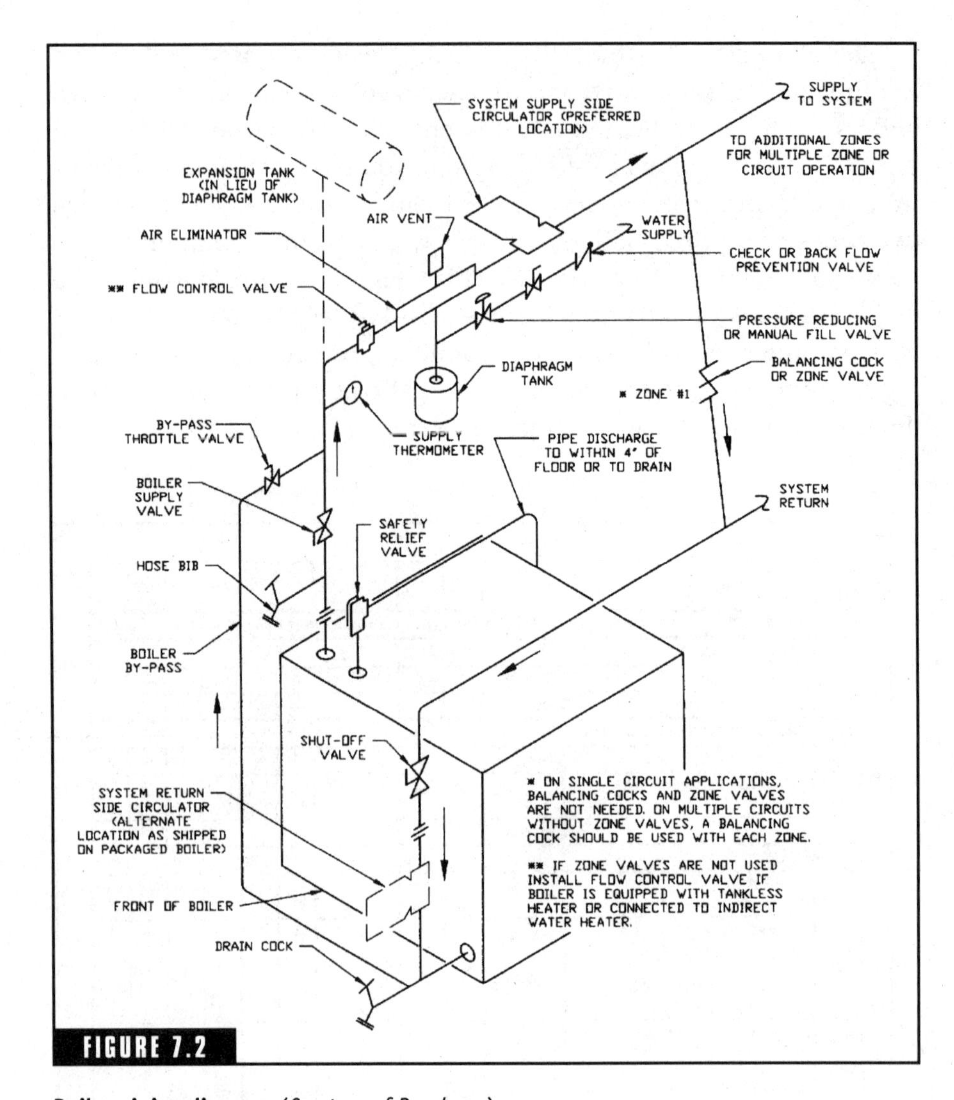

FIGURE 7.2

Boiler piping diagram. (*Courtesy of Burnham.*)

Cast-Iron Boilers

Cast-iron boilers (Fig. 7.3) are the most common type of heat source used in residential hydronic applications. Known as sectional boilers, these boilers can be purchased as individual components and put together on a job. However, most of the boilers sold are already assembled. The boilers are heavy. They can easily weigh 500 lb. This makes

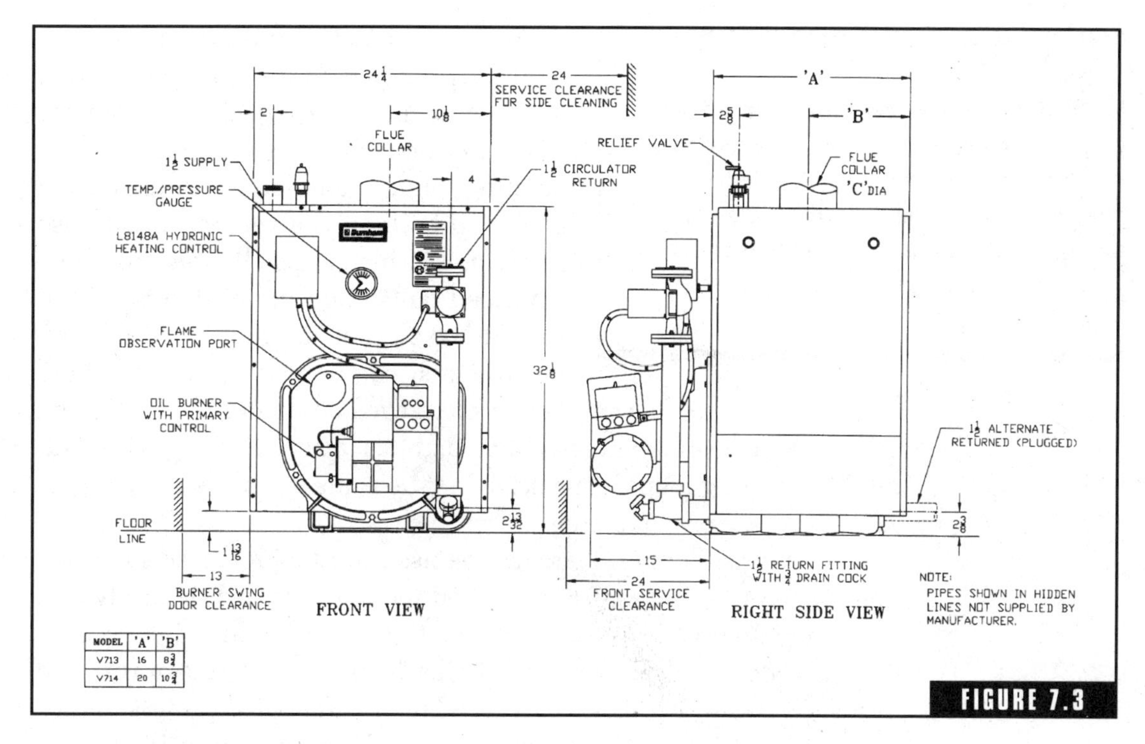

Packaged boiler, without domestic coil. (*Courtesy of Burnham.*)

getting the boilers set in place difficult at times. When conditions allow the use of an appliance dolly, the job of moving a cast-iron boiler is not nearly as intimidating as it is when the boilers must be carried by workers.

Modern boilers are much smaller than their predecessors. This is due to increased efficiency and new designs. Even though the boilers appear small, don't underestimate their weight. Even little boilers are heavy. Wet-based, sectional boilers are assembled by bolting the sections together. When assembled, the sections are known as a boiler blocks. The amount of heat output available from a cast-iron boiler is directly proportional to the number of sections used to create the boiler block. The more sections there are, the more heat potential there is (Fig. 7.4).

MATERIALS

Cast-iron boilers (Fig. 7.3) are the most common type of heat source used in residential hydronic applications. Known as sectional boilers, these boilers can be purchased as individual components and put together on a job.

MATERIALS

Steel boilers are often sold with a 20-year warranty. This is the same warranty that is typically offered with cast-iron boilers. Even though both boilers usually come with the same warranty, there is market resistance in many regions against steel boilers.

Some manufacturers offer their boilers as fully packaged units. This means that the heating unit is shipped complete with the boiler block, a burner assembly, a circulating pump, and controls. Buying a packaged boiler is convenient. It can also save time during an installation. However, packaged boilers are not customized, and this can be a problem. It is sometimes better to match components to a boiler on a job-by-job basis. For example, a prepackaged circulator that will work fine for a baseboard heating system might be too small to use with radiant floor heating loops. If packaged systems are to be used, they must be assessed for proper sizing.

Cast-iron sectional boilers can be used only with closed-loop systems. When installed properly, closed-loop systems rid themselves of excess air after just a few days of operation. This is important, because air in a system where a cast-iron boiler is used can result in rust and corrosion. Assuming that a cast-iron boiler is installed professionally, it should last for decades. It's very possible for a cast-iron boiler to continue working after 30 years of service (Fig. 7.5).

Steel Boilers

Steel boilers are often sold with a 20-year warranty. This is the same warranty that is typically offered with cast-iron boilers. Even though both boilers usually come with the same warranty, there is market resistance in many regions against steel boilers. Many old-time installers feel that steel boilers can't compete with cast-iron boilers in durability. Steel boilers are lighter than cast-iron boilers. The price of a steel boiler is usually several hundred dollars less than a comparable cast-iron boiler. With the many advantages of a steel boiler, it's surprising that more contractors are not using them.

Steel fire-tube boilers surround steel tubes with water and allow hot combustion gases to pass through the tubes. Spiral baffles are inserted in the fire tubes to increase heat transfer. This is done by inducing turbulence and slowing the passage of the exhaust gases. The baffles are called turbulators. The fire tubes may be installed vertically

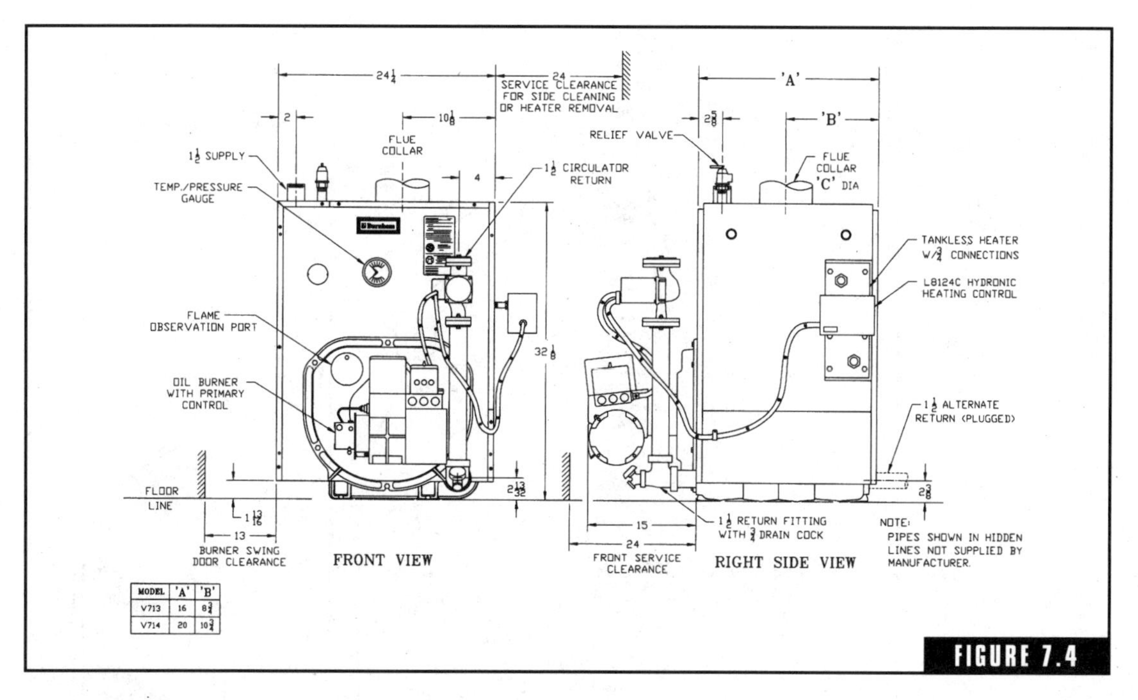

MODEL	'A'	'B'
V713	16	8 3/4
V714	20	10 3/4

Packaged boiler, with tankless coil. (*Courtesy of Burnham.*)

or horizontally, depending on the brand of the boiler. Some boilers are built to have combustion gases pass through multiple fire tubes to increase heat. Generally speaking, a steel boiler that has horizontal fire tubes and where gases are passed through multiple tubes will be more efficient than a boiler that is equipped with single-pass vertical tubes.

Just as cast-iron boilers are suitable only for closed-loop systems, so are steel boilers. Corrosion can be a problem with steel boilers if dissolved air is present in the hydronic system. Since closed-loop systems rid themselves of air quickly, if they are designed and installed properly, corrosion is not usually a problem with steel boilers.

Fuel Oil

The choice of a fuel oil for an oil-burning boilers is not complicated. There are two common choices. A number 1 fuel oil is a suitable choice for an oil burner. A sometimes more expensive, but better choice, is a number 2 fuel oil. The number 2 fuel gives more heat per

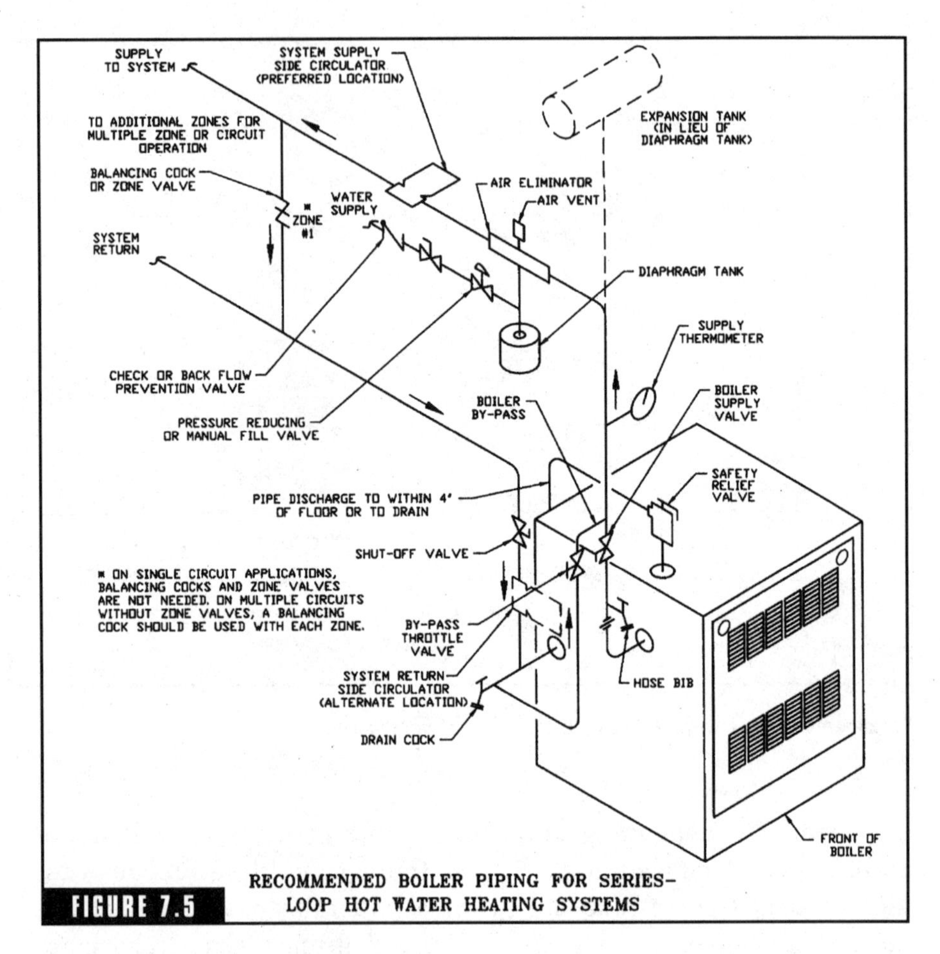

Typical boiler setup. (*Courtesy of Burnham.*)

gallon than number 1 fuel does. It is common for number 1 fuel to be used in vaporizing or pot-type oil burners. Number 2 fuel is a more common choice for standard oil burners. Check the manufacturer's specifications to determine which type of fuel should be used in the oil burner that you are working with.

The storage of fuel oil is usually handled with standard metal oil tanks (Figs. 7.6 and 7.7). The tanks are available in both vertical and horizontal models. Vertical tanks are the most common. It's best when the tanks can be installed in a basement or cellar. Tanks installed outdoors in extremely cold climates can be affected by condensation. Water can build up in a tank and dilute the fuel oil. The tanks must be equipped with both a fill pipe and a vent pipe. When tanks are installed inside a home or building, both the fill and vent pipes are piped to the outside.

Fill pipes should have a minimum diameter of 2 in. Vent pipes should have a diameter of 1¼ in. Fill pipes must be equipped with watertight metal caps. Vent pipes have to be covered with a weatherproof hood or cap. It's common for vents to terminate with whistler-type caps so that people filling systems can tell when the tank is getting full. Vent pipes should be installed so that they are at least 2 ft from the outside wall of a building, to allow good air circulation.

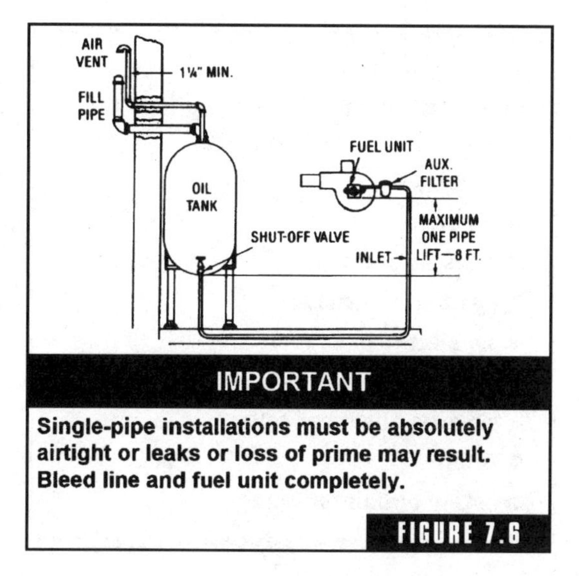

Standard oil tank setup. (*Courtesy of Burnham.*)

Operating Fundamentals

The operating fundamentals of oil burners are not very difficult to understand. High-pressure atomizing oil burners contain nine major parts. They are a nozzle, a nozzle tube, a nozzle strainer, ignition electrodes, an electrodes bracket, an air entrance, an air-adjustment collar, a fan, and rotary turbulator vanes. This type of burner is sometimes called a sprayer. This is because fuel oil is sprayed rather than vaporized. The burner unit is often called a gun. Basically, oil is forced under pressure through a special gunlike atomizing nozzle. The fuel is broken into very small liquid particles to form a fuel spray.

High-pressure atomizing oil burners require pressure to operate. Residential applications usually need 80 to 125 psi of pressure. A commercial burner may operate at pressures ranging from 100 to 300 psi. The burner process starts with oil being delivered to and through a nozzle.

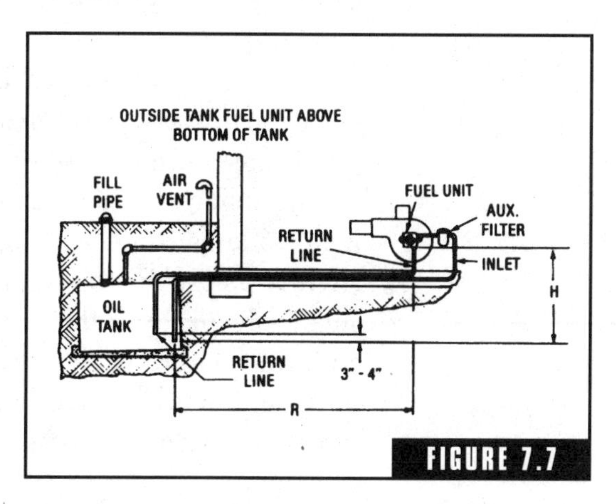

Underground oil tank setup. (*Courtesy of Burnham.*)

MATERIALS

High-pressure atomizing oil burners contain nine major parts. They are

- **a nozzle**
- **a nozzle tube**
- **a nozzle strainer**
- **ignition electrodes**
- **an electrodes bracket**
- **an air entrance**
- **an air-adjustment collar**
- **a fan**
- **rotary turbulator vanes**

The oil is broken into small particles and sprayed into the ignition area. An air supply is brought in through a case opening and forced through a draft tube with the help of a fan. The air mixes with the fuel mist. A turbulator turns to mix the air and fuel completely. This mixture is ignited with a spark that is provided by a transformer, which converts electric current and feeds it to electrodes, where a spark is produced.

Gas-Fired Boilers

Gas-fired boilers are very popular in areas where natural gas is available. These boilers can be very efficient to operate. Keeping operating costs down is something anyone who is paying to run a boiler can appreciate. Basic boiler construction is the same for both gas-fired and oil-fired boilers. There are, however, some differences, so let's discuss them now.

Gas burners can operate with either an atmospheric yellow flame or a power burner. There are gas burners designed for use with natural gas and others that are designed for use with bottled gas. Most of the bottled gas used is liquefied petroleum gas. It's very important to match the burner unit to the exact type of gas to be used for ignition (Fig. 7.8).

A majority of all gas burners used for residential purposes are designed for use as atmospheric yellow flame burners. The atmospheric injection type of burner works on a principle similar to that of a bunsen burner. The burner is nothing more than a small tube placed inside a larger tube. The bigger tube has holes in it that are positioned slightly below the top of the smaller inner tube. Gas coming out of the small tube pulls air through the holes of the larger tube. This produces an induced current of air in the larger tube. The air enters through the holes and is mixed with gas in the tube. The mixture is burned at the top of the larger tube. This type of flame produces little light but extensive heat.

Air supply for an atmospheric injection burner is either primary or secondary air. It is commonly introduced and mixed with the gas in the throat of a mixing tube. Gas passes through a small orifice in a mixer head, which is shaped to produce a straight-flowing jet moving at high velocity. The throat of the mixing tube can be called a venturi. Gas flowing through the venturi spreads and induces air in through an opening at an adjustable air shutter. Because of force from the energy of the gas stream, the mixture of gas and air is pushed into the burner manifold casting. At this point, the mixture goes through ports where additional air is added to the flame to complete combustion. Primary air is the air coming in through the venturi. Secondary air is the air that is supplied around the flame.

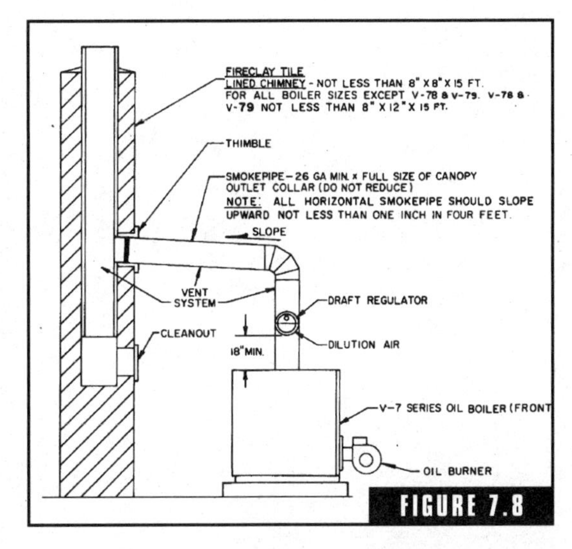

Standard flue setup. (*Courtesy of Burnham.*)

Primary air is admitted at a ratio of 10 parts air to 1 part gas when natural gas is used. When manufactured gas is used, the ratio changes to 5 parts air to 1 part gas. These ratios are just theoretical values. It's common for gas burners to operate well with 40 to 60 percent of the theoretical values. Some elements that affect the amount of gas needed include the uniformity of air distribution and mixing, the direction of travel for gas from the burner, and the height and temperature of the combustion chamber.

Natural drafts provide secondary air. This air should not exceed 35 percent. Draft hoods can be used to control secondary air. The hood can also control backdrafts, which might blow out a flame. Ideally, the flame of a gas burner should burn as blue as possible. This occurs when the mixture of primary and secondary air is controlled properly.

Boiler Selection

Boiler selection (Figs. 7.9 and 7.10) depends on many factors. The first consideration is the type of fuel to be used to operate the boiler. It is possible to obtain a boiler that runs on electricity, but gas-fired and oil-fired boilers are, by far, much more common and more popular. Boilers that are designed for operation with natural gas can be used only where natural gas is offered to customers of utility providers. When

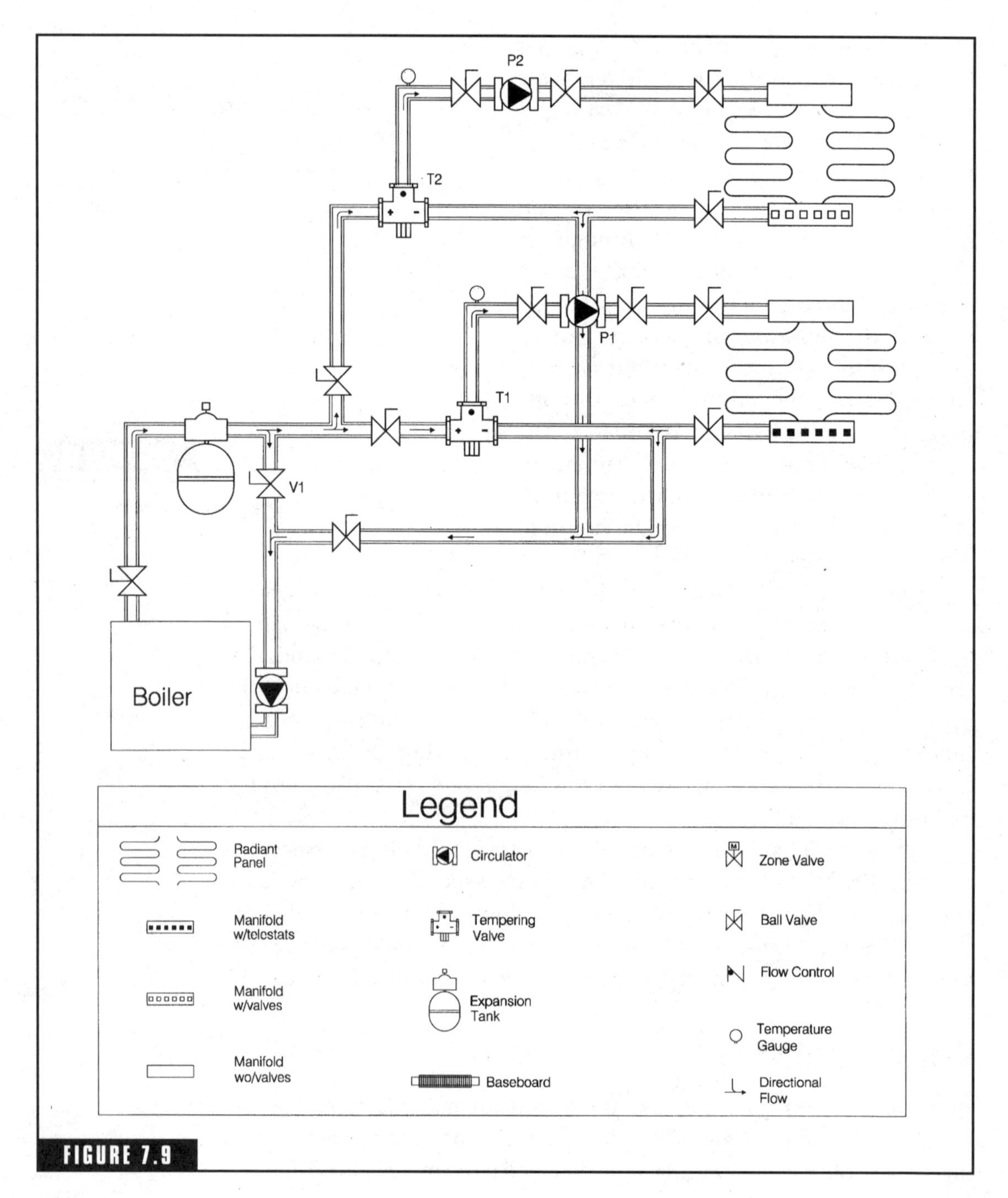

FIGURE 7.9

Noncondensing boiler setup. (*Courtesy of Wirsbo.*)

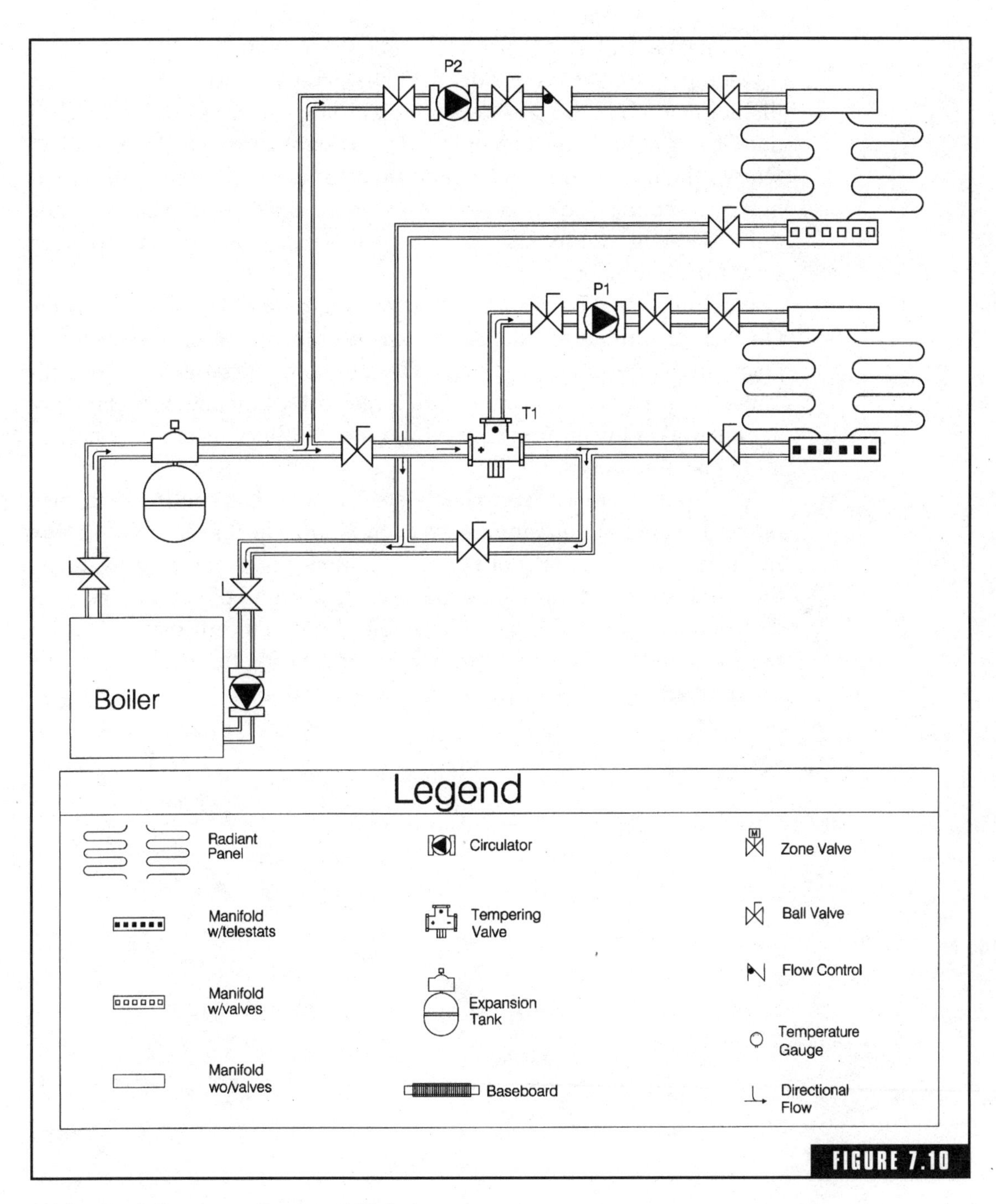

Condensing boiler setup. (*Courtesy of Wirsbo.*)

available, natural gas is an excellent choice as a fuel.

Manufactured gas is usually available in any location. Some people dislike manufactured gas, owing to the large storage bottles used to hold the gas and the perceived risk of explosion. Oil-fired boilers require the use of large tanks, but fuel oil is not explosive in nature. It burns, of course, but it does not explode. Bottled gas can explode. Such accidents rarely happen, but it is a risk that some people are unwilling to accept.

Fuel oil is available in nearly every region and it's a good choice. Oil can be purchased in the off season, during warm weather, and stockpiled for winter use. Most codes allow the installation of two oil tanks for residential use. With two large tanks, homeowners can store summer oil at lower prices for winter use. Oil burners are dependable and make fine heat sources.

The debate of cast iron versus steel is one that continues to rage onward. There are advantages to both, and either is suitable for any residential application. Both types of boilers share similar warranties. Steel boilers tend to be less expensive. Cast-iron boilers may be a bit more efficient, but this is a debatable point. The ultimate decision between the two is up to the person paying for the unit. A quality boiler of either type will serve well for years to come.

CHAPTER

8

Material Selection

Material selection for a radiant floor heating system is pretty simple. There are a number of types of materials that could be used, but there are two particular types of tubing that are used most often. Both of these types of tubing are plastic. Copper is rarely used in new radiant floor heating systems. While copper is still the number one choice of materials for radiant baseboard heating systems, polybutylene and PEX tubing are the leaders in the radiant floor heating field. Polybutylene held a prominent place for quite some time, but PEX is the current leader. This is probably due to the bad press that polybutylene has gotten recently.

Personally, I've used polybutylene in both plumbing and heating systems for many years, and I've never experienced any problems with the material. However, there are a number of documented cases that blame polybutylene (PB) for all sorts of leaks and trouble. I tend to believe that the problems are a result of workmanship and possibly the selection of connection units. Regardless of what I think, the plumbing and heating industry is moving aggressively toward PEX and away from PB. I do like PEX tubing and use it. In terms of working with PEX, it's about the same as PB.

MATERIALS

PEX is a premier tubing for radiant floor heating. The technical name for the material is cross-linked polyethylene.

PEX and PB are quite similar in many ways, but the fittings used with the two types of tubing are not interchangeable. In other words, don't use PEX fittings with PB or PB fittings with PEX tubing. The tools used to cut the tubing are the same. Crimpers used to make connections can be used with both types of tubing, but the crimp rings must be matched to the type of material that you are working with. Support clips and clamps that are designed for PB tubing can be used with PEX tubing. Certainly, there are some crossover areas, but don't mix fittings and crimp rings with the two types of tubing.

PEX tubing is by far the leader in modern radiant floor heating systems. With this in mind, we will concentrate our efforts here on PEX tubing. Copper is almost never used in modern radiant floor systems, so it's hardly worth talking about. However, copper might be used to deliver supply water to a manifold. It could also be used to convey return water from a manifold to a heat source. Of course, copper could be used for an entire system, but the cost is prohibitive, and copper tubing is not as easy to work with as plastic tubing is. For most applications, PEX tubing is the material of choice.

What Is PEX?

PEX is a premier tubing for radiant floor heating. The technical name for the material is cross-linked polyethylene. Unlike PB tubing, PEX tubing is cross-linked and stronger. Talking about molecular chains and the process used to create PEX could get boring. It's enough to know that the product exists and performs well in both plumbing and heating applications. PEX has a three-dimensional network of molecular chains that makes it extremely durable. The tubing can be used with a wide range of temperatures and pressures. This type of tubing is leaping to the front of the list of material choices for radiant heating and plumbing water distribution, but the product is not new. In fact, PEX has been around for many years.

Polyethylene can be cross-linked in a number of ways. Depending on which method is used, the properties of the finished product differ. One cross-linking method occurs during production of the product. A second method of cross linking is performed after tubing has been manufactured. Without getting too technical, the first method of cross linking is done above the crystal melting temperature. In common terms, this is called hot cross linking. The second method, called cold cross linking, is done after tubing has been manufactured. There is a difference, and many contractors feel that tubing which has been cross-linked using the hot method is better.

The Engel method of cross linking is a hot cross-linking method. This is the method used on the tubing that I use most frequently. With the Engel method, the cross linking takes place during the extrusion process of the base polyethylene. This creates a tubing that is knowing within the industry as PEX-A tubing. When the Engel method of cross linking is used, tubing is created that is a product of more precise control over the degree, consistency, and uniformity of cross linking. Since the tubing is cross-linked in a uniform method, the result is a stronger tubing that has no weak links in its molecular chains.

A secondary cross-linking method is known as either the electronic or radiation process. This creates PEX-C tubing. In doing this, high-density polyethylene (HDPE) tubing is extruded. Once the tubing is made, it is subjected to high voltages of electricity. The molecular structure of the polyethylene is cross-linked during this process. This type of process is a cold procedure. PEX created with a cold process is not as strong as PEX made with a hot process. The electronic process produces tubing that has a less consistent level of cross linking.

Another cold cross-linking method is the silane method. With this method, a special premixed polyethylene-based resin is extruded into HDPE tubing. Once the tubing is made, it is placed in a sauna. The heat and moisture from the sauna complete the cross-linking process. As is typical with cold cross linking, the silane method produces cross-linked tubing that is inferior to tubing that is cross-linked with a hot method. Personally, I prefer to work with tubing that has been cross-linked with a hot process.

Stress

A lot of stress can be applied to tubing used for radiant floor heating systems. The same is true of tubing installed in ceilings. Almost everyone knows that buildings settle over time. This means that floors and ceilings can move. The movement experienced during the settling process can damage tubing that is not installed and protected properly. Tubing installed in concrete is, of course, subject to stress. This is true, too, of tubing that is installed in wood members of construction elements. The settling process is not the only stress that tubing might be subjected to.

Concrete expands and contracts when it is subjected to significant temperature changes. This can put a lot of pressure on tubing that is installed in the concrete. Some contractors give very little thought to this fact. It's not unusual for people to think that the tubing used for radiant heating in a floor is installed in a manner similar to that for plumbing materials. Typical plumbing and electrical installations in a concrete floor are installed below the concrete. This is not the case with radiant floor tubing. Standard procedure for radiant floor tubing is to install it in the concrete. Yes, it is embedded in the concrete rather than under it. This increases the stress on the tubing. When the concrete is moving, contracting, and expanding, the tubing must stand up to the stress.

Another consideration when tubing is installed in concrete is the abrasion factor. Concrete has a rough surface. When the concrete moves, the rough texture can eventually rub holes in tubing or puncture the tubing. Copper tubing often has a chemical reaction with concrete that will create pinholes in the copper. There is no chemical reaction between concrete and plastic tubing, but the cuts, rubs, and punctures are still possible. Unlike plumbing pipes that can be encased in foam insulation for protection from concrete, tubing used for radiant heating systems in a floor cannot be covered with insulation. Doing so would defeat the heat output of the tubing for making living space comfortable.

Shearing and stretching are other concerns when tubing for a heating system is installed in a concrete slab. Installing tubing in a floor to avoid problems requires good selection of materials and a professional installation. Heat tubing is subjected to temperature extremes from season to season. There are times when the tubing is hot and other times when it is cold. This type of stress can also take its toll on tubing. One manufacturer is trying to

MATERIALS

One manufacturer is trying to capture a world record for its PEX tubing. After 25 years of being exposed to temperatures of 203°, at 175 psi, its tubing is still holding up to the intense test.

capture a world record for its PEX tubing. After 25 years of being exposed to temperatures of 203°, at 175 psi, its tubing is still holding up to the intense test. It appears that this PEX tubing is going to hold the world's record, but it won't be official until the tubing fails, and that has not yet happened.

Oxygen Diffusion

Chemical reactions and oxygen diffusion can be problems with some heating materials. Cross-linked polyethylene is very resistant to chemical reactions with any chemicals that are commonly found in plumbing and heating applications. Owing to the unique molecular structure of PEX, the tubing is stable and inert. Acidic conditions can deteriorate copper tubing. It's very common for acidic water to create pinholes in copper tubing. This happens in both plumbing and heating systems when the water contains too much acid. Fortunately, PEX and PB tubing are not affected by acid commonly found in plumbing and heating systems.

Oxygen diffusion can corrode a heating system and the system's components. Plastic tubing is very durable, but it can be broken down with oxygen diffusion. This can be a big problem, but it's not hard to deal with. Plastic tubing is permeable to the passage of dissolved oxygen molecules through the walls of the tubing. If this happens, the dissolved oxygen can enter a "closed" system. When you are dealing with a closed system, the fact that it is closed is needed to protect the system from corrosion. If the system is penetrated, damage can occur. The technical term for the invasion is oxygen diffusion.

By definition, oxygen diffusion is the transfer of oxygen molecules through the walls of nonmetallic tubing. All new heating systems contain dissolved oxygen molecules. These molecules are present in all fresh water. Large bubbles of air are purged from new systems. This is done before the system is ever started into operation. Even when the air bubbles are gone, the dissolved oxygen remains present. Air vents and scoops are not capable of removing the dissolved oxygen.

Once a heating system is activated, the water in the system comes up to heating temperature. When this happens, the dissolved oxygen modules come out of solution and become more aggressive. The

QUICK»TIP **Oxygen diffusion can corrode a heating system and the system's components.**

molecules look for ferrous components to attack. If ferrous objects are accessible to the molecules, rust and corrosion will not be far behind. This is a standard procedure for typical hydronic heating systems.

When dealing with a cast-iron boiler, metallic pipe, and baseboard heat, the dissolved oxygen finds its targets and attaches to them. This creates a layer of rust. But, once the rust layer has been created, the dissolved oxygen is pretty much used up. In other words, the heating system reaches a point where there is a rust level and then the rust problem ceases. This has been going on for decades and is completely acceptable. There is, however, a difference when dissolved oxygen is present in a heating system that is made up of nonmetallic piping, such as a radiant heating system that is created with PEX or PB tubing.

So, what's the difference between a metallic heating system and a nonmetallic system in terms of dissolved oxygen? Why does the rust reach a saturation point and stop in a metallic system and continue to grow in a nonmetallic system? Dissolved oxygen in a metallic system is limited. Once the dissolved oxygen is all accounted for in a layer of rust, the process stops. This is not true of a nonmetallic system. The reason is that dissolved oxygen can continue to enter a nonmetallic system through the walls of nonmetallic tubing. This means that the corrosion process is ongoing with a nonmetallic system that is not protected.

How bad can the corrosion process get? Bad enough. If a cast-iron boiler and ferrous components are used with a nonmetallic piping system, a number of problems can occur. The ultimate failure of the system will come when the boiler fails from the corrosion. Other problems include circulator failures, holes in metallic tanks, and sludgy buildups that can reduce water flow. It's enough to know that corrosion cannot be allowed to fester in a non-metallic heating system that uses ferrous components.

Stopping Oxygen Diffusion

Stopping oxygen diffusion is not very difficult. In fact, it's easy enough that there is no good excuse for allowing corrosion to exist in a nonmetallic system. Some methods of protection are more expensive than others. Within the heating industry, there are four typical means of controlling oxygen diffusion. Knowing this, let's examine each of the options.

Some contractors isolate ferrous elements of a heating system from destructive water with the use of a heat exchanger. A nonferrous heat exchanger (Fig. 8.1) can be installed to keep system water from corrosive parts, such as cast-iron boilers, expansion tanks, and pumps or circulators. The heat exchanger protects components on the boiler side of the heating system but does not protect components on the heating side of the system. This means that all components on the heating side of a system should be made of nonferrous materials.

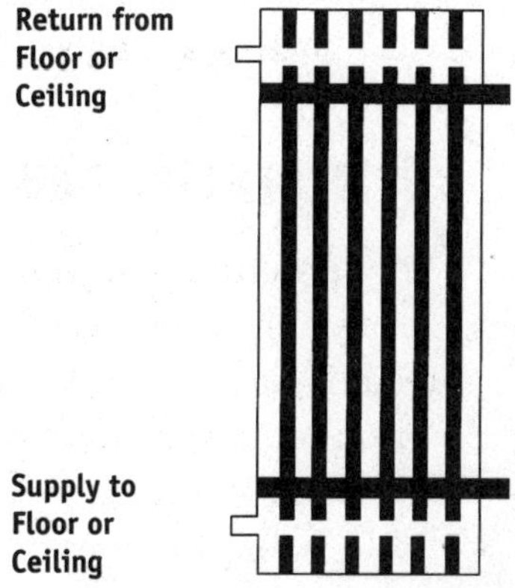

Return from Floor or Ceiling

Hot from Boiler

Supply to Floor or Ceiling

Return to Boiler

FIGURE 8.1

Heat exchanger. (*Courtesy of Wirsbo.*)

A heating system can be designed and installed to eliminate all ferrous materials. This is one of the best ways to deal with the problem of oxygen diffusion. For example, bronze pumps and glass-lined tanks can be installed to prevent rust buildups. The cost of a full nonferrous system can be discouraging. To overcome this, some people use corrosion inhibitors with all heat transfer fluid. When used properly, corrosion inhibitors do work, but the process is labor-intensive. It is not something where a solution is added to the heating liquids and forgotten. Someone has to monitor the mix of heat fluids and corrosion inhibitors. If the mixture of corrosion inhibitors falls below desired levels, corrosion can occur. For this reason, not many people favor the use of corrosion inhibitors.

One easy cost-effective way to deal with oxygen diffusion is to install heat tubing that has an oxygen diffuser barrier. This type of PEX tubing limits the diffusion of oxygen into the heat-transfer fluids. Basically the PEX tubing that is made with a barrier is similar in function to copper tubing. The barrier allows the PEX system to build up a base level of rust with existing oxygen in the water but prohibits the ongoing corrosion problem.

Choosing a method to deal with oxygen diffusion is a personal decision. A lot of contractors opt for PEX tubing with a barrier, but heat exchangers are also used often. Then there are contractors who build

TRADE SECRET

One easy cost-effective way to deal with oxygen diffusion is to install heat tubing that has an oxygen diffuser barrier.

noncorrosive systems by eliminating ferrous components. You can use any method that you are comfortable with, but make sure that you protect heating systems effectively.

Temperature and Pressure Ratings

Temperature and pressure ratings are important considerations when choosing materials for a heating system. PEX tubing is available in different ratings for all types of jobs. For example, you can buy a PEX product that is rated for a temperature of 200° and a pressure rating of 80 psi. Another type of PEX tubing is rated for temperatures of up to 180° with pressures of up to 100 psi. A third type of PEX tubing carries a temperature rating of 73.4° and a pressure limit of 160 psi. You have to match the right tubing to each job. The ratings assigned to PEX tubing are issued by the Hydrostatic Design Stress Board of the Plastic Pipe Institute.

Tubing Size and Capacity

Tubing size and capacity are also factors when choosing a tubing for a heating system. One type of PEX tubing that has a protective barrier to prevent dissolved oxygen from entering a closed system is available in four sizes. Tubing with a nominal inside diameter of 3/8 in. is available in this form. The 3/8-in. tubing will contain 0.491 gal of liquid in a section of tubing that is 100 ft long. The same type of tubing with a 1/2-in. nominal inside diameter will hold 0.920 gal of liquid for every 100 ft of tubing installed. Tubing with a nominal diameter of 5/8 in. will house 1.387 gal of liquid per 100 ft of tubing. The number of gallons of liquid contained in 100 ft of tubing with a nominal inside diameter of 3/4 in. will be 1.89 gal.

One of the big selling points of plastic tubing is the length of the coils available. Obviously it is best to minimize connections for piping that is concealed. Copper tubing is available in coils, but it can be difficult to work with, and the length of the coils is limited. I suppose the length of any coil is limited, but plastic tubing is available in much longer coils than copper tubing is. For example, PEX tubing with a nominal inside diameter of 3/8 in. can be purchased in coils that range from 400 to 1000 ft. Tubing with an inside diameter of 1/2 or 5/8 in. can be purchased in coils running from 300 to 1000 ft in length. Larger tubing like the 3/4-in. size is available in lengths from 300 to 500 ft. PEX tubing is available in sizes of up to 2 in. in diameter.

Most people don't imagine using plastic tubing in typical hot-water baseboard heating jobs. This may be because PEX is rarely used for this purpose, but some types of plastic tubing can be used in such applications. For example, one PEX product is rated for a temperature of up to 200°. This fits the criteria for a hot-water baseboard heating job, and the PEX tubing is protected from oxygen diffusion with a barrier, so it can be used with ferrous components. When the cost of PEX is compared to that of copper, and the advantages of plastic tubing are fully considered, you can probably expect to see more and more systems being installed with PEX tubing. If you didn't know that PEX could be used with standard baseboard systems, now you do. Get the jump on your competition and be a trendsetter. The lower cost of PEX tubing makes using it in a baseboard system very attractive.

Manifold Systems

Manifold systems are an essential component of a radiant floor or ceiling heating system. You can build your own manifold stations, but it is often easier to just buy prefabricated units. Manufacturers of tubing generally offer a full array of system components. It's a good idea to keep systems as compatible as possible. With this in mind, it makes sense to purchase manifolds and their related components from the manufacturer of the tubing to be used in a system, whenever possible.

Assuming that you will be buying preassembled manifolds, look for products that offer options in mounting arrangements and configurations. You should also seek manifolds that provide options for zoning and balancing. If you decide to build your own manifolds, you will have complete control over design issues. And many manufacturers will custom-build manifolds upon request.

Manifolds are only a part of a manifold station. The materials used to mount manifolds must also be compatible with the overall system. A number of mounting options are available to contractors. In addition to being functional, the materials used to mount manifolds may need to be attractive. Sometimes it's not practical to put manifolds in out-of-the-way places. When this is the case, an effort must be made for appearance purposes. Hinged compartments that conceal manifolds are an ideal choice for situations where manifolds must be installed in areas of open sight.

Fittings

The fittings used with PEX and PB tubing are similar but are not interchangeable. Different types of fittings are used in heating systems. Most contractors use copper insert fittings and copper crimp rings. But sometimes compression fittings are used. Most contractors prefer crimp fittings to compression fittings. The risk of leaks is usually much lower when crimp fittings are used. Selection and installation of fittings is a major factor in how well a system functions. Never cut corners on fittings or in making connections. A little mistake in this area can result in major trouble down the road.

Most leaks in a radiant heating system occur at connection points. It is often said that fitting connections are the weakest link in a radiant heating system. Most contractors assume that if connections are pressure tested before concealment the risk of leaks is eliminated. This is not entirely true. Of course, pressure testing at the time of installation should be done, and it will expose leaks that are present at the time of testing. But leaks can come later.

It's not uncommon for fitting connections to fail long after they are made. A connection that tests fine at the time of installation can still leak later. Why is this? Because fitting connections frequently fail as a result of long-term heating and cooling. Natural expansion and contraction associated with changing temperatures can take their toll on connections. When heating and cooling under pressure occurs, it is called thermocycling.

Products such as fittings can be given thermocycle tests. Listing ratings for fittings can help to assure quality in an installation. Third-party certification is another indicator of a quality fitting. When you are selecting materials for a heating system, look for products that have been tested and rated. The selection process for materials is very important. If you use inferior products, you are likely to run into trouble at some point. You may have to pay a little more at the time of installation for top-notch products, but the end result should be much better and can save you a lot of money and trouble in the future.

Controls

The controls used on a heating system can vary a great deal. Controls that are perfect for one system may not be suitable for another system. Customers are usually the ones who should choose control options. Unfortunately, most customers don't have any idea what the difference is between a zone valve and a tempering valve. Ideally, contractors should educate their customers about the many control options available. It's unreasonable to assume that average people can make viable decisions about what type of control to install with a heating system. Even some heating contractors have trouble deciding which type of control is best suited to a system.

Injection mixing controls are one option for controlling a heating system. Three-way tempering valves are a popular means of control for radiant heating systems, and so are telestats and zone valves. Proportional valve controls and zone control modules are also available for controlling a heating system. Every system should be evaluated on its individual basis to determine which type of control will best serve the heating system.

Other Materials

Other materials that are used when installing a radiant heating system include clips, sleeves, staples, and so forth. Many contractors use staples and staple guns to secure their heating tubing. Some contractors use plastic J-hooks to secure tubing. Clips and cable ties are also used to secure PEX and PB tubing. Plastic rails are sold that allow tubing to be snapped into channels in the rail for securing the tubing. Sleeves are used under sawn joints in concrete. Bend supports are used to bring tubing out of floors and into walls without crimping the tubing. All materials used in a heating system should be chosen carefully.

It's tempting to buy whatever products are being offered for the lowest price. Many contractors buy products at the lowest possible prices. This can be a mistake. There is nothing wrong with bargaining for the best price, but make sure that the materials you are buying are worth your time and that they will not hurt your reputation. Bottom line: Get the best price you can on the best products that you can find, but don't settle for inferior products just because they are cheap.

CHAPTER

9

Circulating Pumps

It could be said that a circulating pump is the heart of a heating system. If a heating system is compared to a human body, the circulating pump might well be compared to a heart. Both the heart and the circulator are responsible for pumping fluids through their respective systems. Even with the fact that circulators are installed in a majority of hydronic heating systems, a lot of contractors and installers would be hard-pressed to explain the differences between various types of pumps. There are distinct differences between circulating pumps, and that's part of what we are about to explore.

Circulating pumps are vital elements to a hydronic heating system, and this includes radiant floor heating systems. Choosing a pump for a system and picking a place for the installation can have a lot to do with how effective a heating system will be. When used with a closed heating system, circulating pumps are responsible for moving water throughout the heat-distribution system. Unlike an open system, where pumps are required to lift fluids, lifting is not required in a closed system.

Centrifugal pumps are used most often in hydronic heating systems. The interior part of the pump that rotates is called an impeller. A center portion of the impeller is called the eye, and this is where fluid is accelerated rapidly and pushed through the vanes of the impeller. Fluid is spun and pushed between the impeller disks as it heads for the volute. When the fluid reaches the volute of the system, it is converted from a velocity head to a pressure head. In other words, fluid comes to the volute with speed, and when it reaches the volute the speed is changed to pressure. After this occurs, the fluid flows around the contoured volute and goes out through the discharge port.

It's important that the process of fluid coming to the volute remain constant. This is not to say that a pump must continuously suck water into the eye of the impeller. Actually fluid entering a centrifugal pump has to be pushed into the pump by system pressure. This comes from upstream of the inlet port. Many people, including professionals, are confused by this fact. As important as the data is, it needs to be understood. Misconceptions on how the pump functions can result in a heating system that just doesn't work very well.

Pump Design

Pump design (Fig. 9.1) can vary greatly and still maintain similar operation. For example, it's common for different circulating pumps to be made with volutes that don't share the same shape. Even when this is the case, the pump works on the same principle. However, the shape of the volute does determine how the pump should be connected to the piping for a heating system. Again, this is a fact that all professionals should be aware of, but not all pros are clear on the differences.

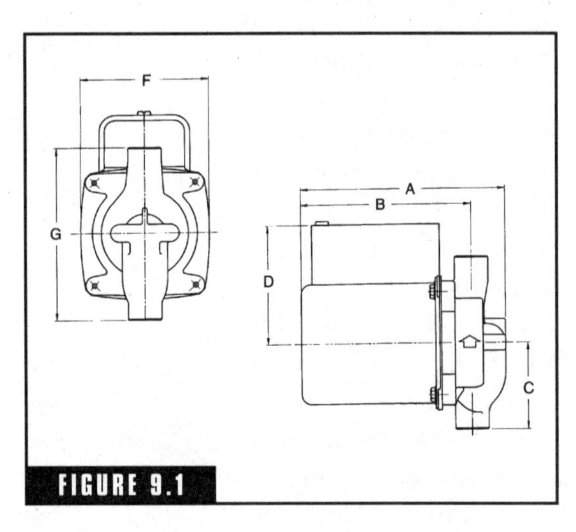

FIGURE 9.1

Typical circulating pump. (*Courtesy of Taco.*)

One common type of circulating pump is an inline pump. This type of circulator has the inlet and discharge ports along the same centerline. This is the most common type of circulator used for residential heating applications. Inline circulators are also used in light commercial jobs. Owing to the design of an inline circulator, there is no need for a lateral offset between the inlet and discharge ports.

Large heating systems generally use end-suction pumps. These pumps have a modified volute design. An end-suction pump creates a 90° turn within the piping system. It's common practice for an offset to be installed between the centerlines of the inlet and discharge piping when a right turn is made in the piping. End-suction pumps are common in large heating systems, such as those used in hotels and large stores, but inline circulators are by far much more common when working with residential properties.

QUICK»TIP **Large heating systems generally use end-suction pumps.**

Another type of circulator that is used with small to medium circulators is the wet-rotor circulator. This type of circulator has been refined over the last few decades to make it a sensible choice for smaller systems. A wet-rotor circulator combines the motor, the shaft, and the impeller in a single assembly that is housed in a chamber filled with system fluid. Since the motor of a wet-rotor circulator is housed in system fluid, it is cooled and lubricated by the system fluid. This means that there is no need for a cooling fan or oiling. Since the rotor assembly sits on bushings in the rotor housing, it can ride on a film of system fluid. With this being the case, there is no requirement for oiling the unit. Many advantages are gained when a wet-rotor circulator is installed and only a few disadvantages.

We've already discussed the fact that wet-rotor circulators don't require oiling. This is certainly an advantage, but there are more advantages to consider. Because of the design and installation of a wet-rotor pump there is no leakage of system fluid due to worn seals. The diminutive size of wet-rotor pumps is another advantage. Their size allows them to be installed easily, located in many fashions, and supported without the need for heavy-duty materials. Wet-rotor circulators are very quiet during operation. This is because no cooling fan is spinning around. Contractors have a broad selection of circulators available to them with multiple-speed motors. This adds diversity to the list of advantages associated with wet-rotor circulators.

Cost is usually a consideration in any heating system. Since wet-rotor circulators have few operating parts, the cost of the pumps is usually lower. When a system requires limited flow rates and head rates, a wet-rotor circulator can be an ideal choice. Close-coupled installations for series pump applications are another possibility that adds to the list of advantages. Most wet-rotor circulators are compatible with a wide range of electronic controls. This is because permanent split capacitor motors are used with wet-rotor circulators.

Very few disadvantages are associated with wet-rotor circulators, but there are a couple. It's possible that a pump that has been shut down for a long period of time will not have enough power to free a stuck impeller. The low starting torque of the permanent split capacitor motor doesn't always have enough power to turn an impeller that has been static for a long time. The only other real disadvantage to consider is what happens when there is a problem with the circulator. If the problem cannot be corrected through the external junction box used with a wet-rotor circulator, access to the pump must be gained. This means opening the heating system and probably losing some system fluid. Air often enters the heating system when a system is opened, so this all adds up to a potential disadvantage.

When you add up the pros of wet-rotor circulators and compare them to the cons, it's easy to see that these pumps can prove very desirable in small heating systems. The pumps can be purchased with cast-iron volutes to be used with closed systems, or you can get the pumps with bronze or stainless-steel volutes when an open system is used. A majority of wet-rotor circulators are designed for use with open systems. Many small models of the pumps are designed for use in multizone systems. They can be ideal for use in systems where every zone of a heating system will have an independent circulator.

Another type of circulator that can be used with small to medium-sized heating systems is the three-piece circulator. As the name implies, there are three elements of this type of pump. There is a body assembly, a coupling assembly, and a motor assembly. Wet-rotor circulators are housed within system fluid. This is not the case, at least not entirely, with a three-piece circulator. The motor of a three-piece circulator is kept out of the wetted portion of the pump. The great advantage to this is that the motor can be serviced without any need for entering the wet portion of the heating system.

Spring assemblies used with three-piece circulators absorb vibration or high torque between the two shafts as the motor starts. This is only one type of design, but it is a common one. An impeller shaft penetrates the volute through shaft seals that must limit leakage of system fluid. This happens even when high pressure is involved. The shaft seals frequently lose some fluid, but the loss is minimal. Most of the fluid evaporates quickly and is never even seen. There are certain advantages to three-piece circulators, but there are also some substantial disadvantages.

One of the strongest advantages of a three-piece circulator is the ease with which the motor can be serviced. This is a considerable advantage over a wet-rotor circulator. As long as bearings are lubricated properly, three-piece circulators offer the potential for a longer operating life than some other types of circulators. And the strong starting torque power of a three-piece circulator eliminates the fear of having a stuck impeller that cannot be freed by the motor. Advantages exist, but you must weigh them against the disadvantages.

Direction of flow for a circulator is generally indicated by an arrow located somewhere on the body of the circulator. If the shaft of the pump is horizontal, the direction of flow can be horizontal, upward, or downward. The key is keeping the shaft horizontal. In practice, the preferred direction of flow is upward. This is because air bubbles can be cleared more quickly with an upward flow.

Three-piece circulators must be oiled regularly, and this can be a pain in the neck. If the oiling is not done, system failure can occur. The heavy construction of a three-piece circulator is not always desirable. The external motor and coupling assembly required by a three-piece circulator makes noise when the pump is in operation. This can be a distinct disadvantage when the pump is located within hearing distance of living or work space. The routine maintenance of this type of pump is another disadvantage. If shaft seals become worn or defective, they can leak system fluid. This rounds out the routine disadvantages associated with three-piece circulators. Three-piece circulators are available for both open- and closed-loop systems.

Pump Position

The pump position (Fig. 9.2) for circulators used in smaller heating systems is usually horizontal. The horizontal position removes thrust load on the bushings, since the weight of the rotor and impeller is balanced off by the horizontal placement. Direction of flow for a circulator is generally indicated by an arrow located somewhere on the body of the circulator. If the shaft of the pump is horizontal, the direction of flow can be horizontal, upward, or downward. The key is keeping the shaft horizontal. In practice, the preferred direction of flow is upward. This is because air bubbles can be cleared more quickly with an upward flow.

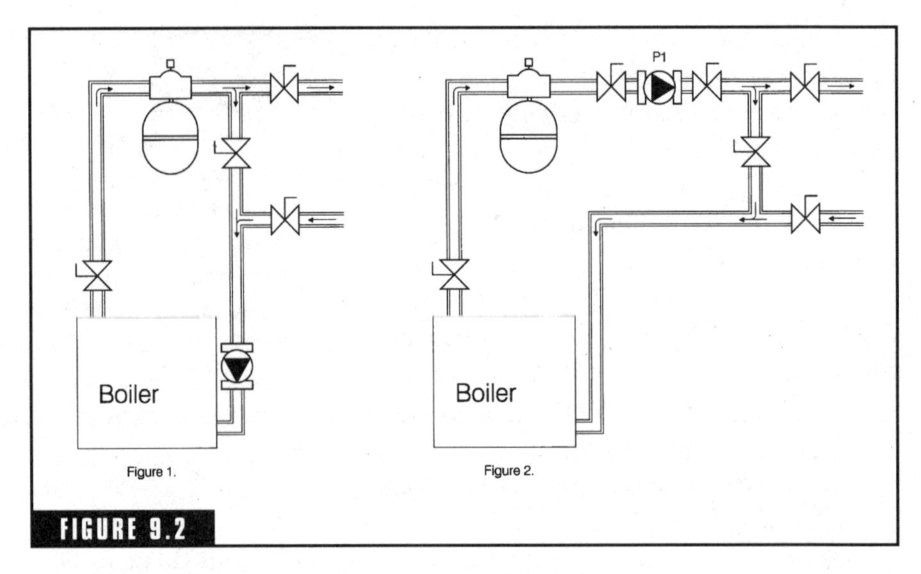

FIGURE 9.2

Possible pump placement positions. (*Courtesy of Wirsbo.*)

Some circulating pumps are light enough to be supported by the pipes that they are installed on. It's wise, however, to provide individual support for each circulator installed. This is especially important when a pump is particularly heavy or generates a lot of torque when starting. It is acceptable to support the piping on each side of the circulator, at a point as close to the connections as possible, to create support for the pump while the actual support is on the piping. Three-piece circulators require even more support. This is due to their offset design and the weight of the motor. It is highly recommended that shock-absorbing hangers be used with three-piece circulators. If a three-piece circulator is attached to piping that is secured rigidly to a wall, vibration from the pump may transfer through the piping and supports to a point that will interfere with comfort in the living or work space.

Residential heating systems often involve the use of pump headers. When several circulators are used with a hydronic heating system, the use of a header is extremely effective. Headers that come off a boiler close to the floor are sometimes supported by concrete blocks. When headers feed off of the upper section of a boiler, they are generally supported by all-thread rod that is secured at a floor joist. Threaded steel or black-iron pipe is the material most often used to construct a header to hold circulators. When a multiple-circulator header is used, the far end of the piping must be supported very well. If it's not, too much stress is placed on the connection where the header piping connects to the heat source.

> **QUICK»TIP** **For ease of removal, circulators are normally installed with flanges. Circulators that are installed with flanges and double valves are very easy to remove.**

Mounting Circulators

When mounting circulators on a piping system, you have to consider the fact that the circulator might have to be removed for servicing at some point in the future. Many contractors install valves both before and after a circulator to make it easier to remove the pumps when needed. Circulators should not be installed with fixed threaded connections. For ease of removal, circulators are normally installed with flanges. Circulators that are installed with flanges and double valves are very easy to remove.

One side of a circulator flange is an integral part of a pump's volute. The matching mate of the flange is threaded onto system piping. When the circulator is put in place, the two flanges are mated together and bolted tight. An O-ring or gasket compresses between the flanges to prevent leakage. Most small circulators are equipped with a two-bolt flange. The flanges are often made of cast iron, for use in closed systems, but they are also available in bronze and brass for open systems.

Isolation flanges are very nice. They contain a built-in ball valve that can be opened or closed with a screwdriver. When the isolation flange is closed, the circulator can be removed from the system without disrupting the rest of the system. The same effect can be achieved with valves installed on each side of a circulator, but the isolation flange is a little neater and works very well.

Flanges are normally used to install circulators (Fig. 9.3), but half unions can be used in some cases. When you are working with very small circulators, you might find that half unions are your connection method of choice. A half union is threaded directly onto a pump's volute. The half unions can be bought with or without an integral ball valve that allows for isolation. Extremely small circulators are sometimes designed to be soldered directly to copper tubing. As long as the circulator is isolated with valves on both sides of it, the copper tubing can be cut to remove the small circulators if needed.

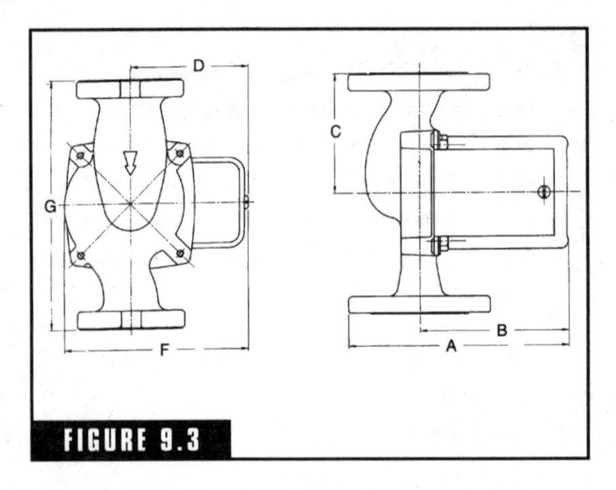

FIGURE 9.3

Flange-type circulating pump. (*Courtesy of Taco.*)

Circulator Location

Circulator location is an important factor in the design of a heating system. A bad circulator location can result in frequent problems and aggravation. To ensure quiet operation that is reliable, careful consideration must be given to circulator placement. Old-time professionals know that circulators should be located so that their inlet ports are close to the connection point of a system's expansion tank. This principle applies to both residential and commercial heating systems. While the work has been done extensively in commercial applications, it seems to slip between the cracks with residential jobs. The point is, a rule of thumb for circulator location is close to an expansion tank.

The amount of liquid in a closed-loop system doesn't change. It doesn't change in the piping or the expansion tank. Regardless of whether a circulator is on or off, the volume of fluid simply doesn't fluctuate. When a closed system is operating normally, it is at what is known as the point of no pressure change. When a circulator is located with its inlet port near an expansion tank, there is less change in the heating system. A horizontal piping system that is filled with fluid and pressurized to a set pressure stays the same when the circulator is turned off. But, when the circulator is turned on, it creates a pressure difference between its inlet and discharge points. The pressure in the expansion tank remains static. A combination of things begin to happen. There is a pressure difference at the circulator. A pressure drop due to head loss occurs in the piping, and the point of no pressure change plays a role in the change. Essentially, pressure in most of the system increases when a circulator cuts on. This helps to get air out of a system through the air vents installed in the system. Since there is a short section of pipe between the expansion tank and the inlet port of the circulator, there is only a very slight drop in head loss within the system piping.

If a circulator were installed without its inlet port being close to an expansion tank, a substantial pressure drop would be experienced in system piping when the circulator cut on. One drawback to this is that the heating system has difficulty in ridding itself of air when pressure drops. This can lead to pump cavitation. In some cases, the pressure drop can be enough to actually pull air into the heating system through the air vents. It's not unusual for some systems that rely on circulators with their inlet ports away from the expansion tank to work fine, but many of them don't. A variety of factors are involved.

Circulator location is an important factor in the design of a heating system. A bad circulator location can result in frequent problems and aggravation. To ensure quiet operation that is reliable, careful consideration must be given to circulator placement.

Factors affecting a system's performance include pressure distribution, system height, pressurization of the system, pressure drop, and fluid temperature. Problems with systems occur most often when there are high fluid temperatures, low static pressure, low system height, and high pressure drops. There is no viable reason to gamble on problems that can be avoided with circulator placement.

Conventional sectional boilers that operate with low head loss characteristics can accommodate circulators that are installed with their return ports close to an expansion tank. This is due to the low head loss, and a circulator should not be installed in this manner when the heat source has a high head loss. Since you can't go wrong by putting the inlet port close to an expansion tank, just do it.

Understanding Head Factors

If you start a conversation in a supply house about head factors, you are likely to hear a lot of different points of view. Contractors and installers all have their definition of what the head produced by a pump is. Many people will say that the head of a pump is the height to which a pump can lift and maintain a column of water. Others will claim that it is the pressure difference that the pump can produce between inlet and outlet ports. Which answer is correct? Well, they are both partly right, but neither is completely correct. It is generally accepted that the proper definition of the head created with a pump is an indication of the mechanical energy transferred to the fluid by the

QUICK»TIP **If a circulator were installed without its inlet port being close to an expansion tank, a substantial pressure drop would be experienced in system piping when the circulator cut on.**

pump. The energy can be in the form of increased flow rate or increased pressure. This involves either kinetic energy or potential energy.

Since the fluid used in hydronic heating systems is incompressible, the flow rate and flow velocity entering a pump is the same as it will be when it leaves the pump. The increased head for this type of situation results in a pressure increase. How much the pressure increases depends on the flow rate. Also the density of the fluid being processed through the pump can affect the pressure changes. The measurement of the head is in mechanical energy and is rated as an element of "feet of head."

Circulator performance is sometimes judged by the pressure difference between the circulator's inlet and outlet ports. In fact, some circulators are equipped with tapped openings that allow pressure gauges to be installed for monitoring purposes. The density of fluid being moved by a circulator must be known to convert head and pressure differences. To arrive at the head, you must divide the pressure differential between inlet and discharge ports or a circulator by the density of the fluid being moved.

Performance Curves

Performance curves (Fig. 9.4) are used to show a pump's ability to add head to a fluid. The curves are drawn on graphs that indicate peak performance. A performance curve shows how much head a pump can add to a fluid as a function of the flow rate passing through the pump. The creation of a pump curve originates from test data based on water in temperature ranges of from 60 to 80°. When antifreeze is used in a system, it can change the performance factors. There is usually a small decrease in head and flow rate capacity at the pump. Typically, the variations in small heating systems tend to be minimal and don't usually require any alteration of a system design.

Performance curves are used to match pumps with the needs of a heating system. The data available for a performance curve is instrumental in choosing a circulator of the proper size. Any manufacturer of a circulator should be happy to supply a performance curve for its product. Don't hesitate to ask for this valuable tool in sizing a heating system's components.

The Difference

What is the difference between high-head pumps and low-head pumps? Some circulators produce high heads at lower flow rates while other pumps produce lower heads over a wide range of flow rates. The pumps producing the lower heads tend to create a stable head for operating purposes. What makes the difference? It is the design of the pump that determines the performance experienced. The diameter and width of an impeller can be changed to create different performance from a pump. A pump made to produce a low head will likely have an impeller with a small diameter and widely separated disks. This is in contrast to a pump set up for a high head, where the impeller will have a large diameter and very small separation between its disks. Even though two circulators might share identical specifications in terms of horsepower and operate at the same speed, you can achieve the different head characteristics with the impeller and disk sizing.

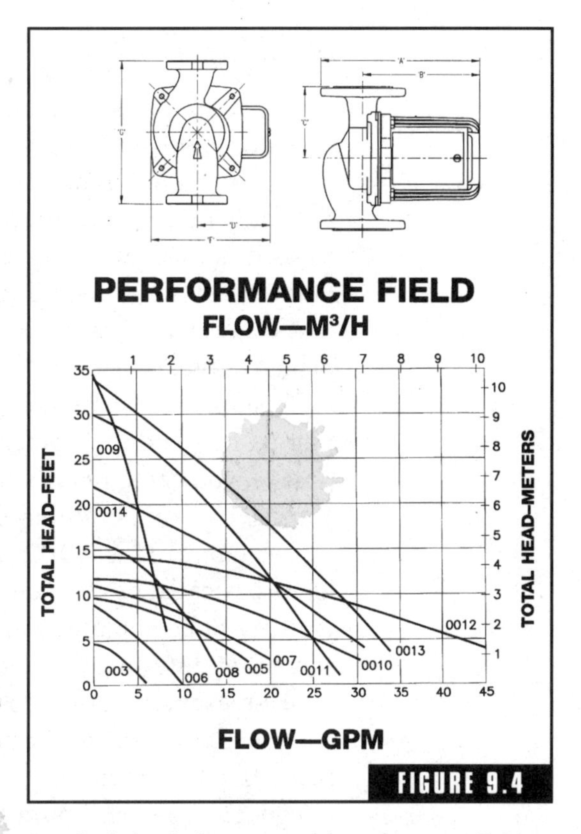

Standard circulating pump with performance field noted. (*Courtesy of Taco.*)

Pumps that have a steep pump curve are meant for use in systems where high head losses at modest flow rates are expected. A circulator that offers a flat pump curve would be used when there is a need to maintain a steady pressure differential across a given distribution system with a wide range of flow rates. Matching a circulating pump to a heating system can be complicated and it is important, so you must invest the time to do it right. Pumps with a flat curve are good for heating systems where there are numerous zone valves that may be opening and closing at various times. To avoid drops when the zone valves are open, the flat-curve circulators compensate for the changes.

Stacking Them Up

When contractors and installers talk about stacking circulators up, they are referring to what is more commonly called a series installation. This type of pump installation is required when a system design demands more circulation than a single pump can produce. In most cases it is possible to opt for one large circulator, but the cost can be more than that of installing two smaller circulators in a series system. Stacking circulators up can make it confusing when judging a performance curve. But this doesn't have to be a problem.

Assume that you are going to install two identical circulating pumps in a series. If you have the performance curve for one of the pumps, you can quickly calculate the total performance of the pair. All you have to do is double the performance of the first pump's performance curve. This works only if both pumps are identical. What you have, basically, are two circulators acting as a multistage pump. There is likely to be some variation on performance, but most professionals agree that doubling the head produced by a single circulator will provide a reasonable estimate of performance expectations (Fig. 9.5).

When circulating pumps are installed in a close-coupled arrangement, the discharge flange of one pump is attached directly to the inlet flange of a second pump. It is usually small circulators that are installed in a close-coupled manner. Typically, the pumps will be operated at the same time whenever either of the pumps is running. Because of the high heads and the associated pressure differential that comes from installing close-coupled dual pumps, it's very important that the installation be done so that the inlet port of the pumping system is close to the expansion tank used with the system. Cavitation is a real risk if the pumping station is not installed with its inlet near the expansion tank.

When ground-source heat pumps or heat exchangers are used, contractors may turn to what is called a push-pull installation. This involves the installation of two circulating pumps, but they are not joined at their flange connections. A section of pipe separates the two pumps. In some cases, the heat exchanger or a coil for a heat pump creates the separation. Having more space between the two pumps cuts down on increased pressure. More accurately, it distributes the pressure more evenly. In doing this, the risk of cavitation is reduced.

Lining Them Up

Another way of installing multiple pumps is lining them up. This is called a parallel installation. In this type of arrangement, the inlet ports of the pumps are connected to a common header pipe. The discharge points are also connected to a common pipe. This type of installation is appropriate for systems where a high flow is required at a modest head. You might find this to be the case when working with a multizone parallel system that is controlled by a number of individual zones. The performance curve of pumps installed in a parallel manner can be arrived at by doubling the flow rate of a single circulator at each head value.

SPECIFICATIONS	UL & CSA LISTED
CASING	Cast Iron or Bronze
IMPELLER	Non-Metallic
SHAFT	Ceramic
BEARINGS	Carbon
CONNECTIONS	Flanged ¾", 1", 1¼", 1½"
PRESSURE RATING	125 psi Maximum
TEMPERATURE RATING	230°F Maximum
MOTOR TYPE	Permanent Split Capacitor
HP	⅛ HP @ 3250 RPM
ELECTRICAL CHAR:	115/60/1 1.40 Amp Rating Impedance Protected

FIGURE 9.5

Sample specifications used to compute performance from a circulating pump. (*Courtesy of Taco.*)

It's difficult to predict the exact performance of pumps installed in a parallel fashion. System resistance has a lot to do with the performance of the pumps. Some people feel that putting two pumps in parallel will double the flow rate of a system. This is not the case. Two pumps do not double the flow. If two pumps are installed in a parallel system, they can be manifolded together. A check valve is required if both pumps will not be running simultaneously. Check valves must be installed in the discharge pipe of each pump. Failure to install the check valves can result in much of the flow created by the operating pump being backwashed through the second pump that is not running while the primary pump is.

Cavitation

Liquids boil when they are subjected to specific combinations of pressure and temperature. A liquid can be prevented from boiling at a particular temperature if there is a certain amount of pressure on the liquid. This process is called vapor pressure. For a liquid to boil there must be a formation of vapor pockets in the liquid. Vapor pockets appear to be bubbles, and they do look like bubbles, but they are not air bubbles. Water that has no air in it can still exhibit vapor pockets that appear to be bubbles. The vapor pockets form when the pressure of the liquid drops below the vapor pressure for the current temperature.

Vapor pockets forming in water create a situation where the density of the water is considerably lower than that of the liquid surrounding the vapor pocket. It is estimated that the density inside the vapor pocket can be as much as 1.5 times lower than the density of the surrounding water. When the pressure of a liquid increases to a point above the vapor pressure, vapor pockets implode. This can be a very destructive process. The action can be strong enough to literally rip away materials from surrounding surfaces.

Cavitation occurs when a liquid at a given temperature has a pressure that drops below the vapor pressure. This cavitation process is most likely to occur in the impeller of a circulating pump or a partially closed valve in a heating system. Valves subjected to cavitation can be damaged, but pump impellers are much more likely to sustain more substantial damage. It doesn't take long for a pump's impeller to be badly damaged from cavitation. In fact, an impeller can be eroded to extreme levels in a matter of only a few weeks if a pump is suffering from cavitation.

What causes the damage to an impeller? When water comes into the eye of the impeller, it goes through a rapid drop in pressure. If the pressure at the eye of the impeller drops below the vapor pressure, vapor pockets form very quickly. These pockets are sent out from the impeller and reach the system fluid. When this happens, the pressure of the liquid increases above the vapor pressure. This causes the vapor pockets to implode, which is what damages the impeller. It is usually the edges of the impeller that are most affected. At this same time there is a drop in the pump's flow rate and head. This is due to the water being partially compressed. The long-term effect of cavitation is a worn impeller that must be replaced.

Gaseous cavitation occurs when there is air in a liquid. The air could be in the form of air bubbles or could be found inside vapor pockets that are caused by vaporous cavitation. When pressure is lowered, it can allow some dissolved air to come out of solution. For this to occur in an established heating system is rare. However, it is not uncommon for gaseous cavitation to take place when a new heating system is first put into service. Compared to vaporous cavitation, gaseous cavitation is not very destructive, and it is usually a short-term condition that does not require any real concern.

Cavitation in a heating system should be avoided, and it can be. By keeping static pressure for a system high and temperatures for the system low, the risk of cavitation is greatly reduced. Keeping circulators installed with their inlet ports close to an expansion tank also lessens the risk of cavitation. Circulator placement should be low in a system to maximize static pressure at the pump's inlet. Avoid putting flow-regulating valves near the inlet of the circulator. Consider installing a deaerating device in the heating system to control air within the system. The installation of a straight piece of pipe that is at least as long as 10 pipe diameters on the inlet port of the pump can help to control cavitation. And make sure that circulating pumps are installed with their inlets well below the static water level of a system. These few guidelines can make a major difference in the performance of a heating system.

Picking a Pump

Picking a pump for a heating system isn't difficult, but the selection process is important to the overall performance of the heating system. Before you can choose a pump for a heating system, you need to know the flow rate at which the system will perform well. You should also consider the thermal requirements of the system and the heat-distribution units that will be used. In the case of radiant floor heating systems, you may not have much to consider in terms of heat emitters. It is not as though you will be working with baseboard heating units, radiators, or heating panels.

You will of course need to know the head loss of the system at the flow rate which is required for the system. Working with pump curves and flow rates and head loss is all part of the pump selection process. Any circulator you choose should have a little power to spare. Installing a circulator that barely meets the system requirements is a mistake. Oversize the circulator a little to ensure good success with the heating system. A rule-of-thumb formula is to install a circulator that is rated at between 10 and 20 percent more powerful than what the system is expected to need. Any additional oversizing will result in wasted energy and cost. Plus, a pump that is too powerful can increase the flow rate of a system to a point where noise or other problems might become present within the system. You should strive to pick a pump that has a performance curve that will intersect with the system curve for maximum efficiency.

If you do your homework on circulating pumps, you will find out just how important the selection of the pumps is. Whether installed in a series or parallel manner, the pumps used with a system must be matched properly to the system to avoid problems and excessive operating costs. It doesn't take a lot of time to define what the right pump is, but it can be quite costly to install an improper pump and later have to warranty the problems. A contractor's reputation is often on the line when a pump is being chosen. Don't take this important element of system design lightly.

CHAPTER 10

Controls and Control Systems

The controls and control systems used with heating systems are essential elements of the heating systems. Some people consider the controls to be the brains of the heating system. Without proper controls, the heating systems could not function very well. If controls for a heating system are chosen and set up properly, a heating system can be very effective and efficient. Many parts of a heating system are affected by the controls of the system. For example, circulating pumps, burners, and mixing valves are all affected by the controls of a heating system. In order for a heating system to maintain a stable level of operation, the controls of the system must be installed correctly.

The controls for a heating system can range from extremely simple devices to very complex controls. Many control systems involve a variety of control types. There are many factors to consider when designing a control system. What works well with one heating system may not perform well with other systems. Installing controls requires a good working knowledge of various types of controls. For example, some controls are set manually while others work dependent on temperature. A designer must know which types of control to use for different types of heating needs. Some heating systems can work successfully with a simple on-off switch. This is as easy as it gets. But many heating systems require a much more complex system of controls to function at optimum performance.

The design of control systems can cover a wide range of options. Both electronic and microprocessor-based controls can be used in hydronic heating systems. The array of controls available to today's designers are, to say the least, vast. Modern controls are capable of performing many tasks that some years back would have required some human effort. The accuracy and ability of modern controls is sometimes hard for people to believe. Technology has produced a long line of products which can improve a heating system; so let's take a look at some of them.

On-Off Controls

On-off controls are one of four general types of controls. The other three types of controls are staged controls, modulating controls, and outdoor reset controls. Of the four, the on-off control is the simplest. Sometimes only one type of control is used, and then there are times when more than one type is used in a single system. The controls can regulate a number of things in a number of ways. For example, a circulator might be made to start, stop, or run at a given speed. Any element controlled may have a range of control, and having a wide range of controls to choose from makes this process easier. A major reason for having good controls is that the controls provide a heating system with a means of operating at peak performance.

Some people think of on-off controls as something like toggle switches. A toggle switch could be used as an on-off control. Having such a switch in an open position would keep a heating system from receiving electrical power. Closing the switch would complete the circuit and enable the heating system to run. But some types of on-off controls are automatic in nature. An example of this would be a standard room thermostat. The thermostat is technically an on-off switch, but after an initial setting, it does not require manual manipulation to activate a heating system. This concept confuses some people. How can an automatic thermostat be considered an on-off switch? Well, let me explain.

A control, like a room thermostat, is capable of making a heating system run or shut off. But, the control cannot vary the rate at which heat is delivered. In other words, the heating system is either on or off, and this is why the thermostat is an on-off control. The fact that the thermostat can have the heating system cut on when room temperature drops to a certain point doesn't make it any less an on-off control. Once a room comes up to the set temperature of a thermostat, the thermostat cuts the heating system off. This is done automatically, but it still keeps the thermostat in an on-off category. If the control could force a heating system to run at a higher heating rate, that would break the control out of the on-off category. For general purposes, any control that merely turns a heating system on or off is an on-off control, even when it is capable of working automatically.

Thermal mass is a factor in determining what type of control to use. A forced hot-air heating system has less thermal mass than a hydronic baseboard heating system. If an on-off switch is used with each of these types of heating system the hydronic system will produce more overall comfort. This is due to the added mass that the hydronic system has over the hot-air furnace. But radiant floor heating that is installed in a concrete floor has tremendous mass when it is compared to the mass of other types of heating systems. Since concrete gathers and holds heat from a radiant system, the concrete becomes a part of the heating mass. It's been said that the thermal mass of a heated concrete slab can be as much as 70 times greater than the mass obtained when heating air.

In some heating systems, such as a hot-air system, sudden bursts of heating are quite noticeable. For example, a furnace cuts on and runs for a few minutes to meet the demands of a room thermostat. As soon as the furnace cuts off, air infiltration from windows, doors, and walls can become more noticeable. The battle between warm and cold air rages on in spurts. These spurts can be discomforting. Hot-water baseboard heating systems experience spurts in their heat output, but occupants of heating space don't notice the fluctuation as much. This is due to the added mass of the hydronic heating system. When radiant floor heat is used, the fluctuations that occur are even less noticeable. This is due to the fairly constant temperature given off from the concrete as a massive heat mass.

Surges in heat output are more noticeable when there is less thermal mass. Larger storage capacity of a heat mass is needed to dampen or lessen temperature swings during fluctuations which occur with on-off switches. Radiant floor heating has a large mass and tends to produce even comfortable heat, even when spikes in heat output are present. When more control is needed to minimize strong swings in room temperatures, a staged control can be used.

Staged Controls

Staged controls differ from on-off controls in that the staged controls have an ability to regulate heat capacity. The heat capacity can be controlled in fixed intervals as the heat load increases. Since staged controls can do this, they can eliminate much of the discomfort experienced when a heating system spikes with an on-off switch. Residential applications of staged controls tend to have only a few stages of operation. But larger jobs such as commercial buildings may have a dozen or more stages set up with the control system. The advantage of a staged control is comfort, but efficiency is usually better, also, when a staged control is installed.

As the name implies, staged controls work in stages. A heating system starts off at a low setting or stage. When there is enough demand for more heat, a second stage is entered. This process continues for as many stages as necessary to reach maximum heat output. When the first stage of heating is being used, it will operate intermittently. As heating demand escalates, the system's first stage or lower stages will begin to run continuously to meet the demand while the higher-level stage runs intermittently. Once all stages are running continuously, the system is up to maximum output.

The use of a staged control generally reduces operating costs. It also provides a more stable level of comfort, since there are not such extreme drops and jumps in heat output. To gain the best match between heat load and heat output, many stages are required. This is due to the need to define output in smaller increments. However, most residential applications don't require more than two or three stages. Many oil-fired boilers used in homes are set up for only a single-stage heating system, so this rules out the staged control system. Gas boilers are more likely to accept staged controls in residential applications. The reason for the difference is the oil nozzle in an oil-fired boiler. The

QUICK»TIP The use of a staged control generally reduces operating costs.

size limitations on small oil-fired boilers are not compatible with a staged system. There is more cost involved with the installation of a staged system, but if a job is large enough to warrant such a system, the added cost can usually be recovered through lower operating costs.

Modulating Controls

Modulating controls are the cat's meow when it comes to nearly ideal control over heat load and heat output. This is because a modulating control makes continuous shifts in the heat output to match the heat load. Usually modulating controls work by controlling the water temperature being supplied to a heat system. By raising and lowering the temperature of supply water, the modulating control can maintain a constant favorable heat. A motorized mixing value is normally used to accomplish the moderation between temperatures. Sometimes, however, a variable-speed pump is used to bring hot water into a constantly circulating distribution system. There are a few other ways to accomplish modulating control, but the two methods discussed above are the most common.

Outdoor Reset Controls

Outdoor reset controls are extremely effective in controlling stable heat temperatures. This type of control reacts to outside temperatures. When the temperature outside becomes colder, the outdoor reset control increases the temperature of water being supplied to a heating system. In contrast, as the outdoor temperature rises, the temperature of supply water is lowered. This process produces very constant heat while cutting back on operating costs. There is no question that an outdoor reset control is efficient and effective, but relatively few of them are used in average homes.

Electrical Functions

Electrical functions for heating systems are controlled by switches and relays. These devices are classified by the number of poles and throws that they have. Most switches have between one and three poles. The poles refer to the number of independent paths through which elec-

ch that is electrically

tricity can pass. When switches and relays are classified by the number of throws that they have, the consideration is how many settings exist where electricity can pass through the device. Electrical devices chosen for a system must meet the criteria required by the system. For example, the current and voltage ratings of a switch must match the current and voltage specifications for the heating system's application. The ratings for the switch must meet or exceed the amount of current and voltage that will be required to operate the heating system.

Some people are not sure what a relay is. Simply put, a relay is a switch that is electrically operated. A relay can be operated remotely. The two major parts of a relay are its coil and its contacts. There is also a spring, a pendulum, and the relay terminals. If any of these components fail, a relay can become useless. Contacts are held in their open position by the spring. If the right amount of voltage reaches the coil, a magnetic force is created. This process causes the contacts to come together. When this happens, the circuit is complete. The relay contacts will stay in touch with each other until the voltage load on the coil is reduced, at which time the contacts will come apart and return to their open position.

There are time-delay relays and general-purpose relays that might be used in a heating system. A general-purpose relay is normally used when a relay is required to be installed as a separate component. These relays can be plugged into a relay socket where external wires are attached to screw terminals or quick-connect tabs. When a delay is needed between two control events, a time-delay relay is used. Since time-delay relays have adjustable timing circuits built into them, they can be set to different modes. Time-delay relays plug into sockets just as general-purpose relays do. The purpose of a time delay is to protect a compressor against unequalized pressure starts and to prevent short-cycle operation.

Thermostats

Thermostats are well-known controls associated with heating systems. To most people outside of the heating industry, a thermostat is thought of as the only control needed or used with a heating system. Homeowners, for example, know that turning the dial of a thermostat up or down will cut their heating systems on or off. This is true, of course, but the thermostat is not the only control involved.

QUICK»TIP **Thermostats used with heating systems have what is called a setpoint temperature. This is simply the temperature which the thermostat is set to maintain.**

A standard room thermostat is essentially a temperature-operated switch. The thermostat indicator is placed on a particular temperature rating. If the air temperature that the thermostat is exposed to is colder than the temperature that the thermostat indicator is set on, the heating system will come on to warm the room. Once the room temperature is equal to the setting on the thermostat, the heating system will be turned off by the thermostat. Turning a single-stage room thermostat to a higher temperature once a heating system is running will not increase the rate of the heat output. The thermostat only controls the on-off function of the system and does not regulate the rate at which heat is produced. Many people don't understand this.

Thermostats used with heating systems have what is called a setpoint temperature. This is simply the temperature which the thermostat is set to maintain. In a perfect world, a heating system would maintain the setpoint temperature at all times, but this isn't practical. In order to keep a room at its setpoint temperature constantly, a heating system would be forced to cut on and off very frequently. This would put the heat source under extreme stress and would shorten its working life. To avoid this problem, thermostats are designed with a differential.

The differential of a thermostat is sometimes adjustable, but some thermostats have factory-set differentials that are not adjustable. Most thermostats are set to have a differential of up to 5°. This means that a thermostat will not cut on until room temperature drops to a point 5° below the setting of the thermostat. It's common to find a thermostat with a differential of 3°. The lower the differential is, the more often a heating system will cycle on and off. This is good for comfort but hard on the heating system.

Thermostats should never be mounted on exterior walls or near sources of heat or cold. A thermostat should be placed on an interior wall in a central location. Most thermostats are installed about chest high, but this can vary with individual needs.

Thermostats should never be mounted on exterior walls or near sources of heat or cold. A thermostat should be placed on an interior wall in a central location. Most thermostats are installed about chest high, but this can vary with individual needs. For example, a person who is confined to a wheelchair would benefit from a thermostat that is mounted at a lower height. This is not only a reach issue, but it's that the thermostat should be installed at a height where comfort is most often sought. Since a person in a wheelchair will not normally be as high off the floor as a person walking, the lower mounting level will allow for a better comfort zone.

Aquastats

Aquastats are controls that monitor and control the temperature of liquids, such as water in a hydronic heating system. An aquastat fulfills two purposes. It acts as a safety device as well as a control. In the case of a radiant heating system, an aquastat might be used to monitor and maintain the temperature of supply water going into the heating tubing. Some aquastats are made to strap onto a pipe. Others have a copper capillary tube that is actually inserted into the liquid being monitored (Fig. 10.1).

When a strap-on aquastat is used, its sensing bulb is pressed tightly against the pipe that the aquastat is attached to. The bulb contains a fluid that increases in pressure as the fluid temperature rises. If the bulb expands, pressure is placed on a diaphragm or bellows assembly.

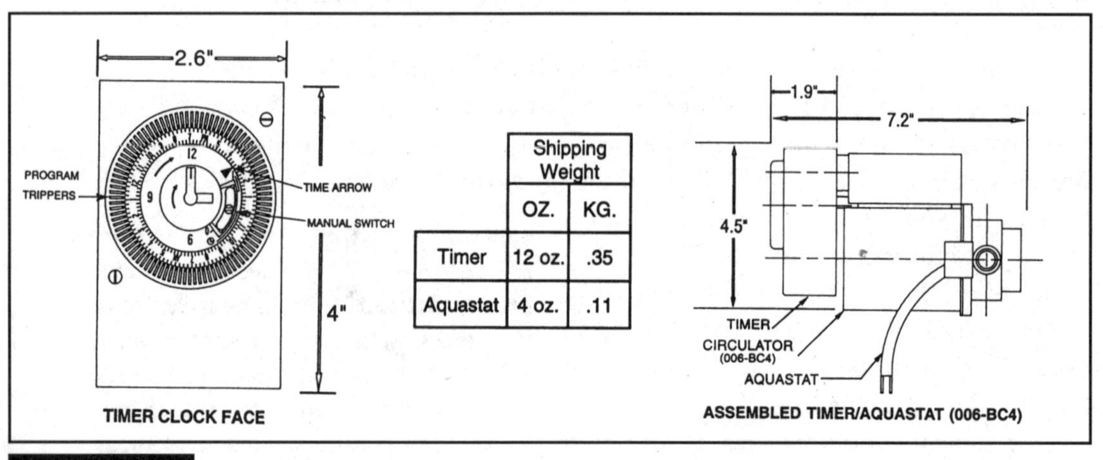

	Shipping Weight	
	OZ.	KG.
Timer	12 oz.	.35
Aquastat	4 oz.	.11

FIGURE 10.1

Aquastat. (*Courtesy of Taco.*)

Expansion occurs and either opens or closes the electrical contacts of the control. For maximum control, the sensing bulb of an aquastat should be immersed in the liquid that is being monitored. Special fittings are available that allow aquastats to be mounted on regular threaded fittings with the sensing bulb in the pipe where liquid will be monitored.

High-Limit Controls

High-limit controls are often used on small boilers, such as those used in residential applications. This type of control is often packaged with boilers that are being sold for residential use. In this case, a single control contains several components. There can be a transformer, a relay, and an aquastat all contained in a small package. A sensing bulb comes out of the aquastat and goes into the rear case of the boiler block. When a high-limit control is used, it is both simple and inexpensive. The one device fulfills several functions with good efficiency on small heat sources.

Relay Centers

Relay centers are often set up for heating systems where multiple zones are to be used. The relay center keeps wiring and control operation together in a central location. A relay center can be set up so that a domestic water heating device can take priority over the heating zones if hot water for domestic use is called for. Typical relay centers are designed to handle anywhere from three to six heating zones. It's possible to have a relay center that is designed for use with 24-V ac zone valves or line voltage circulators. Contractors often install relay centers that are larger than what is immediately required. This makes expanding the number of zones on a system easier. If the demand for zones exceeds six zones, multiple relay centers can be installed.

Zone Valves

Zone valves are used to control the flow of water in hydronic heating systems. Each circuit of a heating system needs some form of control. The control can be independent circulating pumps or zone valves.

MATERIALS

Zone valves are used to control the flow of water in hydronic heating systems. Each circuit of a heating system needs some form of control. The control can be independent circulating pumps or zone valves. Since zone valves are less expensive than circulating pumps, zone valves are most often use to control the flow in heating zones.

Since zone valves are less expensive than circulating pumps, zone valves are most often use to control the flow in heating zones. There are many different types of zone valves, but they all share common characteristics. All zone valves have a valve body and an actuator. It is the actuator that moves to allow liquids to flow or to stop them from flowing. Electrical voltage is applied to an actuator to make the valve shaft of the zone valve move.

Valves that open or close too quickly can cause water hammer in a piping system. The loud banging that comes from a water hammer is disturbing and should be avoided. Some zone valves protect against water hammer by using small electric motors with gears to open and close the valve in a zone valve. The small motors turn the shaft of the valve fast enough to open the valve fully in time for rapid head delivery, but slowly enough to prevent water hammer. A second type of zone valve uses heat motor actuators. An internal resistor heats the actuator so that it will operate the valve. This type of zone valve opens more slowly than the type with the motor and gear, but it is adequate for most heating systems.

Controls are essential to the proper operation of a heating system. Every heating system can have individual needs, though most of the systems will be similar in nature and design. Some contractors install systems with somewhat of a cookie-cutter design. This can be a mistake. A standard control system might work fine for eight out of ten heating systems, but what about the other two? Every heating system should be evaluated individually for its control needs. Most big jobs are drawn and specified by experts. This is not usually the case in residential work. It's often the heating contractor or someone at a supply house who comes up with residential designs. You owe it to yourself and your customers to take the design of control systems seriously. After all, if the controls are not selected or installed properly, a heating system is not likely to work well, and this can be a bad reflection on you and your company.

CHAPTER

11

Components for Heating Systems

The components that go into a hydronic heating system are both numerous and important. Many pieces of equipment are needed to make a safe, functional hydronic heating system. People generally think of a boiler and some type of heat emitter, such as baseboard heating units. Rarely is much thought given to circulating pumps, expansion tanks, zone valves, relief valves, and other essential elements of a safe system. The components of a hydronic heating system that may seem inconsequential are not. They are key elements of the system. Failing to install an expansion tank or relief valve could prove disastrous. Without the use of a circulating pump, a modern hydronic system will not function very well. Zone valves are an inexpensive alternative to multiple circulating pumps when creating multiple heating zones with a system. A number of factors come into play when designing and installing a hydronic heating system. Learning what is needed, where it is needed, and how to install it is essential if you wish to install effective hydronic heating systems.

Modern Piping Materials

Modern piping materials in hydronic heating systems include, most often, copper tubing and cross-linked polyethylene tubing (PEX).

MATERIALS

PEX tubing is a polymer (plastic) material. It is extremely flexible, sold in long coils, and suitable for many hydronic applications. One of the most effective applications for PEX tubing is found in radiant floor heating. Standard PEX tubing can handle water with a temperature of 180° at 100 psi. If the pressure is reduced to 80 psi, the tubing can handle water temperatures up to 200°.

Another type of tubing sometimes used is polybutylene (PB) tubing. Copper tubing is the most popular type of piping used for general convection heating, such as systems utilizing baseboard heating elements, kick-space heaters, and space heaters. PEX tubing is most often used for radiant floor heating systems. PB tubing was being used before PEX tubing, and it is still in use, but PEX is replacing it quickly as a prime choice. This is due in no small part to recent problems with PB tubing and pending lawsuits arising from the problems.

The copper tubing used most in heating systems is type M copper in a hard-drawn (rigid) form. Type L copper is sometimes used in tight spots when a flexible rolled copper tubing is more feasible. Distribution tubing typically has a diameter of 3/4-in. Since copper expands and contracts with temperature variations, the tubing must be supported properly to maintain a quiet heating system.

PEX tubing is a polymer (plastic) material. It is extremely flexible, sold in long coils, and suitable for many hydronic applications. One of the most effective applications for PEX tubing is found in radiant floor heating. Standard PEX tubing can handle water with a temperature of 180° at 100 psi. If the pressure is reduced to 80 psi, the tubing can handle water temperatures up to 200°.

Both copper and PEX tubing have their places in heating systems. Matching the proper tubing to the job is important. A general rule of thumb is to use copper tubing for general heating applications and PEX for radiant floor heating. There are exceptions to this, of course, but in general, the formula works.

Fittings

The fittings used in heating systems are often the same as those used in plumbing jobs. There are, however, some special fittings that are used most frequently with heating systems. Typical generic fittings include couplings, slip couplings, reducing couplings, 45° elbows, 90° elbows, male adapters, female adapters, unions, and tees of both full

TACO HY-VENT

The Taco Hy-Vent is available in 1/8-inch, 1/4-inch and 3/4-inch sizes. They automatically vent air from the system when installed at the highest points. The Hy-Vent is a high-capacity float type design. The Taco Hy-Vent is ideal for those high operating pressure jobs.

SIZE & DIMENSIONS

PRODUCT NUMBER	SIZE CONNS.	MAXIMUM WORK. PRESS.	MAXIMUM OPERAT. TEMP.	DIMENSIONS	APPROX. SHIP. WT. LBS.	
					EACH	CARTON
417	1/8" MALE	125 psig.	240°F	3/4" x 1"	.1	1
400	1/8" MALE	150 psig.	240°F	1 29/32" x 3 1/2"	.4	4.5
426	1/4" MALE	150 psig.	240°F	2 1/4" x 4 1/2"	.6	7
418	3/4" MALE	150 psig.	240°F	2 1/4" x 4 1/2"	.6	7

FIGURE 11.1

Air vent. (*Courtesy of Taco.*)

size and reducing sizes. All of these are basic plumbing fittings. Now let's look at the fittings that are primarily used with heating systems.

Baseboard tees, also called baseboard ells, are fittings that are shaped like a standard 90° elbow. However, the fitting has a threaded fitting in the bend of the elbow. The threads are there to accept the installation of an air vent/purger. These fittings are often used when copper tubing rises vertically through a floor and is turning on a 90°angle into a section of baseboard heat. The fittings are usually made of wrought copper or cast brass. With radiant floor heating, air vents are installed at supply and return manifolds (Figs. 11.1 and 11.2).

Diverter tees (Fig. 11.3) create a flow through a branch piping path that passes through one or more heat emitters before reconnecting to a primary piping circuit. The use of diverter tees can involve pushing or pulling water through a piping path. It can be difficult to tell a diverter tee from a regular tee when relying only on outward appearance. However, a peek inside will reveal the diversion section of the tee. When diverter tees are used, they must be installed in the proper location and direction. Diverter tees are equipped with arrows to indicate proper positioning for water flow. Some installers in the trade call the diverter tees venturi tees. Most commonly they are known by the registered trademark name of MonoFlo tees, which is a product and trademark of the Bell & Gossett Company.

417 HOT WATER AND STEAM VENT

The 417 is available in a 1/8-inch connection. It can be used as an automatic or manual vent. The 417's compact design makes it suitable for installation on baseboard, steam or hot water radiators and convectors. This low profile design allows the 417 to fit underneath the cover of the baseboard.

FIGURE 11.2

Slotted air vent. (*Courtesy of Taco.*)

Dielectric unions are often used in plumbing applications. They are often installed on electric water heaters. These same unions are sometimes used with heating systems. A dielectric union mates together with two dissimilar metals. This breaks the continuity of a conductive reaction. In turn, it avoids galvanic corrosion. The unions are used to fit copper materials to steel materials.

Heating Valves

Valves used in heating systems can be of the same type used in plumbing systems, but this is not always the case. Selecting a valve for a heating system may seem like a simple task, but it can be more important than you might think. Valves are typically used for either component isolation or flow regulation. Gate valves and globe valves are two common types of valves used in heating systems. These same valves are also used in plumbing systems. Understanding which valves to use, why they should be used, and when they should be used will be of value to you as you install hydronic heating systems. So let's explore the different types of valves that may be of interest to you in your heating jobs.

Gate Valves

Gate valves are used frequently in both plumbing and heating. These valves are intended for use as isolation valves. They are not meant to be used as flow regulators. This means that gate valves should be either fully open or completely closed. When open, a gate valve does not affect flow velocity very much. Don't install gate valves where flow regulation is required. If the gate is partially down in a gate valve,

VENTURI FITTINGS:

The Taco Venturi Fittings are designed to divert water flow from a given zone to the by-pass loop. The flow from the zone is partially diverted through the by-pass and then returned back into the zone. The Venturi Fitting creates a differential pressure that makes some of the flow want to divert through the by-pass and back to the zone. The Taco Venturi Fitting can be used in upfeed or downfeed applications. The typical application for the Taco Venturi Fitting is to divert heated water through kickspace heaters, convectors, radiators or baseboard.

SIZE & DIMENSIONS

Model No.	Size Inches	Dimensions-Inches		Approx. Ship. Wt. Lbs. Each
		A	B	
VF-075-050	3/4x1/2	2.30	1.00	1/4
VF-100-050	1x1/2	2.75	1.22	1/2
VF-100-075	1x3/4	3.0	1.41	
VF-125-050	1 1/4x1/2	3.0	1.30	1/2
VF-125-075	1 1/4x3/4	3.30	1.50	

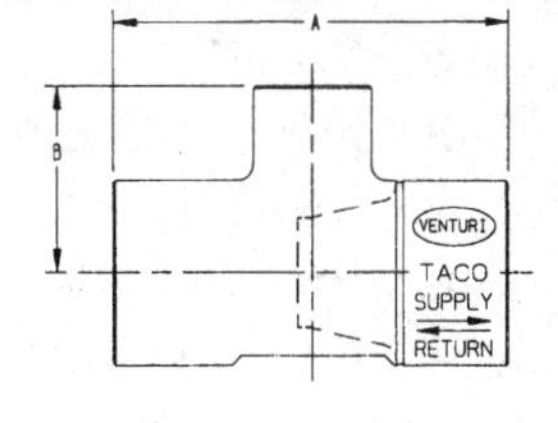

FIGURE 11.3

Diverter tee. (*Courtesy of Taco.*)

to regulate flow, there is a strong likelihood that vibration or chatter will be the result.

Globe Valves

Globe valves can be used to regulate flow. In fact, that's what they are intended to do. There is a right and a wrong way to install a globe valve. Make sure that any globe valve you install is positioned so that water enters the lower body chamber. If water comes in from the other end, noise in the system is a strong possibility. Globe valves should not be used to isolate equipment. The design of a globe valve does not make it ideal for isolation. While globe valves can be used for isolation, they shouldn't be, since they are not the most efficient choice.

Ball Valves

Ball valves are probably one of the most used valves in the heating industry. These valves can be used for isolating components or regulating flow. The positive closing action of a ball valve makes it a fine choice as an isolation valve. While ball valves can be used for flow regulation, it is not considered wise to use ball valves to control flow when the flow must be reduced by more than 25 percent. There is no hard and fast rule on this. It is a recommendation to maintain valve condition.

Check Valves

Check valves are needed to ensure that fluids do not flow in an unwanted direction. Two common types of check valves are used in heating systems. One is a swing check and the other is a spring-loaded check valve. When a swing check valve is installed, it must be installed on a horizontal line with its bonnet pointing straight up. The operation of a swing check is simple. Fluid flowing through the valve in the proper direction holds the flap of a swing check open. If, for any reason, a backflow situation occurs, the flap of the check valve closes, preventing the backflow.

Spring-loaded check valves are not so sensitive to orientation as swing checks are. This is one reason why installers like spring-loaded valves. Because of the spring action, these valves can be use in any orientation. While it is not important that a spring check be installed in a

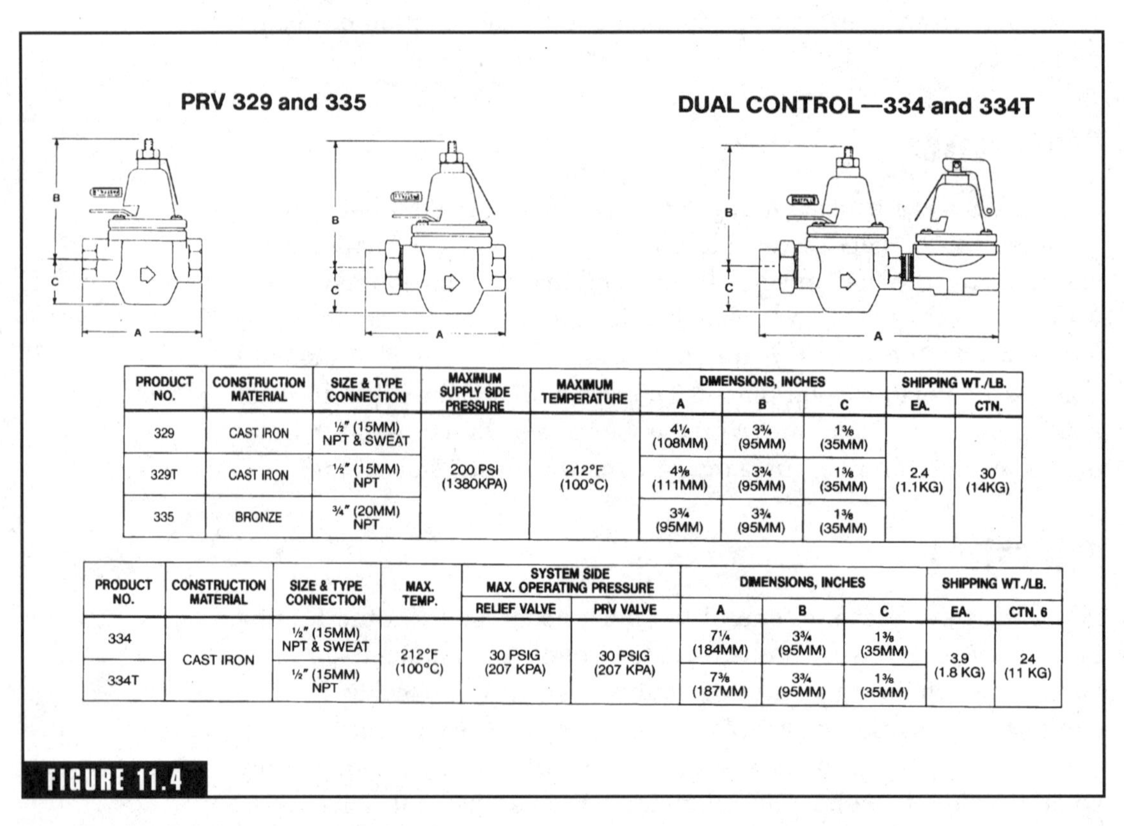

PRODUCT NO.	CONSTRUCTION MATERIAL	SIZE & TYPE CONNECTION	MAXIMUM SUPPLY SIDE PRESSURE	MAXIMUM TEMPERATURE	DIMENSIONS, INCHES			SHIPPING WT./LB.	
					A	B	C	EA.	CTN.
329	CAST IRON	½" (15MM) NPT & SWEAT	200 PSI (1380KPA)	212°F (100°C)	4¼ (108MM)	3¾ (95MM)	1⅜ (35MM)	2.4 (1.1KG)	30 (14KG)
329T	CAST IRON	½" (15MM) NPT			4⅜ (111MM)	3¾ (95MM)	1⅜ (35MM)		
335	BRONZE	¾" (20MM) NPT			3¾ (95MM)	3¾ (95MM)	1⅜ (35MM)		

PRODUCT NO.	CONSTRUCTION MATERIAL	SIZE & TYPE CONNECTION	MAX. TEMP.	SYSTEM SIDE MAX. OPERATING PRESSURE		DIMENSIONS, INCHES			SHIPPING WT./LB.	
				RELIEF VALVE	PRV VALVE	A	B	C	EA.	CTN. 6
334	CAST IRON	½" (15MM) NPT & SWEAT	212°F (100°C)	30 PSIG (207 KPA)	30 PSIG (207 KPA)	7¼ (184MM)	3¾ (95MM)	1⅜ (35MM)	3.9 (1.8 KG)	24 (11 KG)
334T		½" (15MM) NPT				7⅜ (187MM)	3¾ (95MM)	1⅜ (35MM)		

FIGURE 11.4

Pressure-reducing valves. (*Courtesy of Taco.*)

straight-up position, it is essential that the valve be installed with the right direction of flow. An arrow on the valve makes it easy to know which direction to install the valve in.

Pressure-Reducing Valves

Pressure-reducing valves (Fig. 11.4) are used to lower the pressure of water in a system before it enters a boiler or water-distribution system. In the case of heating, the pressure-reducing valve, also known as a boiler feed valve, is used to lower pressure from the water-distribution system prior to its entering a boiler. As with check valves, pressure-reducing valves must be installed in the proper direction. The valves have arrows to point them in the direction of flow. There is usually a lever on the top of a pressure-reducing valve that can be lifted manually to speed up the filling process for a boiler during a setup and start-up procedure.

Pressure-Relief Valves

Pressure-relief valves are important safety valves. When temperature or pressure reaches a risky level, these valve will open to release water that might otherwise cause damage or even an explosion. These valves are required by code on all hydronic heating systems. It's common for new boilers to be shipped with a relief valve already installed. A typical rating for a relief valve in a small boiler is 30 psi. Never install a heating system without a relief valve. Relief valves are equipped with threads to accept a discharge tube. This tube is very important. If a relief valve blows off without a proper safety tube installed, people could be seriously injured from hot water or steam. I see far too many relief valves that are not equipped with discharge tubes. The tube should run from the valve to a point about 6 in. above a floor drain. If a drain is not available, at least extend the pipe to within approximately 6 in. above the finished floor level.

Backflow Prevention

Backflow prevention is an important part of a hydronic heating system. It simply would not do to have boiler water mixing with potable water. Local codes require backflow prevention, and various types of devices are available to achieve code requirements. Most systems utilize in-line devices. These backflow preventers must be installed in the proper direction. They have arrows on the valve bodies to indicate the direction of flow. It is common for this type of device to have a threaded opening about midway on the valve that will act as a vent. Again, a safety tube should be installed on the threaded vent opening to prevent spraying or splashing if the valve discharges.

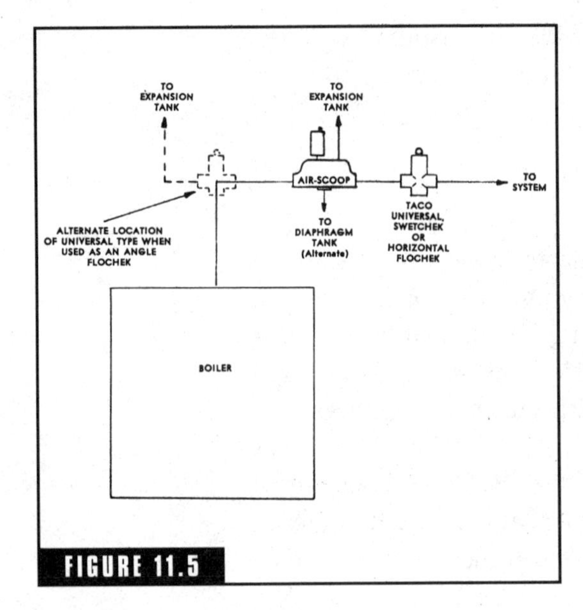

FIGURE 11.5

Flow-check valve. (*Courtesy of Taco.*)

Flow Checks

Flow-check valves (Figs. 11.5 and 11.6) are another type of valve used in heating systems. These valves have weighted internal plugs that are heavy enough to stop thermosiphoning or gravity flow when a system's circulating pumps are not running. Some flow checks have two ports, while others have three. Valves with two ports are intended for use in horizontal piping. When three ports exist, one can be plugged and the valve can be used in a vertical application. A small lever on top of the valves allows the valve to be opened manually in the event of a circulator failure. The lever, however, cannot be used to control flow rate. Again, the direction of flow is important when installing a flow check. An arrow on the body of the valve will indicate the direction of flow as it should be used with the valve.

Mixing Valves

Mixing valves are used to mix cold water with hot water to create a regulated temperature in water being delivered from the mixing valve. This might be the case in a hydronic system that uses both baseboard heat emitter and radiant floor heating. The temperature of water for the floor heating would need to be lower than that of the water used for the

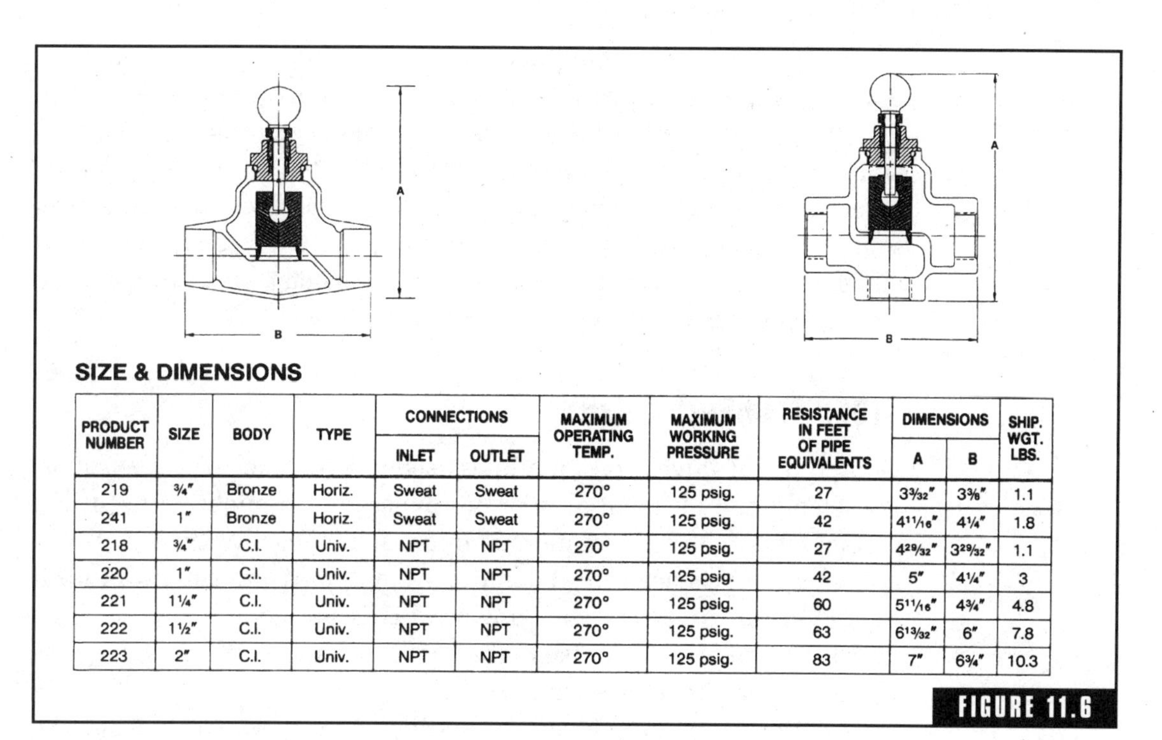

SIZE & DIMENSIONS

PRODUCT NUMBER	SIZE	BODY	TYPE	CONNECTIONS		MAXIMUM OPERATING TEMP.	MAXIMUM WORKING PRESSURE	RESISTANCE IN FEET OF PIPE EQUIVALENTS	DIMENSIONS		SHIP. WGT. LBS.
				INLET	OUTLET				A	B	
219	¾"	Bronze	Horiz.	Sweat	Sweat	270°	125 psig.	27	3³⁄₃₂"	3⅜"	1.1
241	1"	Bronze	Horiz.	Sweat	Sweat	270°	125 psig.	42	4¹¹⁄₁₆"	4¼"	1.8
218	¾"	C.I.	Univ.	NPT	NPT	270°	125 psig.	27	4²⁹⁄₃₂"	3²⁹⁄₃₂"	1.1
220	1"	C.I.	Univ.	NPT	NPT	270°	125 psig.	42	5"	4¼"	3
221	1¼"	C.I.	Univ.	NPT	NPT	270°	125 psig.	60	5¹¹⁄₁₆"	4¾"	4.8
222	1½"	C.I.	Univ.	NPT	NPT	270°	125 psig.	63	6¹³⁄₃₂"	6"	7.8
223	2"	C.I.	Univ.	NPT	NPT	270°	125 psig.	83	7"	6¾"	10.3

Flow control valve. (*Courtesy of Taco.*)

baseboard system. This is possible with the use of a mixing valve. Using knobs or levers on the outside body of a mixing valve is all that is required to regulate the temperature of water being delivered from the valve.

Zone Valves

Zone valves can be used in place of additional circulators when a hydronic heating system is being zoned off into different zones. Circulating pumps cost more than zone valves, so the zone valves are often used in place of additional circulators. Some old-school installers don't like zone valves. A preference between circulators and zone valves is a personal matter. Most installers are comfortable using zone valves, and many contractors use them extensively.

There are two types of zone valves. One type uses a small electric motor combined with gears to produce a rotary motion of a valve shaft.

The other type uses heat motors to produce a linear push-pull motion. Both types consist of a valve body and an actuator. Either type of valve must be operated either fully open or completely closed.

Zone valves for residential systems are located on the supply pipe of each zone circuit. They are generally positioned near the heat source. In most cases, zone valves are equipped with transformers that allow them to be wired with regular thermostat wire. Some zone valves, however, are designed to work with full 110-V power.

Other Types of Valves

Other types of valves are sometimes used in hydronic heating systems. For example, a differential pressure bypass valve might be used in a system where there are numerous individual zones. When a large circulating pump is used for the entire system, pressure can build up if several of the zones are shut down. This is not good, owing to high flow rates and possible noise in the zones that are running. The bypass valve eliminates the problem.

Metered balancing valves are another type of valve that may be encountered in a hydronic heating system. These valves may be used in multizone systems where more than one parallel piping path exists. The flow rates in the pipes must be balanced to produce desirable heating conditions, and metered balancing valves make this possible.

Another type of valve that is sometimes used is the lockshield-balancing valve. This is a valve that allows a system to be isolated, balanced, or even drained at individual heat emitters. These valves can be purchased in a straight in-line fashion or in an angle version. The installation of lockshield-balancing valves is not common in typical residential applications, but they may be installed in such systems.

Circulating Pumps

Circulating pumps are to a heating system what the heart is to the human body. The pumps move fluid through the pipes of a heating system. Closed-loop, fluid-filled, hydronic heating systems may be equipped with a single circulating pump or with many. When zone valves are use, it is common for only a single pump to be installed. Most circulating pumps are centrifugal pumps. The design and types of circulating pumps vary. Many of them are of a three-piece design while others are in-line designs.

Wet-rotor circulators are quite common in small heating systems. This type of pump has a motor, a shaft, and an impeller fitted into a single assembly. The assembly is housed in a chamber that is filled with system fluid, and the motor of the pump is cooled and lubricated by the system fluid. There are no fans or oiling caps. Maintenance of a wet-rotor circulator is minimal. Because of their quiet operation and worry-free maintenance, wet-rotor circulators rule the roost when it comes to residential and light commercial heating systems.

TRADE SECRET

Deciding on where to place circulating pumps is not a job that should be taken lightly. Proper placement has much to do with the quality of service derived from a pump. A rule of thumb to follow is to keep all circulators located in a manner so that their inlet is close to the connection point of the system's expansion tank.

Three-piece circulators are also common in residential and light commercial applications. One advantage to this type of pump is that the motor is not housed within the system fluid. If a problem exists, the motor can be worked on without opening the wet system. The disadvantage of a three-piece circulator is that it must be oiled periodically and more noise is present during operation, since the motor is mounted externally.

Pump Placement

Deciding on where to place circulating pumps is not a job that should be taken lightly. Proper placement has much to do with the quality of service derived from a pump. A rule of thumb to follow is to keep all circulators located in a manner so that their inlet is close to the connection point of the system's expansion tank. The reason for this is that the expansion tank is responsible for controlling the pressure of a system's fluid. By keeping circulators near the part of the system that is considered to be the point of no pressure change, which is the location of the expansion tank, the circulators are always working with a constant pressure. Therefore, the circulators can give better, more uniform service.

How a circulating pump is mounted in a system is also important. It is not wise to hang a circulator on flimsy piping. Many installers build their headers, the section of piping where circulators are mounted, with steel pipe and then switch to copper tubing as they distribute water to heat emitters. This is a good idea. The circulators can, however, be mounted in copper pipe or tubing lines, but the pump should be supported well with some type of hanger or bracket.

Most circulators used in residential heating systems are designed to be installed with their shafts in horizontal positions. In doing this, pressure from the thrust load on bushings, due to the weight of the rotor and impeller, is reduced. Like so many other in-line components of a heating system, circulators must be installed with the proper orientation to water flow. There are arrows on the pump housings to indicate the proper direction of flow for installation.

Circulators are mounted to a system with the use of flanges. The flanges are bolted to the circulator, so that the pump may be removed for repair or replacement. Smart installers place valves on either side of circulators to make the replacement procedure easier if it is ever required. By having isolation valves both above and below a pump, it is a simple matter to remove the pump without draining the entire system.

Expansion Tanks

Any hydronic heating system must be equipped with an expansion tank (Fig. 11.7). When water is heated in a heating system, the liquid expands. Since thermal expansion of this type is unavoidable, it must be given a means to occur without damage to the heating system. This is done with an expansion tank. The air cushion that is provided in an expansion tank allows water to expand and contract naturally, without fear of damage to the heating system or people in its vicinity. Without an expansion tank, a hydronic heating system could rupture or explode, causing severe damage to both property and people.

The concept behind an expansion tank is easy to understand. The tank has a specified amount of air in it. When water is forced into the tank, the air is compressed. The volume of water in the tank increases, but the air cushion creates a buffer for the expanding water. For example, if the water temperature in the expansion tank is 70°, the tank might be half full of water. The same tank holding water at a temperature of 160° might have only one-fourth of its space not filled with water. These are not scientific numbers, just examples. Basically, the cooler the water is, the less water there will be in the tank.

Old-style expansion tanks were basically just metal tanks with an air valve on them where air could be injected. A common problem with this type of tank was related to keeping air in the tank. As air escaped, the tank filled with more water than it should have. This condition is known as waterlogging. It used to be a common problem in both heating and well systems. The problem of waterlogging has been all but eliminated with new technology in the form of diaphragm-type

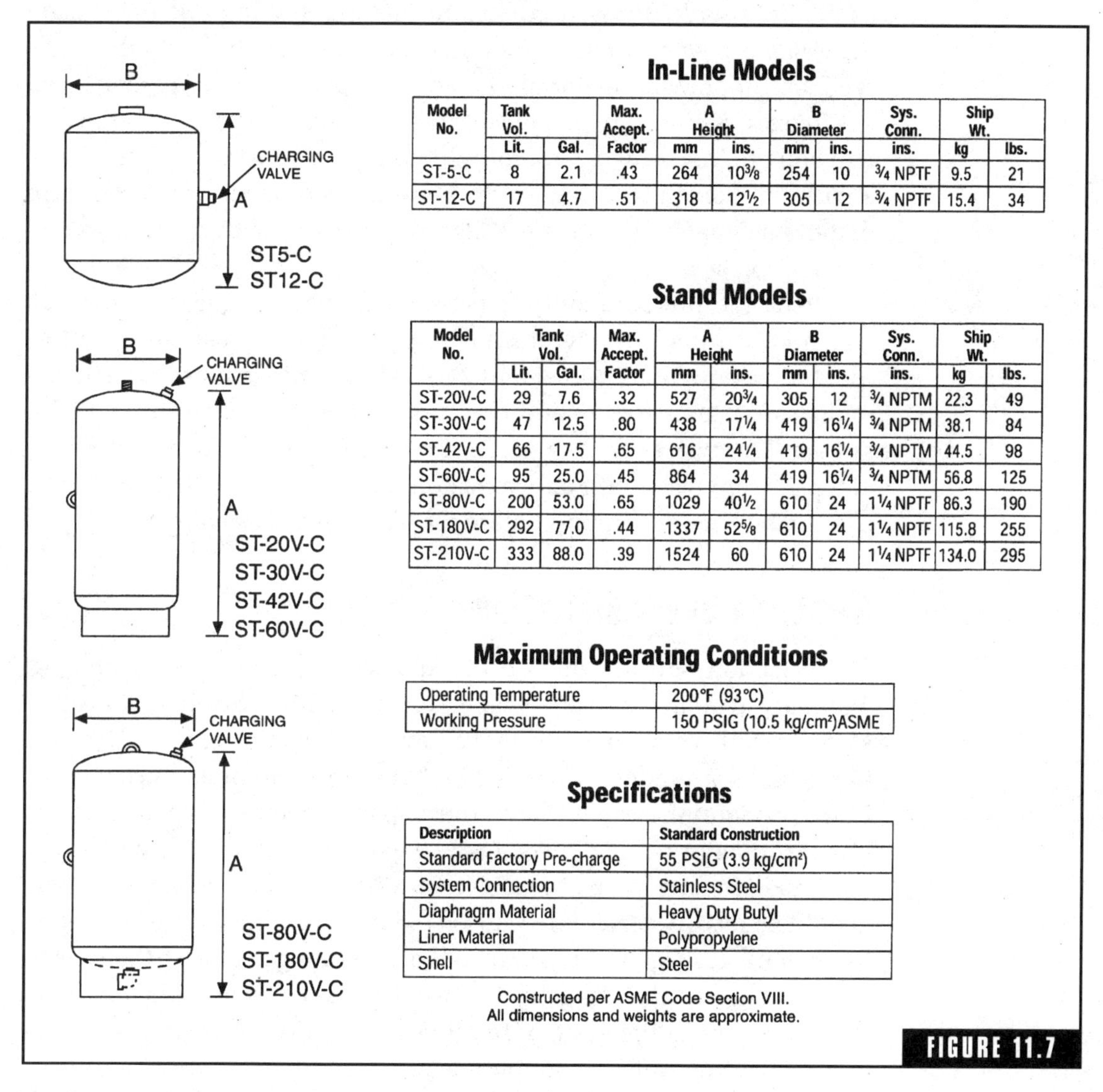

In-Line Models

Model No.	Tank Vol.		Max. Accept.	A Height		B Diameter		Sys. Conn.	Ship Wt.	
	Lit.	Gal.	Factor	mm	ins.	mm	ins.	ins.	kg	lbs.
ST-5-C	8	2.1	.43	264	10 3/8	254	10	3/4 NPTF	9.5	21
ST-12-C	17	4.7	.51	318	12 1/2	305	12	3/4 NPTF	15.4	34

Stand Models

Model No.	Tank Vol.		Max. Accept.	A Height		B Diameter		Sys. Conn.	Ship Wt.	
	Lit.	Gal.	Factor	mm	ins.	mm	ins.	ins.	kg	lbs.
ST-20V-C	29	7.6	.32	527	20 3/4	305	12	3/4 NPTM	22.3	49
ST-30V-C	47	12.5	.80	438	17 1/4	419	16 1/4	3/4 NPTM	38.1	84
ST-42V-C	66	17.5	.65	616	24 1/4	419	16 1/4	3/4 NPTM	44.5	98
ST-60V-C	95	25.0	.45	864	34	419	16 1/4	3/4 NPTM	56.8	125
ST-80V-C	200	53.0	.65	1029	40 1/2	610	24	1 1/4 NPTF	86.3	190
ST-180V-C	292	77.0	.44	1337	52 5/8	610	24	1 1/4 NPTF	115.8	255
ST-210V-C	333	88.0	.39	1524	60	610	24	1 1/4 NPTF	134.0	295

Maximum Operating Conditions

Operating Temperature	200°F (93°C)
Working Pressure	150 PSIG (10.5 kg/cm²)ASME

Specifications

Description	Standard Construction
Standard Factory Pre-charge	55 PSIG (3.9 kg/cm²)
System Connection	Stainless Steel
Diaphragm Material	Heavy Duty Butyl
Liner Material	Polypropylene
Shell	Steel

Constructed per ASME Code Section VIII.
All dimensions and weights are approximate.

FIGURE 11.7

Expansion tank. (*Courtesy of Amtrol.*)

tanks. These tanks have a flexible diaphragm built into them that regulates air charges and greatly reduces or eliminates waterlogging.

How do diaphragm tanks work? They are fitted with a synthetic diaphragm that separates the captive air in the tank from water in the tank. By doing this, air loss is greatly diminished. These tanks are the rule rather than the exception in modern heating and well systems. It is important that the diaphragm material used in an expansion tank be compatible with fluids used in the heating system. All diaphragm materials are safe to use with water, but butyl rubber, which is one type of diaphragm material, is not compatible with glycol-based antifreezes, which may be present in some heating systems. If the system you are installing a tank for will contain glycol-based antifreezes, a tank with an EPDM diaphragm material is a better choice. Hydrin diaphragm materials are most commonly used with solar-powered, closed-loop systems.

Check the pressure and temperature ratings assigned to any tank you are planning to use with a heating system. A typical rating for residential use might be 60 psi and 240°. These ratings must be matched to the safety temperature and pressure relief (T&P) valves used with a system. If a typical T&P valve for a residential water heater, rated for 150 psi, were used with an expansion tank system that is rated for 60 psi, the tank could rupture before the T&P valve discharged.

Selection and Installation

The selection and installation of expansion tanks can be confusing. Tanks are available in many shapes and sizes. Most residential heating systems will work fine with tanks having capacities of 10 gal or less. Some tanks are mounted vertically. Others are mounted horizontally. Most residential systems have the expansion tank suspended from either piping or ceiling joists. Large expansion tanks are usually floor mounted. In any case, the air-inlet valve for any expansion tank should be readily accessible. This is required in case air must be added to the tank. It's a good idea to install a pressure gauge near the inlet of an expansion tank. The gauge makes it possible to monitor the static fluid pressure in the system. In all cases, regardless of tank style, the expansion tank must be supported properly to avoid operation problems.

Controls for Heating Systems

Controls for heating systems are integral parts of a functioning system. Every hydronic heating system relies on controls to work properly. There are four basic categories for the controls to fall into. There are controls that are merely used to turn a system on or off. Other types of controls are staged controls, modulating controls, and outdoor reset controls. The choice and installation of controls is, to a large extent, a matter of the heating system being fitted with the controls. The best advice is to refer to manufacturers' recommendations and follow them when selecting and installing controls for heating systems. There are far too many facets of controls to cover completely in this chapter. However, we will overview some of the controls, so let's do that now.

Thermostats

Thermostats, like those on the walls of homes where heating systems are installed, are examples of on-off controls. Devices used to open or close an electrical contact are the most common type of control used in heating systems. Burner relays and setpoint controls are also examples of on-off controls. These types of devices simply allow a system to cut on or off. The devices do not control or regulate the heat output of a system. This confuses some people. Many people think that a thermostat regulates heat. This is not the case. A thermostat that is set for a high temperature allows a heat source to run longer, but not to burn hotter.

Controlling in Stages

One way to match the output of a heat source with a building's need for heat is to control the heat output in stages. This is known as staged control. It means that the controls used for this type of control cut on and off in stages. The stages can range from zero to maximum heat output. The stages come on and go off in sequence. The use of staged sequences makes it easier to balance heat output with heat need. This type of control is rarely needed in typical residential applications.

Modulating Control

The ultimate in matching output with heat need comes from modulating control. The advantage of modulating control is that heat output is continuously variable over a range from zero to full output. Modulating controls make this possible. Most modulating controls do their job by controlling the temperature of water passing into and through heat emitters. There are several different ways to accomplish modulating control. The most common involves the use of mixing valves or variable-speed pumps. But electrical elements and even thermostatic radiator valves can produce the effects desired with modulating control.

Outdoor Temperatures

Outdoor temperatures affect the effectiveness of a heating system. When a standard hydronic system is designed, it is set up to provide a certain room temperature based on a certain outdoor temperature and the temperature of water in the heating system. An outdoor reset control is ideal for balancing outdoor temperatures with the water temperature in hydronic heating systems. With this type of balancing, a building enjoys near-perfect heating control, even when outside temperatures are uncooperative. When outdoor temperatures plummet, the temperature of water in the heating system is raised through the use of an outdoor rest control. In reverse, if outdoor temperatures warm up considerably, the temperature of water in the heating system is lowered. This balancing act combines to make a more efficient and more comfortable heating system.

All sorts of controls are used in heating systems—switches, relays, time-delay devices, high-limit switches, triple-action controls, multizone relays, resets, and so forth. The safest bet when dealing with controls is to follow the recommendations of manufacturers. The list of potential controls is a long and complicated one.

CHAPTER 12

Expansion Tanks

Expansion tanks are needed on hydronic heating systems. They are a safety device that deals with the expansion of the liquids used in hydronic systems. Without the use of an expansion tank, a heating system could rupture or explode. As important as these tanks are, they are sometimes taken for granted. The sizing of the expansion tank is an integral part of designing a good heating system. You might think that expansion tanks are so common that all installers and contractors would be fully aware of the need for the tanks. Many are, but a lot can't explain why the tanks are needed. Oh, they can tell you that the tank is needed and that it is related to the expansion of liquids in the heating system. But, ask some installers more detailed questions about expansion tanks and you might cringe at their answers. I know I have.

How much do you really know about expansion tanks? Could you pass a quiz on the subject? Is there a required air pressurization for a diaphragm-type expansion tank? How important is the type of material used in making the diaphragm of an expansion tank? These are just some of the questions that are difficult for some installers and contractors to answer. Don't worry, if you don't know the answers now, you soon will.

Hydronic heating systems depend on liquids to deliver heat. Water is usually the medium for this purpose. Any liquid running through a hydronic heating system is subject to expansion. This is due to the heating of the fluid. The process is called thermal expansion. Since thermal expansion cannot be eliminated, it must be controlled. This is the job of an expansion tank. An expansion tank must contain an air pocket to compensate for the expansion of liquids in heating systems. Modern tanks have diaphragms in them. This helps to prevent water-logging, which was a serious problem in older types of expansion tank.

Generally speaking, liquids are incompressible. If a container is filled with liquid and the liquid is later heated, expansion is going to occur. To avoid a rupture of the containing vessel, there must be excess room that allows for the expansion. As simple as this fact is, it is very important. It's possible that a container that is not properly set up or protected can explode violently.

Systems that are open to the atmosphere don't require an expansion tank (Fig. 12.1). Expansion in a nonpressurized system can be dealt with by keeping the fluid level below the rim of the container. This extra space allows the fluid to expand, reaching toward the rim without overflowing. But, in a typical heating system, this sort of setup is not practical. Closed-loop heating systems are used with hydronic

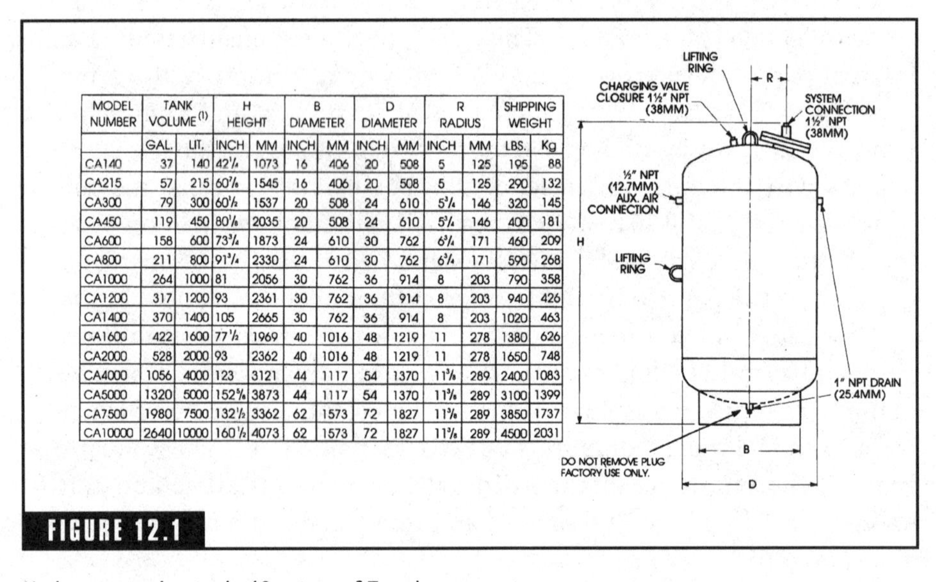

MODEL NUMBER	TANK VOLUME (1)		H HEIGHT		B DIAMETER		D DIAMETER		R RADIUS		SHIPPING WEIGHT	
	GAL.	LIT.	INCH	MM	INCH	MM	INCH	MM	INCH	MM	LBS.	Kg
CA140	37	140	42¼	1073	16	406	20	508	5	125	195	88
CA215	57	215	60⅞	1545	16	406	20	508	5	125	290	132
CA300	79	300	60½	1537	20	508	24	610	5¾	146	320	145
CA450	119	450	80⅛	2035	20	508	24	610	5¾	146	400	181
CA600	158	600	73¾	1873	24	610	30	762	6¾	171	460	209
CA800	211	800	91¾	2330	24	610	30	762	6¾	171	590	268
CA1000	264	1000	81	2056	30	762	36	914	8	203	790	358
CA1200	317	1200	93	2361	30	762	36	914	8	203	940	426
CA1400	370	1400	105	2665	30	762	36	914	8	203	1020	463
CA1600	422	1600	77½	1969	40	1016	48	1219	11	278	1380	626
CA2000	528	2000	93	2362	40	1016	48	1219	11	278	1650	748
CA4000	1056	4000	123	3121	44	1117	54	1370	11⅜	289	2400	1083
CA5000	1320	5000	152⅝	3873	44	1117	54	1370	11⅜	289	3100	1399
CA7500	1980	7500	132½	3362	62	1573	72	1827	11⅜	289	3850	1737
CA10000	2640	10000	160½	4073	62	1573	72	1827	11⅜	289	4500	2031

FIGURE 12.1

Modern expansion tank. (*Courtesy of Taco.*)

heating systems. There should be room for the water to roam, so to speak. This is accomplished with an expansion tank. The size of the expansion tank and its location in the heating system are both important elements of the system design.

Open tanks were used to control the expansion of water in heating systems at one time. It was common for these tanks to be installed in the attics of buildings. There were problems with this arrangement. One of the disadvantages was that fluid was lost through the opening at the top of the tank. As fluid was lost from the tank, it had to be replaced with fresh water. This meant that dissolved oxygen was being introduced into the heating system. As most pros know, the dissolved oxygen generated rust, which is not a good thing.

Another problem with the open tank in an attic was that it limited the pressure capabilities of the heating system. It was not uncommon for the tanks that were placed in attics to freeze during extreme weather conditions. If the liquid in the tank froze, it was not a stretch to expect the tank to split. When the ice thawed, the defective tank would flood living space below it. There had to be a better way, and there was. The result was a closed expansion tank that could be installed in a lower, heated space.

Standard Expansion Tanks

Standard expansion tanks followed the open-tank protection used in early heating systems. The standard expansion tanks were a definite improvement, but they were far from perfect. This same type of situation existed for plumbing systems where well pumps were being used. The big problem with standard expansion tanks was waterlogging. Occasionally air had to be injected with air to maintain a balanced pressure. A waterlogged pressure tank could not perform its function.

When standard expansion tanks were used, they would be installed on a heating system and when the system was filled with water, the tank would trap air in itself. The air created a cushion to deal with the expansion of fluid in the heating system. Normally an expansion tank will reach a pressure of about 5 psi below the setting of a system's pressure-relief valve when the system is working at maximum operating pressure. Standard expansion tanks were common years ago, and many of them are still in use. They exist in both plumbing and heating

MATERIALS

Diaphragm tanks (Fig. 12.2) are the way to go. They are affordable, effective, efficient, and easy to work with.

systems. There were, and are, a number of disadvantages associated with standard expansion tanks.

A typical installation for a standard tank will have the tank hanging above the boiler it is serving. Specials fittings on boilers can allow air bubbles to be driven out of solution by heating and to rise into the tank. A dip tube in the boiler allowed water to pass from the boiler to the standard expansion tank. This process kept the transfer of air to the distribution system at a minimum. Unfortunately, this type of system allows cool water to backfeed from the tank to the boiler. The water brings with it dissolved oxygen, which stimulates rust. All in all, the design is far from perfect.

Waterlogging is another big problem with a standard expansion tank. If the tank fills with liquid, there is no longer an air cushion available to counteract the effects of liquid expansion. A system operating under these conditions is likely to experience some leakage or blow-off from its relief valve. Maintenance on a standard tank should be done at least twice a year. This involves draining the tank completely and then refilling the system. This is usually the only way to maintain a proper air balance in the tank. Modern heating systems overcome the problems of a standard expansion tank with the use of tanks that have built-in diaphragms. There is no longer any need for suffering through the trials and tribulations of older heating systems.

Diaphragm Tanks

Diaphragm tanks (Fig. 12.2) are the way to go. They are affordable, effective, efficient, and easy to work with. Air and water are kept separate in a diaphragm tank. The diaphragm is what creates the separation. You can think of a diaphragm tank as an expansion tank that has a balloon full of air in it. Without any doubt, diaphragm expansion tanks are the only sensible way to go in a modern heating system.

A number of advantages are offered by diaphragm tanks. Air and water do not come into contact with each other in a diaphragm tank. The volume of air in the tank is always protected from reabsorption by the fluid in the heating system. Another advantage is that diaphragm tanks don't require any draining or refilling. Waterlogging is not a

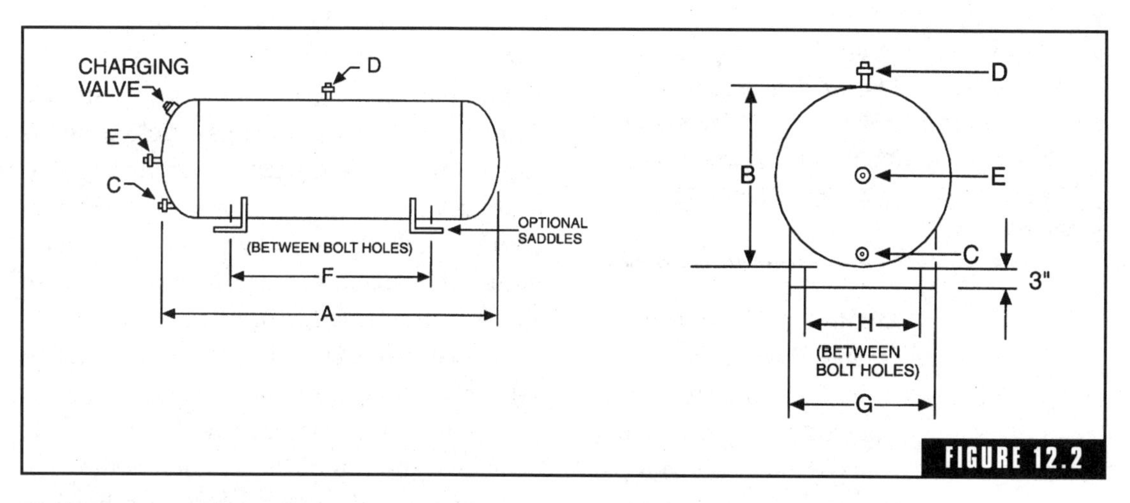

Diaphragm expansion tank. (*Courtesy of Amtrol.*)

problem with an expansion tank that is equipped with a diaphragm. This means that a heating system will work better for longer with less maintenance. Rust is not a problem with diaphragm tanks, which is yet another advantage.

Since diaphragm tanks can be precharged with air (Fig. 12.3), a smaller tank can be used in place of what would have been a much larger standard expansion tank. Most heating systems can be served by a small tank that can be mounted directly to the system piping. Positioning of the expansion tank is not critical when a diaphragm tank is used. This is not to say that the location of the tank is not important,

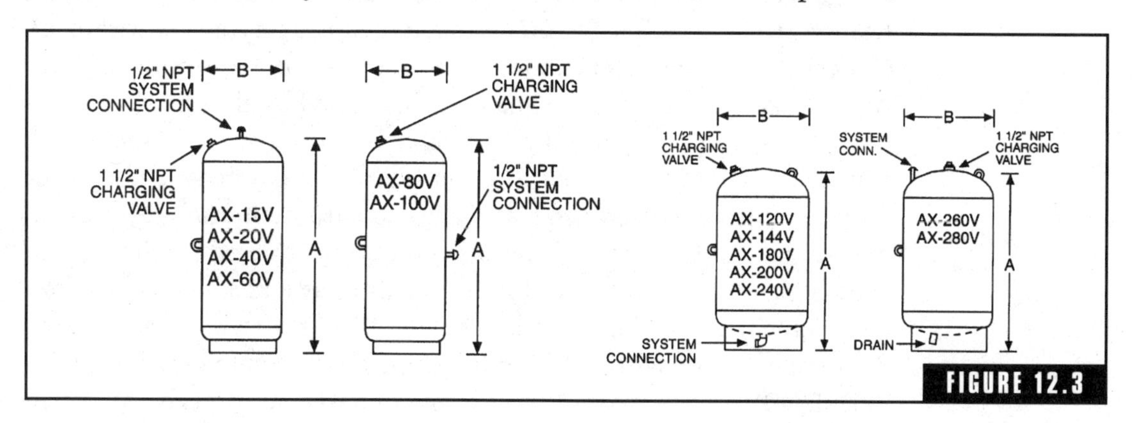

Vertical diaphragm expansion tanks. (*Courtesy of Amtrol.*)

QUICK»TIP A diaphragm-type expansion tank should be sized to assure a pressure of about 5 psi lower than the relief valve setting for the system when the system is at maximum operating pressure.

but the tank can be installed in any position. Since tanks with diaphragms can be installed directly to system piping, there is no need for special fittings, such as those that would be used with a standard expansion tank.

A diaphragm-type expansion tank should be sized to assure a pressure of about 5 psi lower than the relief valve setting for the system when the system is at maximum operating pressure. Anything less can result in a leaking pressure-relief valve. The proper air-side pressurization for an expansion tank is equal to the static fluid pressure at the inlet of the tank, plus an extra 5 psi allowance at the top of the system. This pressure adjustment should be made prior to filling a system with liquids. Adding or removing air is the process involved to balance a system's tank to the proper pressure setting. Once the air-side pressure is set, it ensures that the diaphragm will be expanded fully against the shell of the expansion tank when the system is full of liquid. This is based on cool water. If the diaphragm is not adjusted properly, the diaphragm may be partially compressed before the liquids are even heated.

Matching a Tank

Matching a tank to a heating system requires compatibility between the two (Fig. 12.4). One key consideration is the material that the diaphragm is made of. Not all materials are suitable for use with all heating systems. This is something that not everyone is aware of. Chemical reactions on the diaphragm material can be a problem. Over time, the diaphragm material can be comsumed by the chemical reaction. If this happens, the modern tank is no better than a standard expansion tank. Plus, a sludge can build up in the system. Three major types of materials used for diaphragm construction are butyl rubber, hydrin, and EPDM. Any of these materials are fine when they are in contact with only water. But many heating systems have antifreeze compounds in them. Glycol-based antifreezes will break down diaphragms made with butyl rubber. Typically a diaphragm made with EPDM will be compatible with a glycol-based antifreeze. A hydrin diaphragm is usually reserved for use with closed-loop solar systems when hydrocarbon oils make up the transfer fluid. If a system is being used with fluids that are capable of rusting metallic surfaces, any expansion tank used should be rust-resistant.

Check the Ratings

Check the ratings on any tank that you plan to install in a heating system. Every tank should have a listed rating for pressure and temperature. These ratings are very important. You might find that the ratings are 60 psi and 240°. This is a pretty standard rating. Residential systems will almost always be okay with a tank of this type. Light commercial jobs can usually be fitted with these same tanks. As routine as this may sound, don't take it lightly. If you install a tank that is rated lower than the pressure-relief valve used on a system, the tank could

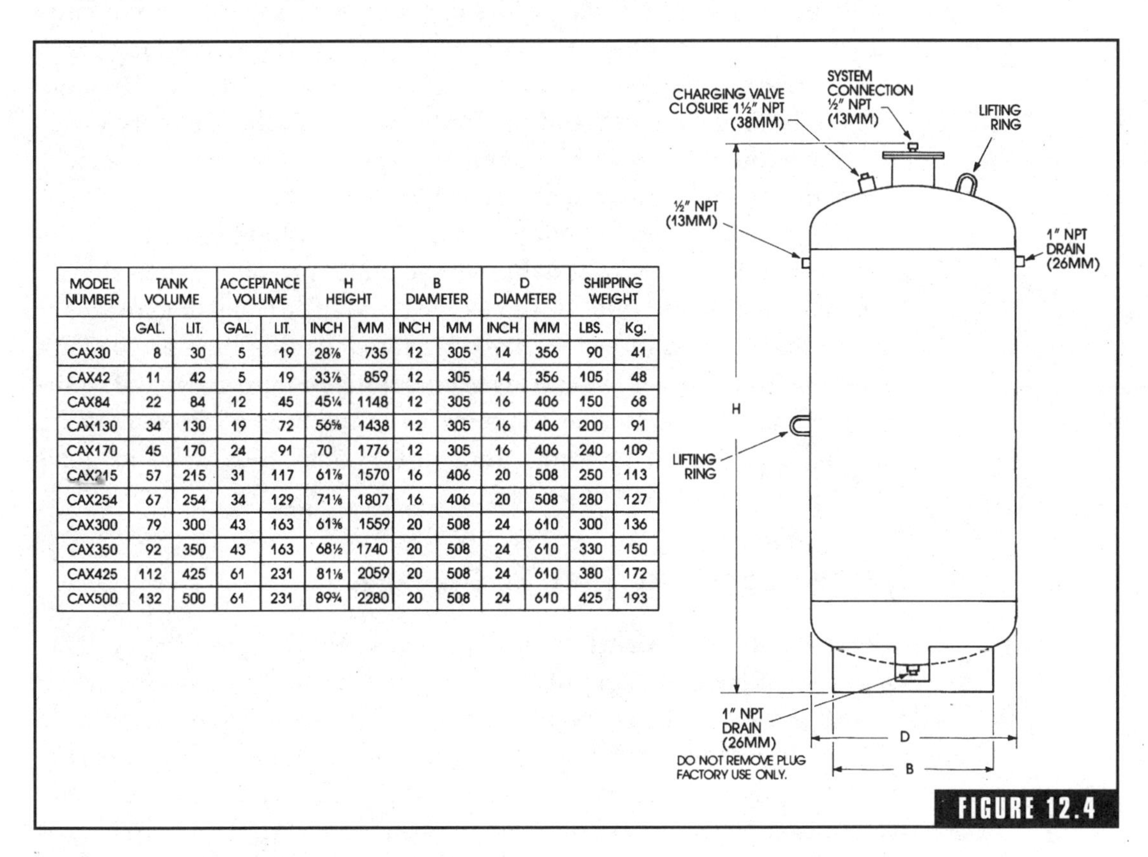

MODEL NUMBER	TANK VOLUME		ACCEPTANCE VOLUME		H HEIGHT		B DIAMETER		D DIAMETER		SHIPPING WEIGHT	
	GAL.	LIT.	GAL.	LIT.	INCH	MM	INCH	MM	INCH	MM	LBS.	Kg.
CAX30	8	30	5	19	28⅞	735	12	305	14	356	90	41
CAX42	11	42	5	19	33⅞	859	12	305	14	356	105	48
CAX84	22	84	12	45	45¼	1148	12	305	16	406	150	68
CAX130	34	130	19	72	56⅝	1438	12	305	16	406	200	91
CAX170	45	170	24	91	70	1776	12	305	16	406	240	109
CAX215	57	215	31	117	61⅞	1570	16	406	20	508	250	113
CAX254	67	254	34	129	71⅛	1807	16	406	20	508	280	127
CAX300	79	300	43	163	61⅜	1559	20	508	24	610	300	136
CAX350	92	350	43	163	68½	1740	20	508	24	610	330	150
CAX425	112	425	61	231	81⅛	2059	20	508	24	610	380	172
CAX500	132	500	61	231	89¾	2280	20	508	24	610	425	193

Expansion tank details. (*Courtesy of Amtrol.*)

rupture. There's not much to talk about here, but you must make sure that the tanks used in systems are compatible with the temperature and pressure ratings pertaining to the rest of the system.

A Wide Selection

A wide selection of expansion tanks is available. The tanks come in a variety of sizes and shapes. Diaphragm expansion tanks for typical heating systems can range in volume from 1 to 10 gal. Tanks of this size will normally meet the need of both residential and light commercial installations. If these tanks are attached to system piping, as they often are, the pipe should be well supported on both sides of the connection point. Some installers mount these tanks horizontally and support the tanks themselves with hangers. Either method is acceptable.

Big heating systems can require larger expansion tanks. For a large system, a floor-mounted tank is ideal. Since the tank sits on a solid surface, there is no need to support it with hangers. Most contractors place these tanks on elevated platforms to protect the tanks from any flooding of the floor area. It's common practice to arrange four cement blocks under such a tank to elevate and support it.

Regardless of the type of tank used, the air-induction valve on the tank should be readily accessible. An accurate pressure gauge should be installed near the inlet of the expansion tank. The gauge makes it easy to monitor pressure in the tank. It's rare for a diaphragm-type tank to become waterlogged, but if this occurs, the pressure gauge will indicate the problem. As long as the pressure at the inlet of the tank is the same as the air-side pressurization, the system should be in good working order, in terms of pressure.

Where Should the Tank Be Placed?

Where should an expansion tank be placed within a heating system? Tank placement can affect the pressure distribution of a system when it's in operation. As a rule of thumb, the expansion tank should be located close to the inlet port of the circulating pump. While this rule has been followed for years in most commercial installations, on many occasions it has not been followed in residential installations. The two types of heating systems differ, but the tank should still be installed near the inlet of the circulator. (Figs. 12.5 to 12.7).

Fluid in a closed-loop heating system is static in volume. This means that the volume doesn't change appreciably. Unlike what some people think, the volume of the fluid doesn't change when the circulating pump is running. Just remember that the liquid volume remains essentially the same at all times in a closed-loop system. Just as the volume of liquid in the heating system is static, so is the amount of air in the system. The concept is known in the trade as the point of no pressure change.

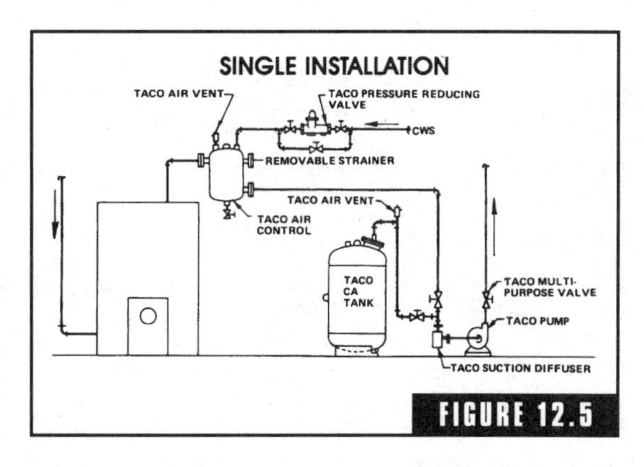

Single installation. (*Courtesy of Taco.*)

When a circulating pump is operating, it boosts system pressure. This is a good thing. It forces air out of vents and cuts down on the risk of having cavitation at the pump. Even so, the volume doesn't change. Pressure changes in the system are balanced between the circulator and the expansion tank when everything is installed properly. If the expansion tank is installed near the discharge port of a circulator, the pressure drop in the system is much more significant, and this can lead to real problems. Air will not be expelled properly and cavitation at the pump is much more likely. Stopping this potential problem is simple; install the expansion tank near the inlet port of the circulator.

There is no big mystery to expansion tanks. The keys to remember are as follows:

- Size the expansion tank properly.
- Use a diaphragm-type expansion tank.
- Make sure the ratings for the tank are compatible with the system requirements.
- Confirm that the diaphragm material is compatible with system fluids.
- Install the tank near the inlet port of the circulating pump.
- Make sure that the air valve for the tank is accessible.

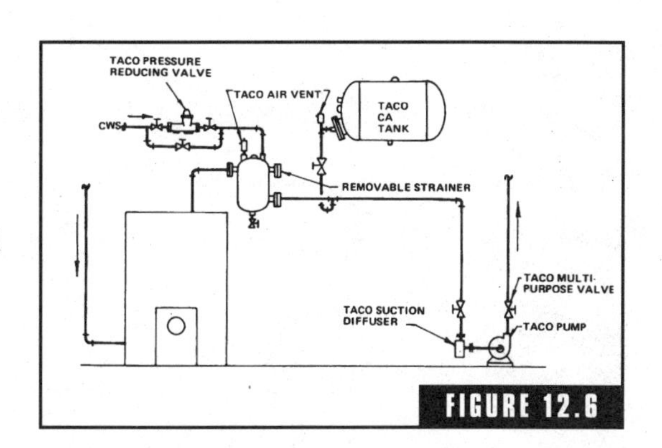

Horizontal mounting method. (*Courtesy of Taco.*)

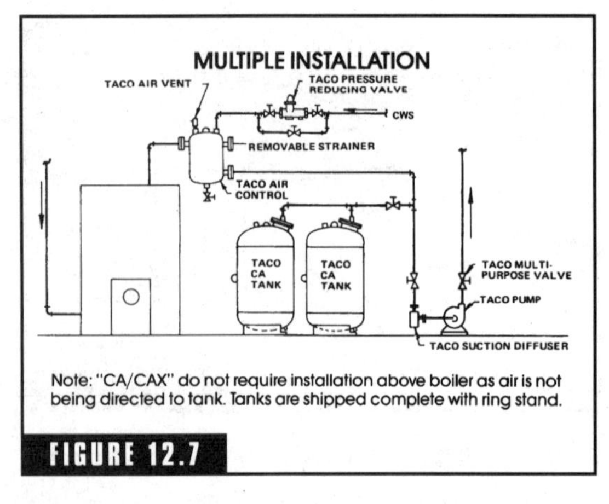

FIGURE 12.7

Installation of multiple expansion tanks. (*Courtesy of Taco.*)

- Install an accurate pressure gauge at the inlet of the expansion tank.
- Support the expansion tank properly.
- Read and heed all manufacturer's recommendations when installing a tank.

If you follow these basic guidelines, the systems that you install should not have problems associated with the expansion tank.

CHAPTER

13

Domestic Water Heating

Domestic water heating is an added benefit available to homeowners who use hydronic boilers to heat their homes. The expense of heating water for domestic use with a boiler that is providing space heating can be considerably less than the cost of a separate water heater. This is no secret. People who heat with boilers have been deriving their domestic hot water from them for years. Some boilers use tankless coils to heat water for domestic use. Other boilers are set up to work in conjunction with a water storage tank. And some heating systems combine a tankless coil with a storage tank. When water is heated for domestic use with a boiler, an additional load is placed on the boiler. This load must be accounted for.

Tankless coils are frequently used to provide domestic hot water from a boiler. These coils are favored for their low cost to operate. But they are not without their problems. Coils can produce nearly endless hot water, but the water temperature may not be as high as some consumers would like. The trade-off of a high volume of hot water in exchange for water of a lower temperature is not always acceptable. Another problem often encountered with tankless coils is their tendency to become clogged after some years of use. Mineral deposits that build up in the coils can reduce water flow considerably. Cleaning a

QUICK»TIP In order for a tankless coil to perform property, the boiler water surrounding the heat exchanger must remain hot at all times. This is accomplished with the use of a triple aquastat control.

coil is not easy. An acid bath will sometimes take care of the problem, but many times it makes more sense to simply replace the coil.

Tankless coils are heat exchangers. They are installed within a hydronic boiler. Since the coil is a heat exchanger, the water within the coil never mixes with the water in the boiler. This is a point that many average people have some trouble understanding. The coils are so named because they consist of coils of copper tubing. Once a tankless coil is installed in a special chamber of a boiler, the coil becomes surrounded by hot boiler water. This is what heats the water within the coil. When there is a demand for hot water to fill a domestic need, cold water passes through the tankless coil. During the trip through the coil, the cold water is warmed. Since the cold water makes only a single pass through the coil, the water is not always warmed to a desirable temperature.

In order for a tankless coil to perform property, the boiler water surrounding the heat exchanger must remain hot at all times. This is accomplished with the use of a triple aquastat control. The aquastat fires the boiler whenever the boiler water drops to a certain temperature. This is usually not a problem during a heating season, but the process becomes somewhat inefficient during warm months. Even though generally heating is not needed during warm spells, the water in the boiler must remain hot to heat domestic water. This too is a drawback of a tankless coil. Owing to the inefficiency of heating domestic water during seasons when general heating is not needed, tankless coils are losing popularity. Some people feel that the use of tankless coils in new installations should be banned, as a means of saving energy.

Combining a Coil with a Tank

Combining a tankless coil with a storage tank is one way of helping to overcome the efficiency problems associated with coils. When a storage tank is added to a tankless coil system, the overall energy usage of a boiler is lowered and the problem of fluctuating water temperature is better controlled. A storage tank is an added expense when installing a system, but the cost is offset by energy savings. In this type of setup, a high limit control is needed rather than an aquastat. In the case of adding a storage tank to an existing system, the existing aquastat may be able to be reconfigured to work as a high limit control. Once the boiler is equipped with a coil, a storage tank, and a high limit control, it will fire only when there is a call for heat or hot water. The boiler will not operate at all times, as it would have to without the storage tank.

When a storage tank (Fig. 13.1) is used with a coil, a small circulating pump is responsible for moving the domestic water. This pump runs when there is a call for domestic hot water. Cooler water that is near the bottom of the storage tank is moved through the tankless coil. After the coil heats the water, it is moved to a separate connection point that is near the top of the storage tank. Mixing water from bottom to top in this fashions helps to ensure a stable water temperature for the domestic water supply. Water continues to circulate in this manner until a thermostat associated with the storage tank is satisfied. When the temperature of water in the storage tanks drops below a predefined level, the circulator cuts on and the process is repeated.

Circulators used with tankless coils and storage tanks come into contact with oxygenated water. This means that the circulator must be made of materials that will not rust. Bronze, stainless steel, and even high-temperature plastics can be used to protect the circulator and the system from rust development. It's common for installers to equip circulators for the purpose of moving domestic hot water to run for several minutes after the tank thermostat is satisfied. When a delay is installed to keep the circulator running, the result is more residual heat being gained from the boiler that would otherwise go off through the chimney.

Independent Tanks

Independent tanks that are indirectly fired are a great alternative to the use of a tankless coil. Most modern hydronic heating systems are being installed with indirectly fired storage water heaters as a part of the system. These tanks have their own built-in heat exchangers. Unlike a tankless coil in a boiler, the heat exchanger in indirectly fired water tanks carries hot boiler water. Fresh water is contained in the tank and surrounds the heat exchanger. This is just the opposite of the way that a tankless coil works. Boiler water passing through the heat exchanger heats the surrounding potable water. These tanks work with an on-demand principle, so the boiler is not working until the temperature of the water in the tank falls below the setpoint temperature. Usually controls on a boiler will give preference to the domestic heating tank.

Specifications and Dimensions

Model No.	Tank Volume Lt.	Tank Volume Gal.	A Height cm	A Height in.	B Diameter cm	B Diameter in.	Dom. Water Conn. Inlet	Dom. Water Conn. Outlet	Boiler Water Conn. Inlet	Boiler Water Conn. Outlet	Ship. Wt. kg	Ship. Wt. lbs.
WH-7C	155	41	122	48	56	22	¾" NPT(F)	¾" NPT(F)	1" NPT(F)	1" NPT(F)	63.8	141
WH-7C-DW	155	41	122	48	56	22	¾" NPT(F)	¾" NPT(F)	1" NPT(F)	1" NPT(F)	68.3	151
WH-60C	227	60	137	54	66	26	1" NPT(F)	1" NPT(F)	1" Cu	1" Cu	75.6	167
WH-60C-DW	227	60	137	54	66	26	1" NPT(F)	1" NPT(F)	1" NPT(F)	1" NPT(F)	80.1	177
WH-80C	302	80	168	66	66	26	1" NPT(F)	1" NPT(F)	1" Cu	1" Cu	801	177
WH-80C-DW	302	80	168	66	66	26	1" NPT(F)	1" NPT(F)	1" NPT(F)	1" NPT(F)	84.6	187
WH-120C	450	119	180	71	76	29¾	1½" NPT(F)	1½" NPT(F)	1" Cu	1" Cu	98.2	217
WH-120C-DW	450	119	180	71	76	29¾	1½" NPT(F)	1½" NPT(F)	1" NPT(F)	1" NPT(F)	102.7	227

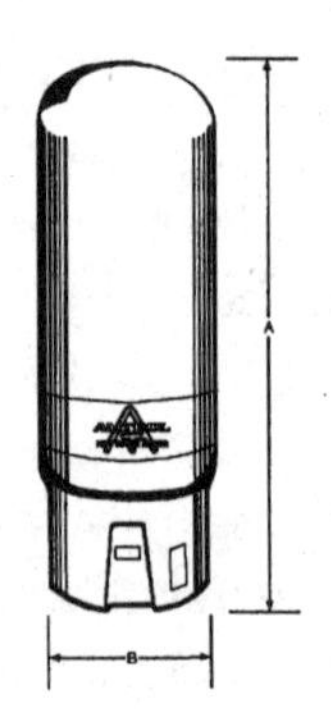

Maximum Operating Conditions

Operating Temperature	160°F
Working Pressure	150 PSIG

Specifications

Description	Standard Construction
Water Side Liner	Polyethylene
Pressure Shell	Steel
Outer Jacket	Polyethylene
Insulation	2" Polyurethane
Heat Exchanger	Finned Copper with Stanoguard Plating
Cold Water Dip Tube	Polypropylene

Performance

Model No.	1st Hour Rating Lt.	1st Hour Rating Gal.	Continuous Flow Lit.	Continuous Flow Gal.	Standby Loss BTU/Hr.	Standby Loss °F/Hr.	Recovery Time(Approx.) Minutes	Boiler Flow Rate Requirements Min.	Boiler Flow Rate Requirements Max.
WH-7C	1183	313	1036	274	400	1.2	6 to 15	7	18
WH-7C-DW	1183	313	1036	274	400	1.2	6 to 15	7	18
WH-60C	1251	331	1036	274	200	0.4	9 to 22	7	18
WH-60C-DW	1251	331	1036	274	200	0.4	9 to 22	7	18
WH-80C	1323	350	1036	274	260	0.4	12 to 29	7	18
WH-80C-DW	1323	350	1036	274	260	0.4	12 to 29	7	18
WH-120C	1463	387	1036	274	400	0.4	18 to 43	7	18
WH-120C-DW	1463	387	1036	274	400	0.4	18 to 43	7	18

All dimensions and weights are approximate.

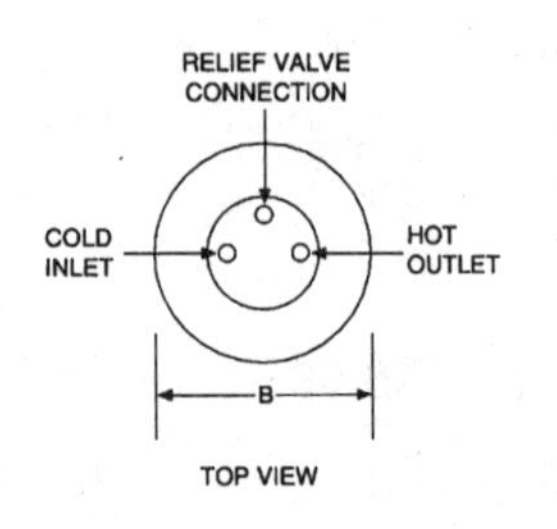

FIGURE 13.1

Hot water maker domestic heating tank. *(Courtesy of Amtrol.)*

In other words, if there is a call for general heating at the same time as a call is generated for domestic hot water, the controls will provide the water heater with hot boiler water prior to fulfilling the general heating need.

Tank construction varies. Some storage tanks are made of stainless steel on the interior. Others have an interior of ceramic-lined steel. A thick insulating jacket surrounds the holding portion of the tank. A well-insulated storage tank reduces the frequency with which a boiler must fire to reheat water. This, of course, increases the energy efficiency of the heating system and reduces operating costs.

There are some trade techniques used to boost system efficiency. It's not uncommon for well-constructed storage tanks to maintain water temperature long enough to give a boiler a couple of hours off during nonpeak demand for domestic hot water. This is pretty good, but it can be better. One way of doing this is to keep the storage tank close to the boiler. Reducing the developed length of the piping between the storage tank and the boiler will reduce heat loss from the piping. A thermostatic mixing valve allows water in a storage tank to be heated to a temperature higher than what is needed at plumbing fixtures. Tying the thermostatic mixing valve into conjunction with the tank thermostat can extend the time between firing cycles for the boiler. All of this reduces operating costs, saves energy, and reduces wear on the heating system.

Auxiliary Loads

Auxiliary loads on a hydronic heating system can range from domestic hot water to heating a swimming pool. By far the most common type of auxiliary load is the one required for domestic hot water. Auxiliary loads can burden a boiler. If a boiler is not sized to handle additional loads, problems will exist. But oversizing a boiler to a point where it can pull dual duty can also be a mistake. In the case of residential heating systems, adding the load of heating domestic hot water should not have a negative effect on a typical boiler. But you can't just take this for granted, especially if there are other auxiliary loads on the heating system.

Reducing the developed length of the piping between the storage tank and the boiler will reduce heat loss from the piping.

Some homeowners heat their garages. The zone for heating the garage can be considered an auxiliary load. If a swimming pool or spa is heated with the use of the same boiler that is used to heat the general living area, the demands on the boiler can be too great. If a single boiler is sized large enough to handle major auxiliary heating loads, it will be far too big to be cost- effective when the auxiliary loads are not demanding heat.

When a single boiler is used to carry auxiliary loads, it is desirable to have the auxiliary loads demanding heat when the primary heating system is not in need of heat. This can work well in some circumstances. For example, if a home uses radiant heat for warmth in the winter and has an auxiliary load to heat a swimming pool in spring and fall, the system might work very well. But, if the auxiliary load was a snow-melting system, the demand for heat from both the primary and secondary heat loads might conflict.

When domestic hot water is the auxiliary load, it's generally fairly easy to avoid conflicts. Most households have only a few peak times for domestic hot water. It's common for there to be a strong demand earlier in the mornings, when people are cooking, bathing, and so forth. Another key demand comes in the early evening when cooking, laundry, and bathing may require hot water. But, during much of the day and night, there is not a significant demand for hot water. When a hot-water storage tank of adequate size is used, it's easy to avoid major conflicts and heavy loads that would unduly burden a boiler.

It Makes Sense

It makes sense to use a boiler to heat domestic hot water. This is especially true with today's modern cold-start boilers and indirectly fired storage tanks. Current controls can divert a boiler's attention from heating living space to heating water. This allows people to enjoy an abundance of hot water, usually without any noticeable loss of heating comfort in the home or office. Adding a storage tank to a boiler makes sense for the consumer and boosts the profit from a job for the contractor.

Selling new customers on the use of domestic hot water from a boiler is not always easy. There are a few reasons for this. I've had customers talk with me about concerns of mixing boiler water with potable water. This doesn't happen, but it can be difficult to explain this properly to potential customers. I've found that pictures speak volumes when it comes to this issue. Showing a customer how a tankless coil is inserted into a boiler may make it easier for the customer to understand how the heat exchanger is protected from contamination.

Another problem to overcome is convincing some people that a hot-water storage tank should be installed. This is especially true when remodeling homes that are owned by older people. For many years, tankless coils were considered all that was needed for domestic hot water. Trying to convince some people that the expense of adding a storage tank will be recovered in lower operating costs can be a real challenge. Gathering statistics from local energy sources and area equipment suppliers can help in this regard. Testimonials from satisfied customers are your best ammunition when selling in this type of situation.

A lot of contractors don't work hard enough to add value to the systems that they are selling with storage tanks. Not explaining the benefits of storage tanks to customers is unfair to the customers. When the features and benefits are explained properly, most consumers will see the value of investing in the additional equipment. It's easy to find plenty of jobs where customers will benefit from adding a storage tank, and this should mean more money in the contractor's bank account. Everyone wins. The contractor makes more money and the customer gets better satisfaction and reduced operating costs. Don't overlook the opportunities that exist with storage tanks.

CHAPTER

14

Slab-on-Grade Piping Systems

Installing radiant floor heating systems in concrete slabs is an excellent idea. Of all the places for radiant heating systems to shine, concrete floors make a wonderful stage. This is due largely to the fact that the concrete floor makes an excellent heat-storage mass. Once the concrete is warmed, it will radiate heat for hours and hours. Also, the heat output from radiant heat in concrete is extremely clean and stable. If ever there were a place that cried for radiant floor heating systems, it's concrete slabs.

In the past few years, radiant floor heating has gained a great deal of popularity. Installations in concrete floors probably have had a lot to do with the good press that the heating systems are getting. It's not uncommon for radiant floor heating systems to be installed in a wide array of floor types, but concrete floors are a natural place to install this type of heating system.

As favorable as a concrete slab is for radiant heating, the performance of a system depends heavily on how it is installed. If the heat tubing is placed too closely together, there will be too much heat in the living space. Installing the tubing too far apart will result in a cold room. The design of any heating system is important, and there is no exception when installing radiant floor heating systems in concrete.

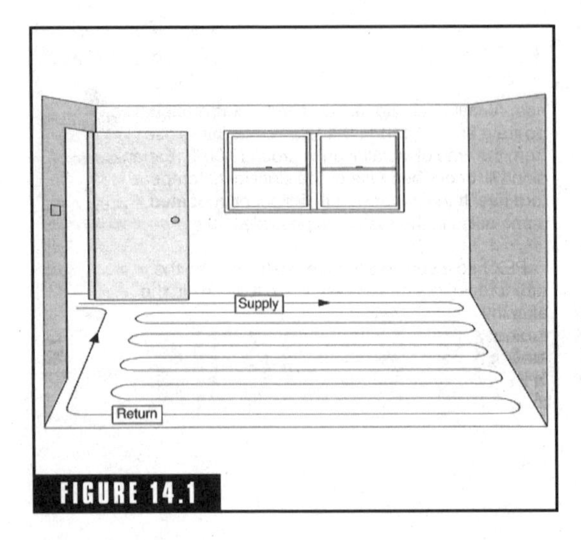

FIGURE 14.1

Single-wall serpentine design. (*Courtesy of Wirsbo.*)

Unfortunately, not all installers understand this. Some people think that since concrete is such a good conductor for a radiant heating system there is little need for expertise in the installation of heat tubing. This simply isn't true. It is just as important to pay full attention to an installation in concrete as it is during any radiant installation.

The first step to a good heating system is a good design (Fig. 14.1). A working design can be drawn on paper or with the help of a computer. Having a solid installation plan is important, but so is the actual installation. When pipes are being installed under or in concrete, they will be extremely difficult to service after the concrete is poured. Some installers get careless when they are installing heat tubing. A sloppy installation can cause any number of problems. Some of the potential problems may show up before concrete is poured. For example, a kinked pipe can be seen. Tubing that is punctured during installation can be discovered with a pressure test prior to pouring concrete. But having heat tubes in a position where they can be rubbed by an abrasive surface after installation can lead to problems long after concrete is poured. If a leak does occur after a slab has been poured, the repair process is both expensive and a real pain in the neck. All this points to the importance of thought and care when designing and installing an in-floor heating system.

QUICK»TIP **Concrete slabs are ideal for radiant floor heating systems. When a slab is a planned part of construction to begin with, a radiant floor heating system is an economical, efficient way to heat the living space.**

Fairly Simple

The installation of heat tubing in a concrete slab is fairly simple when it's drawn on paper (Fig. 14.2). The components of the system are minimal and the positioning of the components is generally the same from one job to another. To illustrate this, consider that your installation will begin with compacted fill as a base for the concrete slab. There are a couple of ways to plan an installation from the ground up. Some contractors used rigid foam insulation boards as the first layer of an installation (Fig. 14.3). Other contractors skip the insulation and begin with a sheet of plastic as a moisture barrier. Most contractors prefer to put plastic down first and then put rigid insulation boards on top of the plastic. The next layer of the system is usually welded wire. It's very common for heat tubing to be installed directly to the wire mesh. This is sometimes done with a tubing tie, but it can also be done with plastic clips. The next layer is the concrete, which in a residential application is usually about 4 in. thick. If a finished floor covering is going to be applied, it completes the total of layers. In a very simplistic way, this is all there is to it.

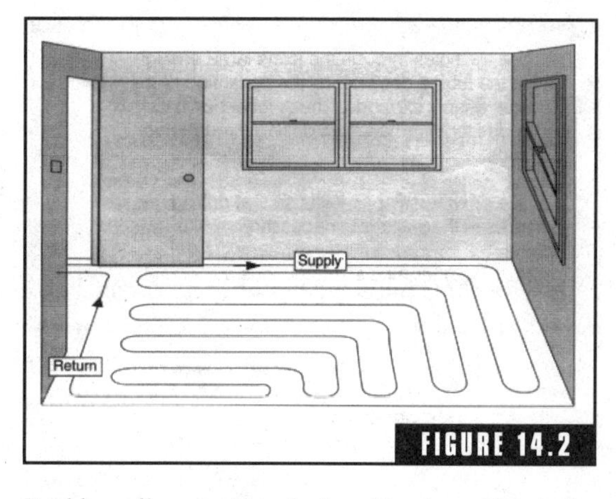

FIGURE 14.2

Double-wall serpentine design. (*Courtesy of Wirsbo.*)

Concrete slabs are ideal for radiant floor heating systems. When a slab is a planned part of construction to begin with, a radiant floor heating system is an economical, efficient way to heat the living space. If you look at an installed system before concrete is poured, most of what you see is loops of plastic tubing. It looks innocent enough. And it doesn't appear to be the sort of installation where a lot of skill or knowledge is needed. In some ways, this is true. Physically, the installation of PEX tubing in a slab-prep area is not difficult. However, the manner of the installation is crucial to the performance of the heating system.

The spacing between tubing varies from job to job, but the spacing of the ties, clips, or clamp rails that secure the tubing remains the same. For example, if nylon cable ties are used to secure PEX tubing to reinforcing wire, the tube ties should be no

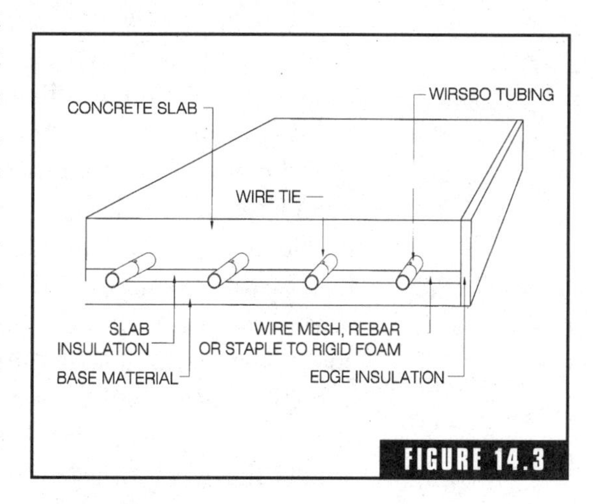

FIGURE 14.3

Foam insulation used with a wet-system design. (*Courtesy of Wirsbo.*)

Not all manufacturers approve of nylon ties being used with their tubing products. Using the wrong method of securement can void a warranty on tubing, so check the manufacturer's recommendations before choosing a means of securement.

more than 30 in. apart. It's better if the supports are not more than 24 in. apart. These recommendations are for tubing that is being installed in straight runs. When the tubing turns, the support ties should be no more than 12 in. apart, and 6 in. is better. Not all manufacturers approve of nylon ties being used with their tubing products. Using the wrong method of securement can void a warranty on tubing, so check the manufacturer's recommendations before choosing a means of securement.

Another device used to secure tubing is a plastic clip. These clips attach to reinforcing wire and then accept tubing to hold it in place. Even though the clips are different from cable ties, the spacing recommendations remain the same. If reinforcing wire is not used in the slab preparation, a different type of clip is needed. This type of clip is designed to be screwed to the foam insulation that is installed below the concrete.

Plastic tubing tracks offer another option for securing tubing in an under-slab installation. These rails attach to the foam insulation via barbed staples. Tubing rails can be purchased in configurations to hold several different sections of tubing. Once the tubing rails are installed, the installation of tubing can go very quickly. Light foot pressure is all that is needed to snap tubing into one of the rails. Contractors have different opinions on which means of attachment is best. The subject could be debated for hours with everyone's opinion having some validity. Since manufacturers may not warranty their products if specific types of attachments are used, you should refer to the manufacturer's recommendations and follow them.

Neatness counts on jobs. It's possible to do a neat installation with any of the means of attachment mentioned above. Personally, I feel that tubing rails give the best image. I also like the rails because they make installations fast. When ties or clips are used, each one must be installed individually. Once a tubing rail is installed, it can accept a lot of tubing at various spacing intervals. This cuts down on installation time. Plus, the tubing is cradled in the rail and should be less prone to rub abrasion.

How far apart should heat tubing be placed? It depends on the individual job (Fig. 14.4). There is no simple rule-of-thumb answer to this issue. If there were, it would probably be that the tubing should be spaced so that each section of tubing is about 12 in. apart from centerline to centerline. This works well for many reasons. One of the reasons is that the reinforcing wire used in slabs is usually 6 by 6 in. This makes any spacing that is in 6-in. intervals practical. Sometimes the tubing is placed closer together, and it can be spaced farther apart. The spacing decision must be made by the designer of the heating system.

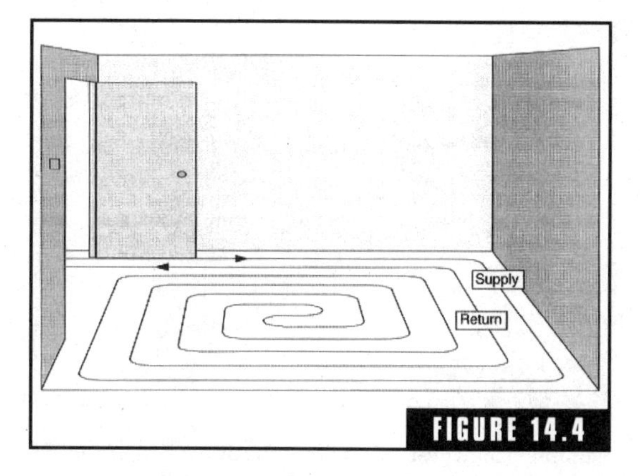

Example of counterflow design. (*Courtesy of Wirsbo.*)

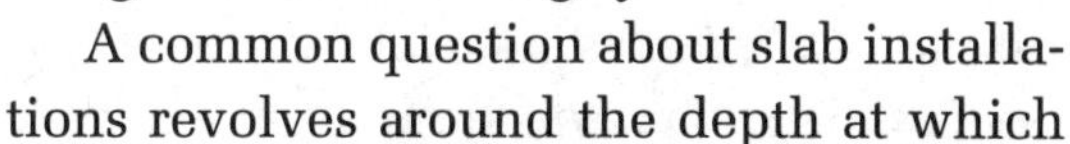

A common question about slab installations revolves around the depth at which the tubing should be placed in the slab. Some of the decision about the depth of coverage with concrete relates to the method used to secure the heat tubing. If the tubing is attached to reinforcing wire, there will be more concrete over the tubing than if the tubing had been placed in a tubing rail. Whether the tubing is at the bottom of the slab or concealed near the center of the slab will not have much effect on the performance of the system once full heat in the living space is obtained. It takes longer for tubing that is buried deep to bring a floor up to temperature, but once the temperature is reached, it can be maintained easily. One way to overcome the slower warmup is to increase the temperature of water supplying the heat tubing. Bottom line: Positioning in terms of depth is rarely a big issue.

How important is foam insulation in a slab-on-grade radiant heating system (Fig. 14.5)? Extremely important if you want the system to be at its best. Early radiant systems didn't have the benefit of rigid foam insulation boards. Modern systems can take advantage of all that the insulation can offer. Additional cost is involved when insulation boards are used, but it's money well spent. Systems installed without insulation cannot keep up with systems that do have insulation when it comes to efficiency. The designer of a sys-

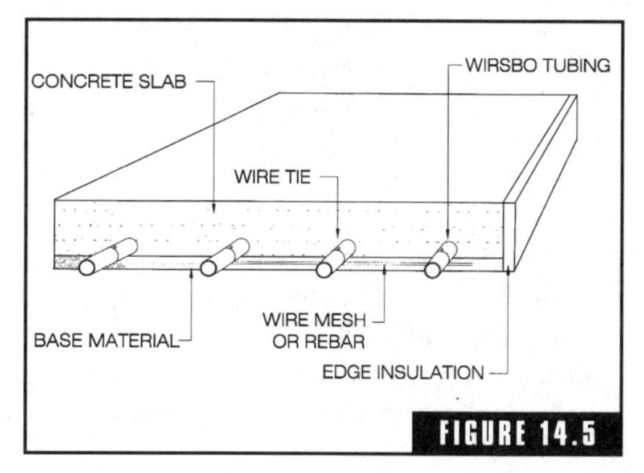

Foam edge insulation used with a wet system. (*Courtesy of Wirsbo.*)

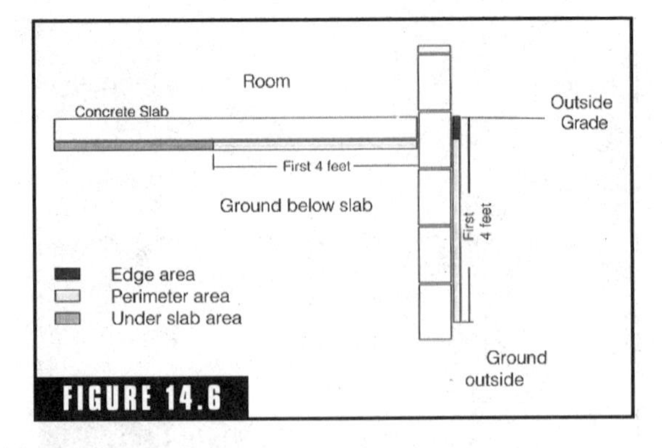

FIGURE 14.6

Example of total slab installation. (*Courtesy of Wirsbo.*)

tem should indicate specifications for insulation, but in case specifications are not available, let's have a little overview of what might be considered standard procedures.

Some installers remember to install foam insulation under a slab but fail to think of installing edge insulation (Fig. 14.6). The edge insulation is a substantial element in creating an efficient radiant heating system under a slab. This edge insulation should be 2 in. thick and placed around the edge of the foundation area. Then, some 2-in. insulations should be laid on the foundation pad area so that it extends toward the center of the floor for about 4 ft. Insulation installed in the center of the pad area can be thinner. Usually a 1-in. thickness is all that's needed. Of course, if you have a heating design to work from, follow it. Another point to note is that insulation should not be installed in pad areas where support piers or bearing walls will be created.

Heating systems that don't employ the use of foam insulation are not as efficient as they should be. Much of the heat produced from heat tubing is lost in a downward transfer to the ground below the slab. Eventually the ground reaches a saturation level that allows heat to pass upward more effectively. However, a lot of heat is lost downward. Much of this heat loss can be stopped by installing foam insulation below the heat tubing. The insulation installed on the inside edges of a foundation wall stops a tremendous amount of heat loss. This is because the exterior air coming into contact with foundation walls and transferring through the walls can reduce heating efficiency. Just as the cold comes in without insulation, heat goes out. All jobs can benefit from insulation on the exterior walls. Heat loss is greatly reduced and efficiency is enhanced when insulation is installed as a part of a radiant floor heating system in a concrete slab.

QUICK»TIP **Heat loss is greatly reduced and efficiency is enhanced when insulation is installed as a part of a radiant floor heating system in a concrete slab.**

A Sample Installation

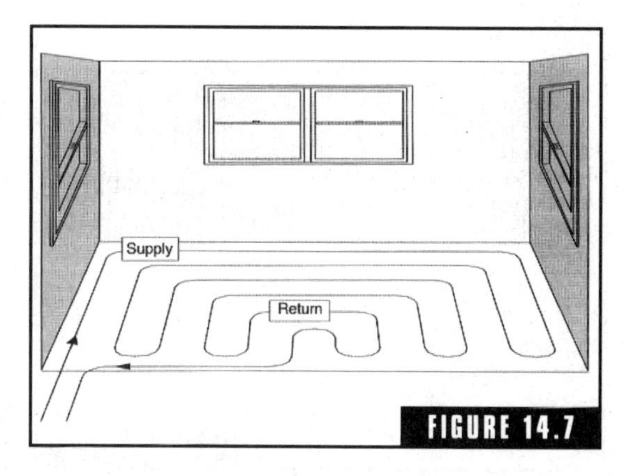

Triple-wall serpentine design. (*Courtesy of Wirsbo.*)

A sample installation can be talked about, but every job is different. Ideally, a heating designer will develop a working drawing for installers to work from (Fig. 14.7). A lot of contractors and installers work without a detailed drawing, but this can be risky. A strong working plan is needed to assure a proper installation. Very few installers can consistently make installations by eyeballing a job. Sure, they can install the tubing and get heat to a building, but the quality of an installed-on-the-run job is often questionable. Good advice is to insist on an approved heating diagram prior to making an installation. Assuming that you have a viable heating plan to work with, making an installation for a slab-on-grade heating system is not very difficult.

Heating contractors have to work closely with other trades when installing heat tubing in an under-slab situation. Plumbing and electrical work should be completed prior to starting any heating installation. The risk of having other trades damage the plastic heat tubing is too great, plus it's not practical to block installation paths needed by other trades. Any work being done in the pad area should be completed before the heating installation is done. At a minimum, a 6-mil polyethylene vapor barrier should be in place before insulation and heat tubing is installed. Insulation boards used should be of a type where a tongue-and-groove joint is made as the boards are installed. There is some risk that the insulation boards will become disturbed until reinforcing wire is installed. The wire should be installed as soon as possible over the insulation. Once the insulation is in place, you can move on to manifold and tubing installation.

The designer of a heating system should indicate all major details of an intended system (Fig. 14.8). This will include potential locations for a manifold system. The manifold should be placed in a central location that can be installed so that it will be accessible. Manifolds are most often positioned to be encased in a cavity in a stud wall. As most installers know, it's critical to turn the heating tubing up in the

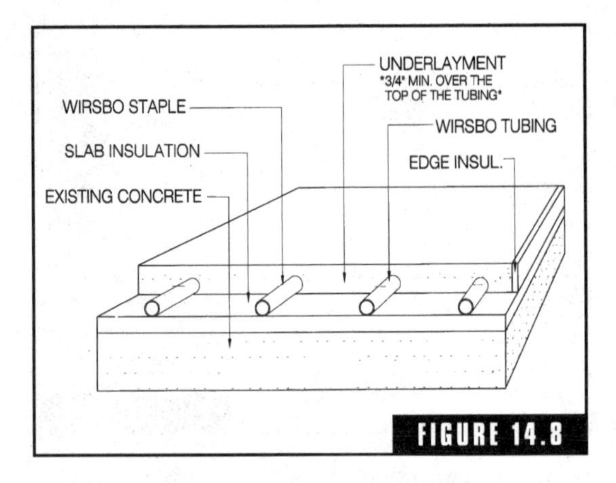

Detailed installation plan. (*Courtesy of Wirsbo.*)

right place during a rough-in. If the tubing is off by even 1/2 in., it can mess up the finished version of a building. The tubing should be turned up out of the concrete so that it will wind up in a planned wall. First of all, the planning of wall location must be precise. Installation of the tubing must also be right on the money. Then the tubing must be secured tightly to ensure that it doesn't move after an installation and before concrete is poured. Special devices are available to accommodate the bends in heat tubing that are needed to bring the tubing out of concrete. These devices should be used during an installation to prevent kinking and to protect the tubing fully.

The tubing used in modern radiant floor heating systems, usually PEX tubing or PB tubing, is very flexible. It can, however, kink if it is not handled properly. Most installers use large rolls of tubing when making a full installation. The tubing is generally placed on a device that is called an uncoiler. Using the uncoiler helps to prevent kinking. Once the uncoiler and the tubing are set up for installation, the process can begin. Assuming that there is a heating diagram to work with, and there should be one, the job is fairly simple.

All tubing installed should be in continuous lengths. Avoid joints at all times. Some installers use cans of spray paint to mark their intended piping path. This can make the job a little easier to keep track of. If you know that you will need 200 ft of tubing to make a run, allow more length to account for the amount of tubing needed to turn the tubing up at a manifold. Plastic heat tubing is pretty inexpensive, so be generous with your turn-up amounts. It can be very frustrating to be nearing the end of a run and find out that you don't have enough tubing to make the installation reach a manifold. If you waste a few feet of tubing, that's okay. It's much better to have more than you need than not enough. Most experienced installers label the ends of tubing that is turned up for a manifold. Personally, I wrap duct tape around the

TRADE SECRET **All tubing installed should be in continuous lengths. Avoid joints at all times.**

tubing and write on it with a permanent marker. You should identify the tubing in some fashion so that you can keep the various zones organized when connecting to a manifold.

Tubing can be unrolled and laid out before being secured, but the tubing should be secured soon to avoid confusion, kinking, and crushing. If any tubing is attempting to bend upward, secure it to take out the bend. Plastic tubing is much more flexible during warm weather than it is during cold temperatures. Keep this in mind as you are installing the tubing, to avoid kinking. The tubing used for radiant floor heating systems is easy to work with, but it does have limitations. When you are making a turn or bend, allow some room for the bend to avoid kinking the tubing.

Another consideration when installing heat tubing is the location of planned expansion joints. These joints are cuts in a concrete floor that give concrete a way to move at prescribed locations in an attempt to prevent uncontrolled cracking in a floor. When these joints are made, a saw is usually involved in cutting the line in the fresh concrete. It's common for the cut to run to a depth of approximately 20 percent of the slab depth. You must make sure that the heat tubing is buried deeply enough in the concrete to avoid being hit by the saw blade (Figs. 14.9 to 14.11).

If you are a cautious installer, you can sleeve heat tubing where it will be passing under planned expansion joints. When a sleeve is used, it should be at least two pipe sizes larger than the tubing that it is protecting. The sleeve can help in terms of risk from cutting sawn joints in the floor for expansion joints. And the sleeves relieve bending stress on any tubing that may be caused as a slab moves with ground movement. The more effort you put into an installation, the fewer problems you will have later in life.

Things to Be Mindful Of

After all tubing is installed and secured, it should be tested for leaks. The test is normally conducted with air pressure. Check with local code authorities to see what testing requirements are in your area, but a pressure of at least 50 psi should be applied to the tubing system and maintained for at least 24 hours. Air temperature can affect the pressure reading a little, but leaks will drop the pressure considerably.

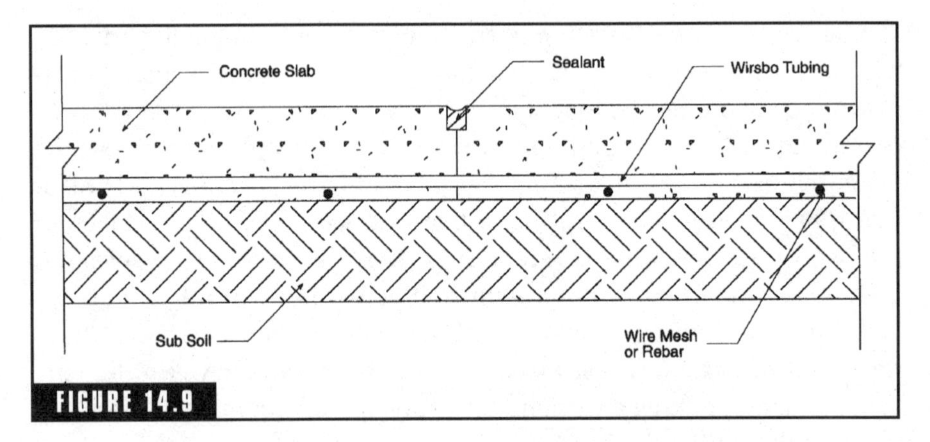

FIGURE 14.9

Cross section of a control joint where heat tubing is installed. (*Courtesy of Wirsbo.*)

Long runs of individual tubing that turn up in a group create a bit of a problem for testing. You could test each tubing circuit one at a time, but it makes more sense to join the tubing to a mutual device so that all tubing can be tested at one time.

There are many ways in which to connect individual tubing to a common device. Some contractors and installers use soft copper tubing and bend it into U-shaped sections that are inserted from tubing circuit to tubing circuit. My way is a bit different. I take a piece of copper tubing, install a number of tee fittings in a line, and cap one end of the device. Each section of heat tubing can be connected to the tee fittings, and the air pump can be connected to the open end of the

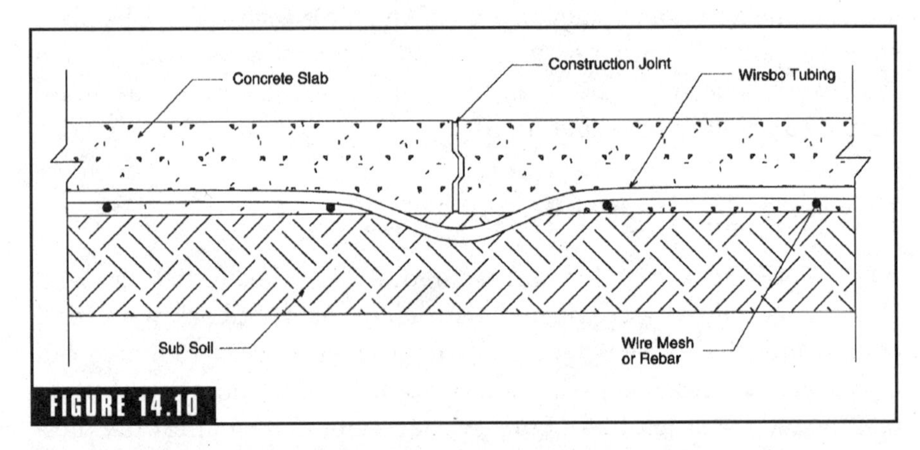

FIGURE 14.10

Heat tubing installed below slab where control joint will be. (*Courtesy of Wirsbo.*)

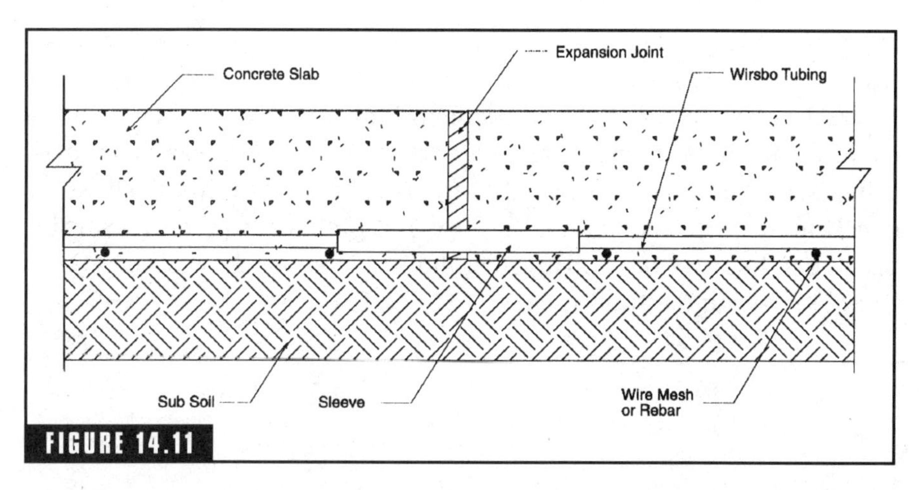

Use of a sleeve with heat tubing that is beneath an expansion joint. (*Courtesy of Wirsbo.*)

device. I like my way best, because I don't have to rely on hose clamps to maintain pressure. With my device, standard crimp rings can be used to attach to branches from my test manifold. Some contractors use water to test their systems, but most modern installers prefer air pressure. If leaks are expected, applying a soapy solution of water to joints will expose even small leaks. The soapy solution will bubble as air from the leak enters the solution. This is the same type of testing that is done for gas piping. If you use hose clamps and U-bends of copper to create your test arrangement, use a soapy solution to test each of the test connections, since they are prone to leakage.

After the slab is poured, you should test the tubing for leaks as soon as it is safe to walk on the slab. This is a step that many contractors skip, since it is not a code requirement. But it is something that many savvy contractors do. The second test eats into profits a little, but it can reduce problems later in a job. If heat tubing is damaged during the pouring of the slab, it can go undetected until a building is completed. If you test your installation while the slab is still green, fixing leaks will be much easier.

It's not common, but heat tubing does get damaged after it's installed. If you don't know about the damage before the slab cures, the concrete is much harder to break through for repairs. Any need for getting under concrete requires work, but the work is easier if the slab has not cured. Taking a little extra time to run a second test, after a slab is poured and before it hardens, can make the effort required for making repairs much less.

Contractors who anticipate problems and prepare for them experience far fewer problems than their competitors who assume everything will go as it should. So many factors are involved with construction and remodeling that it is not practical to plan for all the problems that may occur. However, taking sensible steps to minimize trouble is worth the time spent.

CHAPTER

15

Thin-Slab Piping Systems

Thin-slab piping systems have become more and more popular recently. Not long ago, contractors rarely considered installing radiant floor heating systems in structures where a standard slab was planned. It was generally thought that a standard slab was the only sensible place to install heat tubing. All of this has changed. It is now quite common for heat tubing to be installed in thin slabs and even in dry floor joists. This has opened opportunities for contractors considerably.

No longer is it necessary for a home or office or building to have a thick slab to benefit from radiant floor heating. With today's methods, heat tubing can be secured to a wood subfloor and then have a light layer of concrete poured over it. The thin-slab method requires no more than 1½ in. of concrete. While the slab is thin, the concrete still works as a heat mass that will deliver good equal heat from the heat tubing that is placed in and below it.

Concrete is not the only coverage used for thin-slab systems. There is a gypsum-based product that is also used. The goal is to reduce weight on a subfloor. Most thin-slab systems add between 12 and 15 psf of floor coverage. This may not seem like much weight, but it is enough that it has to be considered in building construction. The added weight is easy to compensate for in new construction, but remodeling jobs offer more complications. Weight is not the only consideration for a thin-slab system. There are many factors to consider, and we will talk about them shortly.

When a thin-slab system is created, heat tubing is secured directly to a wood subfloor. The method of attachment varies, but staples are probably the most common form of securement. Once the tubing is secure, a thin slab is poured over the tubing. When the slab is cured, a finished floor covering is installed over the slab. Floor joists below the subfloor are filled with insulation, usually a fiberglass batt insulation. Some type of containment is installed under the insulation to limit the movement of fibers. The system is simple and easy to install. However, special provisions must be made when a thin-slab system is used.

Floor Loads

Floor loads are the first consideration when planning a thin-slab system. The added weight must be accounted for. In new construction, this can mean increasing the size of floor joists. If this is not feasible, the spacing between the floor joists can be decreased. When remodeling, additional floor joists can be added to make up for lacking structural ability. Bracing supports are another option. A decision has to be made on how best to build a floor that will support the additional weight of a thin-slab system. Structural decisions should be made by qualified professionals. The dead-weight load of a floor is not something to be taken for granted.

Since a professional should determine what needs to be done with a subfloor system, the designer should provide a builder or remodeler with a detailed plan. When this is the case, it's simple enough to comply with the structural drawings and recommendations. There are, of course, times when an engineer or architect will not be involved. Unfortunately, very few remodelers or their customers are willing to pay for a professional evaluation of a bathroom subfloor where a thin-slab heating system is wanted. Adding radiant floor heat in a single bathroom on a remodeling job is usually not the type of job where a big budget exists. Even so, professional structural advice should be sought.

QUICK»TIP **Floor loads are the first consideration when planning a thin-slab system. The added weight must be accounted for.**

Some contractors have enough knowledge to plan their own floor loads. If the contractor has enough knowledge to accomplish the task safely, this is fine. But adding a thin-slab system to an existing subfloor and hoping that the floor can handle the weight is a mistake. In one way or another, confirm the strength of the subfloor being worked with. Add floor joists or other supports as needed. Reject the urge to just install tubing and cover it with concrete. Installing a thin-slab system on a subfloor that is not up to the challenge can result in serious problems later on.

Additional Thickness

The additional thickness of a floor that contains a thin-slab heating system can be a problem. This problem is generally easy to adjust for when the heating system is being installed in new construction. However, making adequate compensation for thin-slab floors in remodeling jobs can be quite a bit more difficult. Several considerations come into play. For example, the minimum amount of headroom required by local codes may be difficult to comply with when an existing floor is raised by 1½ in. Thresholds for doors will be affected by the raised floors used for thin-slab heating systems. Base cabinets in kitchens and bathrooms can be affected by the use of a thin-slab floor. When adding a thin slab in a bathroom, the toilet and the bathing unit can be affected. Most of the problems can be overcome, but the alterations required can be difficult and expensive.

When a thin-slab system is proposed for new construction, adjustments to building procedures are usually easy to work out. Such issues as window height, door height, thresholds, cabinets, plumbing fixtures, and overall ceiling height must all be accounted for. Even when working with new construction, meeting the requirements for a thin-slab system is not always easy.

Working out varying floor heights can be tricky. It's common for radiant floor heat to be used in only portions of a home, such as bathrooms. This means that the bathroom floors will be higher than the hall floor or other floors in the home. If the finished floor level is to be the same throughout a structure, the subflooring for rooms where thin slabs will be used must be lower. In either case, modifications are needed and should be planned for well in advance. Some serious consideration must be applied to the planning of varying floor heights.

Most of the problems encountered with installing thin-slab heating systems come when remodeling buildings and homes. Options for remodeling jobs are much more limited than they are for new construction. This doesn't mean that thin-slab systems should not be considered for remodeling.

The changes in floor heights can create problems for other trades, such as plumbers. Being a master plumber, I'm well aware of how difficult the routing of pipe can be in homes where the joist levels change from spot to spot. A good designer will take the needs of all trades into consideration when working out the best way to accommodate them.

Most of the problems encountered with installing thin-slab heating systems come when remodeling buildings and homes. Options for remodeling jobs are much more limited than they are for new construction. This doesn't mean that thin-slab systems should not be considered for remodeling. If a subfloor has the structural ability to hold a thin-slab system, most other elements of the job can be worked out. However, meeting the minimum headroom requirements of habitable space could create a problem if you will be working with a space where headroom is already minimal. If you can get past the headroom and the weight capacity issues, you should be able to find reasonable solutions for the rest of the problems. To expand on this, let's look at the problems on a one-by-one basis.

Structural Support

Structural support is essential for a good thin-slab installation. When remodeling a structure, a contractor has to consider the effect of additional weight incurred with a thin-slab system. If a floor doesn't have the structural integrity to support a thin-slab system, you will have to find a way to strengthen the subfloor. If there is a basement or crawl space under the room to be equipped with thin slab, you can use timbers and support columns to add strength to the flooring system. The support posts should be installed on concrete pads that are capable of accepting the weight load. Another option is to add floor joists between the existing joists. By reducing the distance between joists, support strength is increased.

Not all rooms have easy access under them. This creates more of a problem, but solutions still exist. If there is no other way to access a joist system, you can remove the existing subflooring to expose the joists. By doing this, you can add additional joists to the system.

Another option would be to remove the ceiling below the floor joists and work from under them. There are, of course, other ways to improve the holding ability of a floor. Regardless of what is done, the procedure should be approved by someone with structural credentials. It may also be necessary to get approval from the local code enforcement office.

Plumbing Fixtures

Plumbing fixtures are affected when a floor level is raised. Flanges for toilets must be raised. This usually isn't a major job, but the work can be difficult on some jobs. In many cases, the existing pipe that the flange is attached to can be cut and extended while a new flange is added. Bathtubs and showers can usually be left alone. The new floor will simply flow around the bathing unit. In the case of a clawfoot tub, the tub will need to be raised, but this is not a big job. Vanity cabinets can be removed and reinstalled after the new thin slab is poured. Normally the plumbing fixtures and cabinets installed in a bathroom don't offer much resistance to raising the floor level of the room.

Door Openings

Door openings between rooms with varying floor heights require a transitional threshold. This threshold is needed to prevent tripping. It is sometimes necessary to make these thresholds on a job-by-job basis. It's not always possible to buy a stock threshold that will make the flow from one floor level to another smooth. It is also possible to form the new slab so that it slopes toward the lower floor level. The risk to this is getting a slab that is too thin to hold up under pressure. The slab may crack if it is too thin.

Headroom

The amount of headroom required in a room is usually enough to allow a thin-slab installation without problems. There are, however, times when an existing room was built to such minimum clearances that the extra 1$^{1}/_{2}$ in. added when a thin slab is installed can be troublesome. Overcoming a headroom problem can cost more than it's worth. You might be able to get a variance for the local code enforcement office. This is the first step that most contractors would take. If a variance is not obtainable, the hard work begins. Raising a ceiling or lowering a floor system can be complicated and expensive. Before this type of work is done, someone must make a decision about how much work and expense is justified to get a thin-slab system. Most contractors would probably recommend a dry-system installation, where the floor level would not have to be raised with concrete covering. Using the dry system would not trigger a change in existing headroom.

Lightweight Concrete

Lightweight concrete (Fig. 15.1) is usually the first type of material considered when planning a thin-slab heating system. Generally speaking, lightweight concrete and gypsum-based underlayments are the only logical choices for thin-slab systems. When lightweight concrete is used to cover heat tubing in a normal installation, the weight added to a subfloor can be as much as 15 psf of floor space. People like the concrete because it is highly resistive to moisture damage. It's possible to leave this type of thin slab uncovered for use as a finished floor. In practice, however, most thin slabs are covered with some type of finished flooring.

When compared with a gypsum-based product, lightweight concrete will usually prove to have a higher thermal conductivity, and this is good. But lightweight concrete does not have as much heat conductivity as concrete where crushed stone aggregate is used, as in full-depth slabs. Concrete used for thin-slab applications is generally made up of portland cement, sand, lightweight coarse aggregate, chopped nylon fibers, and a number of additives that improve flexibility and reduce shrinkage.

Getting lightweight concrete to a floor is not extremely difficult. It's possible to transport the material in wheelbarrows or buckets. A more effective method of transportation involves the use of a hose and a pumping station. Workers will have to work the concrete as it is installed. Screeding is needed as the concrete is poured. It doesn't take much concrete to do a bathroom floor. In fact, 1 sq. yd. of concrete poured to a depth of approximately 1 1/2 in. will cover about 210 sq. ft. of floor area.

MATERIALS

Lightweight concrete (Fig. 15.1) is usually the first type of material considered when planning a thin-slab heating system. Generally speaking, lightweight concrete and gypsum-based underlayments are the only logical choices for thin-slab systems.

Control joints should be planned prior to pouring concrete for a thin-slab system. In thick slabs, the control joints are often sawn after the concrete is poured. This can be done in a thin-slab system if the tubing is below the concrete and the cuts are not made too deep. Since cutting into a thin slab can be risky, some contractors use thin plastic strips to create expansion joints. When the strips are used, they are kept about 3/8 in. below the finished surface of the concrete. By dividing the pour area into small sections with the strips, you can reduce the risk of having the concrete crack later. The strips should be installed anywhere that concrete will come into contact with an immovable object, such as a wall or the edge of a bathtub.

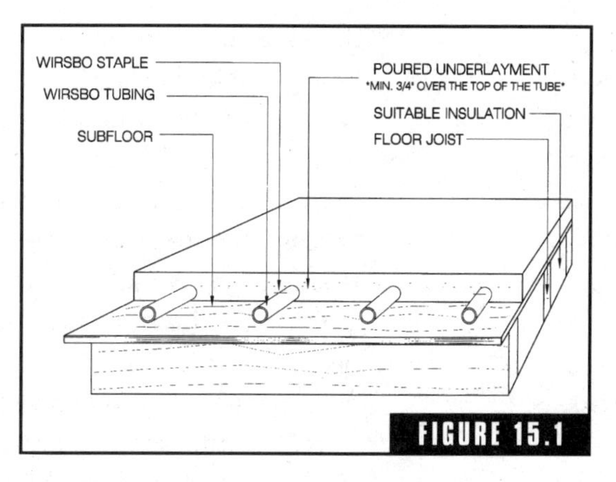

FIGURE 15.1

Thin-slab system detail. (*Courtesy of Wirsbo.*)

Gypsum-Based Systems

Gypsum-based systems are not as well known as lightweight concrete systems, but this may change. The best alternative to a lightweight concrete is a gypsum-based underlayment. This material has been used for years and has proved to be a good product to use with a thin-slab heating system. This type of underlayment is made up of gypsum-based cement, sand, water, and additives to improve flexibility and to reduce shrinkage. The weight of this material comes close to 15 psf. A bonding agent should be used with gypsum-based underlayments. The bonding agent will reduce water problems that might occur with subflooring and will strengthen the bond between the subfloor and the thin slab (Fig. 15.2).

Bonding agents used with gypsum-based underlayments are typically sprayed on the subflooring after all tubing is installed. When the sealing and bonding is done, the gypsum material is prepared in a mixer that is usually placed outside the building. Once the mixture is ready, it is pumped into the job site. When you see

MATERIALS

Gypsum-based systems are not as well known as lightweight concrete systems, but this may change. The best alternative to a lightweight concrete is a gypsum-based underlayment.

the material come out of a hose onto a floor, it appears to be something like pancake batter. Owing to the consistency of the mixture, it flows well and fills in cracks easily. Workers must manipulate the mix to level it out. This is normally done with a wooden float tool. When there is enough material on the subfloor to equal the diameter of the tubing, the pouring is stopped. This step of the process is known as the first stage of the job. Some contractors call it the first lift. It takes about 2 hours for the first stage of the thin slab to cure to a point where it can be walked on. As the mixture cures, it shrinks. This is normal and to be expected. The second stage of the job will correct the shrinkage.

When the second layer of mixture is poured, it is leveled out to maintain a minimum thickness of about 3/4 in. above the heat tubing. Most contractors affix wooden pegs on their wooden float tools to maintain a consistent depth of the mixture. Shrinkage that occurs in the first stage is compensated for with the second pour. Again, the mixture should dry in a couple of hours. However, some mixtures take longer to dry, and site conditions can affect the drying time. Drying time can be reduced by running the heating system during the drying process.

It's critical that the floor be completely dried out before a finished floor covering is installed. Many contractors test their floors by taping sheets of plastic over the poured floor. If water droplets form on the

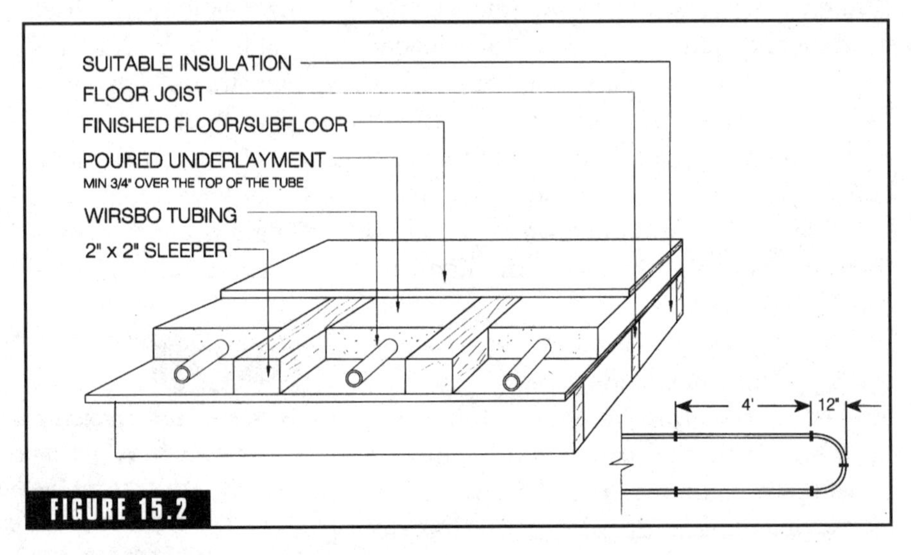

Detail of a poured underlayment system. (*Courtesy of Wirsbo.*)

plastic, the floor is not dry enough to cover with a finished floor covering. When the plastic remains dry, a finished floor covering can be installed.

A thin slab made with gypsum materials can be poured quickly. The material is easy to work with and does not require as much labor to finish as concrete does. Another advantage to the gypsum material is that it is less likely to shrink or crack than concrete is. A thin slab made with gypsum materials has plenty of strength to accept normal foot traffic. However, this type of floor is subject to gouging and should never be used as a finished floor until a suitable floor covering is installed over the thin slab.

A few disadvantages are associated with thin slabs made with gypsum materials. One is that the gypsum-based material doesn't offer as much heat conductivity as concrete does. To overcome this, the heating system should be run at a higher temperature. Otherwise there may be cold spots between heat tubes. Another significant disadvantage is that gypsum materials don't do well when they get wet. Prolonged exposure to water or excessive moisture can damage a thin slab that is made of a gypsum product. The risk of this can be reduced by using a finished floor covering that is water-resistant and caulking all edges and areas where water might seep into the thin slab.

A Little or a Lot

Thin-slab systems can be used to heat a little space or a lot of space. Entire buildings can be fitted with thin-slab heating systems. Selected areas of a home can be heated with thin-slab systems. A lot of people don't think about using thin-slab systems as an alternative to other types of heating systems. But, as people become more familiar with the systems, the demand for them will probably grow. The expense of pouring a thin slab cannot be ignored, but even with the cost of a slab, the overall financial numbers for this type of system can prove competitive with other types of heating systems. An alternative to the slab is the installation of heat tubing beneath subflooring. In this type of installation, no slab is needed. To learn more about this type of heating installation, let's turn to the next chapter.

CHAPTER 16

Dry Piping Systems

Dry piping systems are radiant floor heating systems that don't involve pouring a slab of any type. In these systems, heat tubing is attached directly to subflooring, either above or below it. Instead of pouring concrete or a gypsum-based product over heat tubing that is installed above a subfloor, a deflection system made of aluminum plates is used, although these plates are not always used. Some systems consist only of the heat tubing, but many contractors frown on this practice. One big advantage of a dry system that is installed beneath a subfloor is that there is no need for the floor levels of the building to vary. Dry systems can be used in both new construction and remodeling jobs. The cost of a dry system is low when compared with slab systems, but performance is not as good. Even so, performance is acceptable and the system is a viable consideration.

Many people have little to no confidence in a dry system. Some people feel that a radiant heating system has to be in concrete for it to work well. While it's true that radiant systems in concrete do work extremely well, there is no rule that says concrete must be used with a radiant system. Dry systems are being used frequently in modern construction. Installed properly, these dry systems work very well (Fig. 16.1).

QUICK»TIP **Unlike thin-slab systems, dry systems don't require contractors to beef up flooring supports.**

Most installers use aluminum transfer plates to conduct heat. These plates are not essential to a heating system, but they do make a system more effective. A big advantage to a dry system is that it adds very little weight to a flooring structure. This makes dry systems suitable for both new construction and remodeling without the high cost of flooring reinforcement. The only real weight involved is the plastic tubing and the water in it. This weight is minimal. Unlike thin-slab systems, dry systems don't require contractors to beef up flooring supports.

A dry system is fast and easy to install. Most contractors and installers use staples to hold heat tubing to subflooring. Once a heating plan is drawn, it's a fairly simple matter to place the tubing and staple it to the wood subflooring. There is, however, a risk that the tubing will be damaged by other trades. For example, a carpenter might drive a nail into a floor from above when heat tubing is installed below the floor. A plumber might accidentally cut the heat tubing when sawing a hole for a toilet flange. Or a plumber's drill bit might nick a section of tubing while drilling holes for plumbing pipes. This same sort of accident could happen when an electrician is drilling holes. Any number of possibilities exist for the heat tubing to be damaged. All in all, this is one of the biggest challenges associated with a dry system. It's not practical to install nail plates to protect the tubing, so cooperation must exist between the trades. And it's wise to have as much other work out of the way as possible before installing the heat tubing.

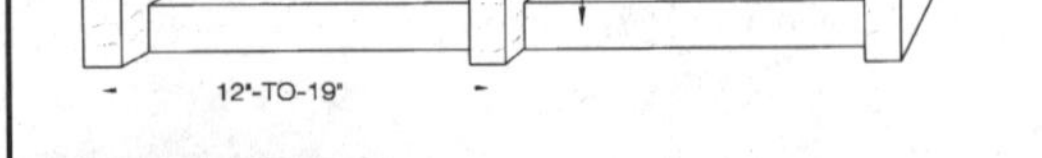

FINISHED FLOOR
"WOOD,TILE,CARPET"
WIRSBO TUBING
WIRSBO STAPLE
SUITABLE INSULATION
FLOOR JOIST
SUBFLOOR
12"-TO-19"

FIGURE 16.1

Dry heating system installed between floor joists. *(Courtesy of Wirsbo.)*

Transfer Plates

Aluminum transfer plates (Fig. 16.2) are often used with dry systems. These plates transfer heat laterally to eliminate cold spots between heating tubes. While the use of aluminum transfer plates can't compete with concrete in heat conductivity, the plates are capable of doing a good job.

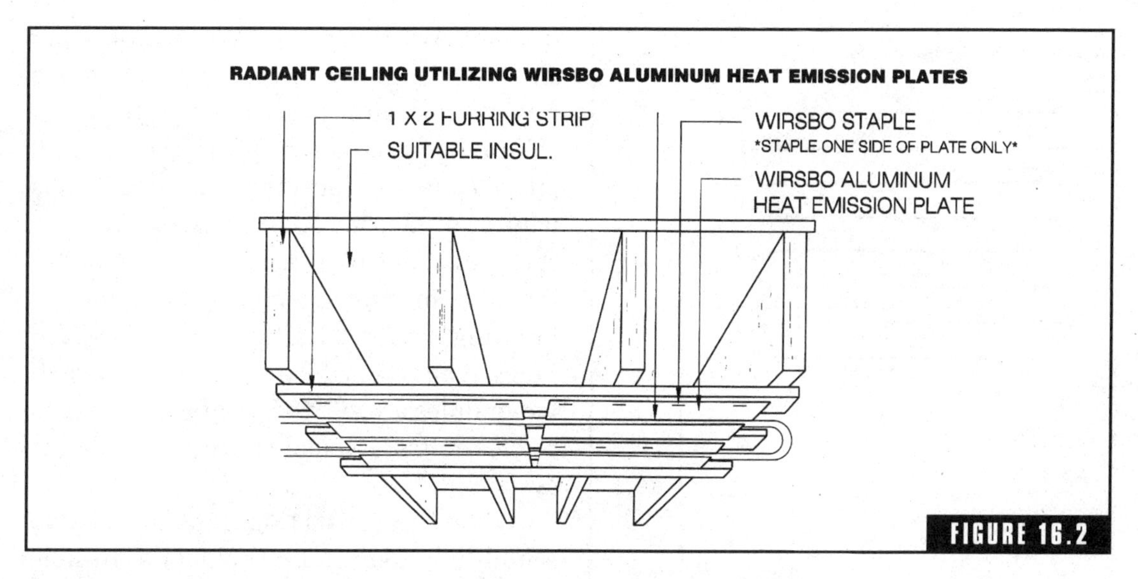

Example of heat-transfer plates in use. (*Courtesy of Wirsbo.*)

Most contractors feel that a dry system should not be installed without the plates, but there are plenty of jobs done that don't include the use of transfer plates.

When aluminum plates are used, they may be installed above or below subflooring. This, of course, depends upon where the heat tubing is located. Insulation is used in conjunction with a dry system. This is the case with, or without, aluminum transfer plates. Installing transfer plates is highly recommended when using a dry system.

Bothersome noise can be associated with aluminum transfer plates. This is due to the thermal expansion of plastic tubing. Most complaints, when they exist, are associated with a ticking noise when the heating system is running. Since building owners don't generally appreciate a noisy heating system, something has to be done to reduce the noise factor. This is not a difficult or expensive proposition.

> **MATERIALS**
>
> **Aluminum transfer plates (Fig. 16.2) are often used with dry systems. These plates transfer heat laterally to eliminate cold spots between heating tubes. While the use of aluminum transfer plates can't compete with concrete in heat conductivity, the plates are capable of doing a good job.**

To avoid noise from aluminum transfer plates, an installer can do one of many things. One of the first considerations is the diameter of holes in joists that heat

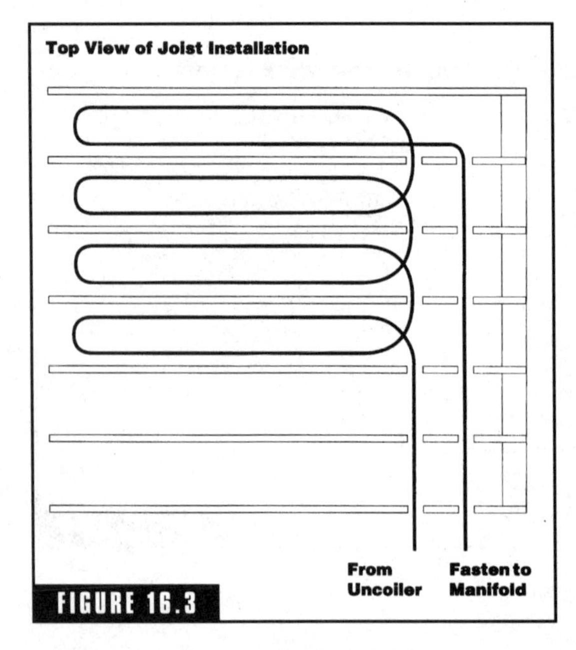

FIGURE 16.3

Routing of heat tubing in floor joists. (*Courtesy of Wirsbo.*)

tubing passes through. When holes are small, tubing that expands can't move freely. This can lead to noise problems. A solution to this is to drill larger holes in joists. However, local building regulations may prevent oversized holes. Many code agencies don't like larger holes, owing to the risk of fire spreading throughout a building. Before you go with oversized holes, check local code requirements. If larger holes are allowable, they can help to reduce noise associated with the heating system.

When truss joists (Fig. 16.3) are used in a building there is no problem with hole sizes. Tubing can be run through the openings in the truss joists. This eliminates any space constrictions that might cause noise in a system. Another noise-reduction tactic is to run tubing in shorter lengths. This creates more loops. These loops help to handle expansion in the tubing. While shorter, more numerous loops seem like extra work, and they are, the benefit of controlling expansion noise makes the additional effort worthwhile. Find a good balance and work with it. One other strategy is to install expansion loops when longer runs are not feasible.

Heat in plastic tubing is what makes the tubing expand. The higher the heat is, the more expansion there will be. Try to keep water temperature in the heat tubing as low as possible. This might best be done by installing a weather-responsive reset control on the heating system. This control will keep water temperature as low as possible whenever possible. It is desirable to keep water temperature low for noise control, but don't compromise heat output in exchange for a quieter system. Higher temperatures are needed in dry systems than would be needed in a slab or thin-slab system. If you reduce the water temperature too much, comfort will not be accomplished.

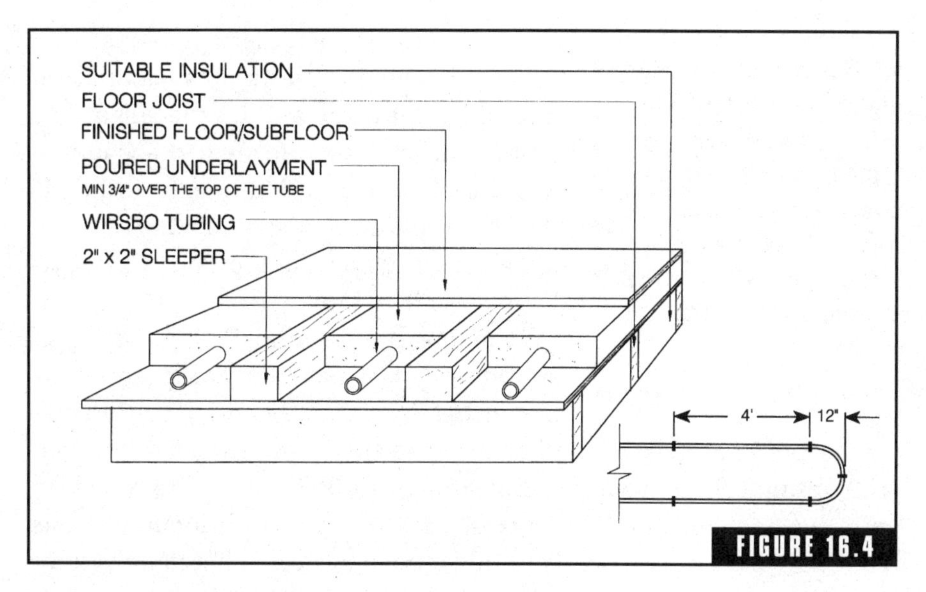

FIGURE 16.4

Sleeper sections in use. (*Courtesy of Wirsbo.*)

On Top

Plastic heat tubing is sometimes placed on top of subflooring. In my experience, this type of installation is not as common as an underfloor installation. When the heat tubing is installed above subflooring, with transfer plates, the process is not complicated. The system starts with subflooring. Sleeper sections (Fig. 16.4) are then installed on the subflooring. The sleeper sections are frequently made from 3/4-in. plywood. Channels are created for the tubing to run through. Then the tubing is installed. However, many contractors run their tubing after the transfer plates are installed. The plates hold the tubing in place. Next the aluminum transfer plates are put into position over the tubing. Then a layer of underlayment is installed. It is usually about 3/8 in. thick. Finished floor covering is then installed over the underlayment to complete the installation on the upper side of the floor. Insulation, usually a glass fiber batt insulation, is installed below the heated floor area. Some form of barrier is installed under the installation to contain any fibers which might become airborne. This completes the installation in the floor areas.

When aluminum transfer plates are installed with a sleeper system, the plates are secured on only one side. This is done to allow the plates

When aluminum transfer plates are installed with a sleeper system, the plates are secured on only one side. This is done to allow the plates free movement as thermal expansion occurs. By having the plates secured on one side, they remain in place but are allowed to move as needed to accommodate expansion.

free movement as thermal expansion occurs. By having the plates secured on one side, they remain in place but are allowed to move as needed to accommodate expansion. Another benefit to this is that the aluminum doesn't buckle or wrinkle as much when the top layer of underlayment is installed.

The installation of heat tubing on top of subflooring is time-consuming and thus expensive (Fig. 16.5) Material used to create sleeper channels is another expense that would not be incurred with an underfloor system. Long straight runs of tubing are preferred with systems installed on top of subflooring. Additional work is required where tubing bends in this type of system. There must either be well-positioned, cut-out sleepers to accommodate the tubing, or a plunge router will have to be used to mill out the long sleepers to allow for turns in the tubing.

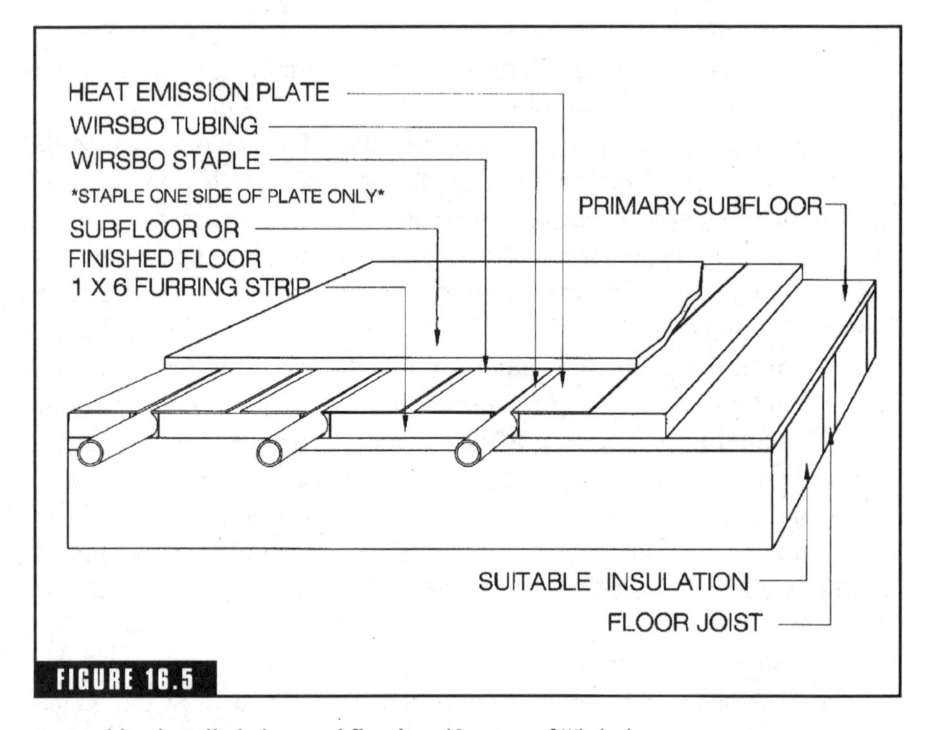

Heat tubing installed above subflooring. (*Courtesy of Wirsbo.*)

Installers should glue as well as nail all sleepers and underlayment to reduce the risk of squeaking floors. It should also be noted that the use of a sleeper system raises the finished floor height. In fact, the floor height is raised about 1 in., nearly as much as it would be with a thin-slab system. This is another reason why a lot of contractors dislike installing heat tubing on top of subflooring. For a number of reasons, installing heat tubing below subflooring is more cost-effective than installing a similar system on top of a subfloor. With this in mind, let's look at the procedure for installing a dry system under subflooring.

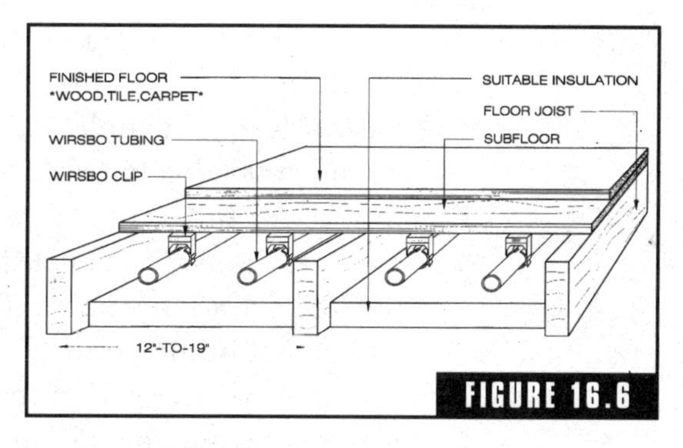

Installation below subflooring using clips to support heat tubing. (*Courtesy of Wirsbo.*)

Under Subflooring

Installing heat tubing under subflooring is simple (Fig. 16.6), cost-effective, and not a bad way to heat a home. Putting heat under a subfloor works well in new construction, and it can do very well in remodeling jobs when access to the underside of the subfloor is available. Since this type of system is installed beneath a floor, it can be installed in a remodeling job without affecting the existing floor covering. This is a substantial advantage.

Tubing that is installed below subflooring may be stapled to the subflooring or supported by heat-transfer plates (Fig. 16.7). Again, the use of transfer plates is recommended. When transfer plates are used, they are stapled to the subflooring and the tubing is run through them. Starting at the top, there is the subflooring. The plates are stapled to the underside of the subflooring. If plates are not used, the tubing is stapled directly to the subflooring. When plates are used, tubing is routed through the plates. Batt insulation is installed below the heat tubing or transfer plates. Once finished floor covering is added over the subflooring, the system is complete in the floor area.

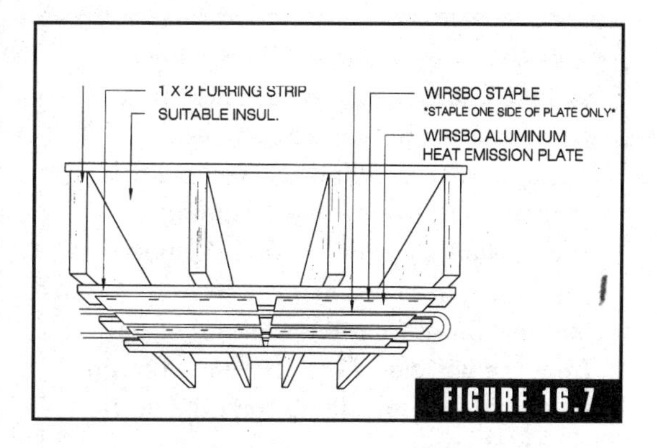

Heat transfer plates can support heat tubing. (*Courtesy of Wirsbo.*)

When installing heat tubing in the cavities of floor joists, you must, naturally, drill the floor joists. Heat tubing must pass through the joists. Ideally, the holes drilled for the tubing should be larger than the diameter needed to pass the tubing through. A rule of thumb is to drill the holes a full 1/2 in. larger than needed. This is not always allowable by local codes, so check the code requirements in your area before drilling extra-large holes. The last thing you need is to be held responsible for ruining a bunch of floor joists. Actually, you could probably correct a code violation for oversized holes by filling them with foam insulation, but there's no reason to take the risk. Check with local authorities before you drill joists.

All holes in the joists should be drilled in a straight line. This can be done easily enough. Most installers use a chalk line to mark their tubing path. Popping a line with a chalk line will provide the horizontal positioning. Then you have to measure down from the subfloor or up from the bottom of the joist to maintain a straight line for tubing from the vertical point of view. Most codes require that joists not be cut too close to either the top or bottom of the joist. Ideally, locate holes for tubing near the center of the joists. Don't even think about notching the bottom of joists to route tubing. This reduces structural integrity and will almost always cause a stir with building inspectors. If you keep your holes near the middle of joists, you shouldn't run into any trouble with code officers or unwanted nails that will dull drill bits and jerk your drill out of your hands.

After holes are drilled for tubing, you are ready to pull the tubing into the joist bays. It is necessary to pull tubing to the farthest joist bay (Fig. 16.8). From here, the tubing can be routed through the joists for heating purposes. When the tubing is in place, it is either stapled to the subflooring or enclosed in a heat transfer plate that is stapled to the subflooring. The heating plan is followed to fill each joist bay with supply tubing. Then, a return loop is installed in the same general manner. It is recommended that a gap of at least 1/4 in. be maintained between all transfer plates. This gaps helps to compensate for thermal expansion.

TRADE SECRET

All holes in the joists should be drilled in a straight line. This can be done easily enough. Most installers use a chalk line to mark their tubing path. Popping a line with a chalk line will provide the horizontal positioning. Then you have to measure down from the subfloor or up from the bottom of the joist to maintain a straight line for tubing from the vertical point of view.

As the joist bays are filled with heat tubing, the tubing typically runs from one end of the joist bay to the other. This requires a U-bend in the tubing. This bend should be made carefully to avoid kinking (Fig. 16.9). It's also essential to avoid binding at any point during the installation. If heat tubing is constricted, it can contribute to noise when the heating system is running. This is why it is advantageous to drill oversized holes, where practical.

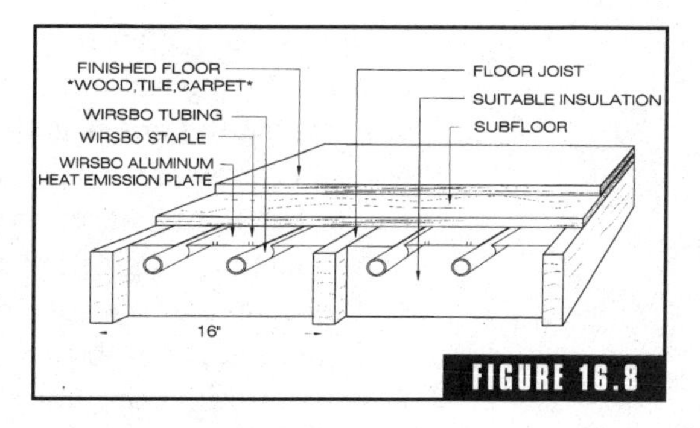

FIGURE 16.8

Heat tubing in floor joists. (*Courtesy of Wirsbo.*)

Obstacles

Obstacles can be a problem when running a dry system. This is especially true when working on existing buildings. All sorts of things can get in the way of tubing paths when you are doing a retrofit job. Even new construction can throw plenty of obstacles at you or your

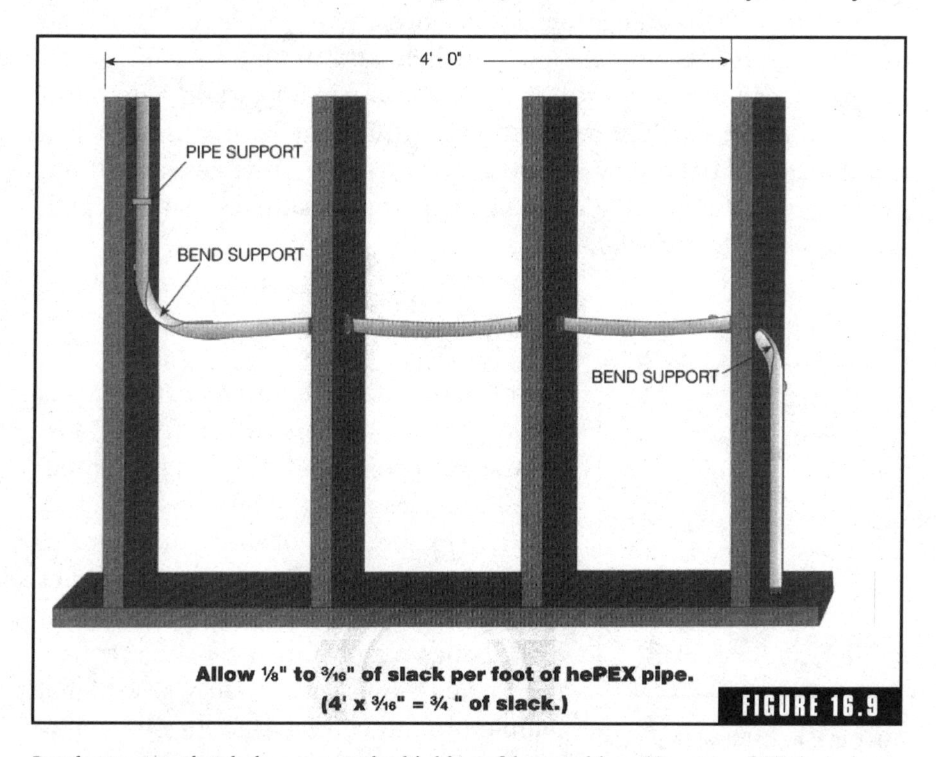

FIGURE 16.9

Bend supports that help prevent the kinking of heat tubing. (*Courtesy of Wirsbo.*)

installers. No matter how well a heating system is planned, expecting the unexpected can only go so far. A key to avoiding problems in this area of installations is good communication with other trades. And it helps to have cooperative trades on the job site. For example, if a carpenter has to cut a section out of a joist and head it off for structural requirements, it's nice to be on good terms with the carpenter.

Obstacles in new construction are far less likely to give you problems than are the obstacles encountered in remodeling jobs. If you have been under the floors of many old homes, you've seen the vast array of objects that exist in the area visited infrequently by anyone. There can be ductwork running all over the place. Electrical wires might be run through joist bays in a way that makes a plate of spaghetti look organized. Plumbers can make a major mess of joist cavities. Larger timbers in old structures can create serious problems when it comes to routing tubing. Steel beams are another potential dead end for the installation of tubing in a straight line. All of these types of potential problems must be taken into account prior to an installation. A detailed site visit is needed before creating a heating plan.

A common misconception is that heat tubing is so small and so flexible that it can be installed anywhere and in any way. While the tubing is easy to work with, installers still need a certain amount of access to make sensible installations. At the very least, obstacles can run the cost of a job way up. In some cases, the obstacles can all but put an end to the installation altogether. You must investigate the path options before you design or bid a job.

No Plates

I said earlier that there are dry systems where no transfer plates are used. Most contractors agree that this is not good practice. Regardless, a lot of jobs are done that don't include the use of aluminum transfer plates (Fig. 16.10). Eliminating the plates will lower the cost of a job. But the financial savings may be lost over time by having to run the heating system at a higher temperature to compensate for the lack of plates. Plus, the floor above the heat tub-

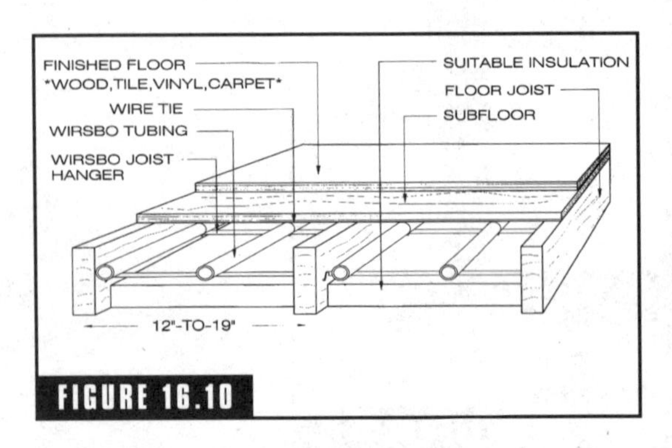

FIGURE 16.10

A dry system that doesn't use heat-transfer plates. *(Courtesy of Wirsbo.)*

ing will probably have cold spots in it if transfer plates are not used.

The transfer plates fill many responsibilities. First of all, they transfer heat laterally. This will not happen if the plates are not installed, which will result in a heating system that is not as efficient. Downward heat loss is increased when transfer plates are omitted. Again, this results in additional operating cost to maintain an equal room temperature. Transfer plates can keep heat tubing in more consistent contact with subflooring. In doing so, the floor above stays warmer. Wood is not nearly the conductor of heat that aluminum is, and without transfer plates, wood is the only conductor available. Tubing that is tight against subflooring during an installation can sag once water is introduced into the tubing. If the tubing is not touching the subflooring, the effectiveness of the heating system is weakened. Raising water temperature within the heat tubing will compensate for a lack of transfer plates, but the increased water temperature results in a more costly operating expense. Plus, high temperatures in radiant floor heat tubing may, over time, degrade the integrity of the tubing to a point of failure. The simple solution is to install transfer plates with every dry system. Not only is this common practice, it is also common sense.

The use of dry systems is on the rise (Fig. 16.11). More and more people are accepting what was once considered a questionable form of heating. Since these systems are capable of delivering good even heat at affordable prices, all contractors and customers should consider dry heating systems. Now, let's move to the next chapter and look at what your options are when installing radiant heat for ice removal.

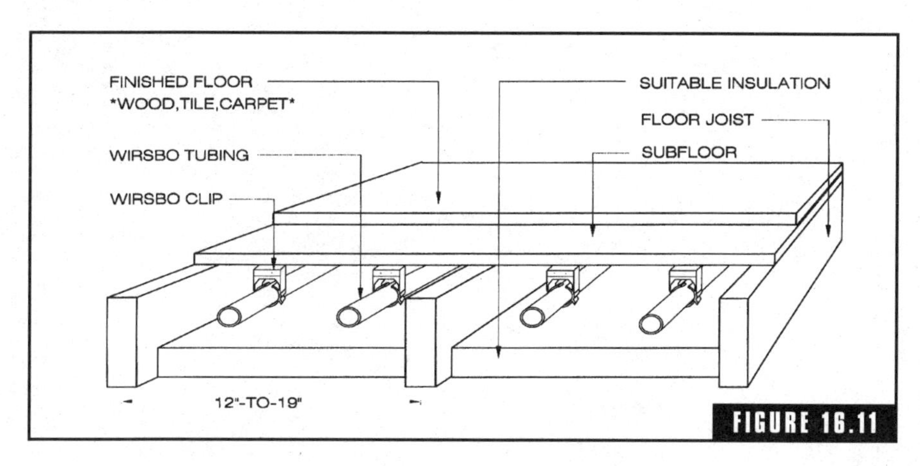

A sandwich method of using heat-transfer plates. (*Courtesy of Wirsbo.*)

CHAPTER 17

Radiant Systems for Ice Removal

Some people believe that installing radiant heating systems for ice removal is an indulgence. There are people who think automatic snow-removal systems are only for the rich. Not so. If a house is being built with plans for a radiant floor heating system, it can make a lot of sense to add snow and ice melting to the duties of the heating system. This will run the cost up, but homeowners may be very happy to pay more when their homes are built if it means never having to shovel snow again. And it's a safety issue. Ice on walkways and driveways causes a lot of slip-and-fall accidents. These accidents can do a lot of damage to a human body. And it's possible that such an accident will result in a costly lawsuit. If this can be avoided by installing a melting system, and it can, some thought should be given to investing in such a system.

Average people don't often think of having an automatic melting system installed along with their home heating system. This is also true of professionals who are having new office buildings built. Large areas are not usually practical to equip with a melting system. But smaller areas, such as parking areas and walkways for a home, are prime candidates for radiant snow melting systems.

There is sometimes argument from opposition that claims an automatic melting system costs too much to operate. This group of people may say that it's cheaper to shovel or plow snow. Cost depends on many factors, but small areas are not generally expensive to clear with automatic melting systems. And shoveling and plowing do little to prevent ice buildup. Sure, a person can shovel a walkway and sand it to make ice more maneuverable, but why bother with this when an automatic system can keep the walkway clear and dry?

Many advantages become apparent when you consider all aspects of an automatic melting system. For example, it's possible to install a system that will melt snow as fast as it falls. However, such a system may cost more to install and operate than its benefits are worth. If you are willing to allow a little snow to build up as most of the snow is being melted, the cost of installation and operation is much less. Various factors come into play when designing a snow-removal system. For example, the designer will have to figure heat capacity, the probable peak snowfall rate, and the budget allowed for system installation and operation.

Modern melting systems can be controlled with smart controls that will turn a heating system on when snow is present and cut off the system when snow and ice have been cleared. This is a big improvement over older controls. Not too long ago, a person had to turn a melting system on and off manually. This could be inconvenient and fairly inefficient. If a snowstorm came in late at night, a substantial buildup of snow could accumulate before it was noticed. Then turning on a melting system would force the system to work hard, which isn't the most efficient way for a system to operate. With automatic controls, snow is never allowed to pile up before the melting system goes to work. Since the smart controls have the melting system working only when it's needed and always when it's needed, the melting system works efficiently and cost-effectively.

QUICK»TIP **It's possible to install a system that will melt snow as fast as it falls. However, such a system may cost more to install and operate than its benefits are worth.**

People who live in areas where heavy snow falls frequently know all too well how much of a job it can be to keep walkways and driveways clear. It's common for people to wait until a storm is over to shovel, blow, or plow snow. Doing snow removal during a storm is sometimes necessary just to keep up with the snow accumulation. When snow is removed before a storm is over, additional removal will be needed by the time the storm is over. When an automatic melting system is used, there is no need to fuss with multiple removal attempts. The automatic system will melt snow continuously, as needed, to keep walkways and driveways clear.

Some of the advantages to automatic melting go beyond the most practical aspects. Many of the benefits come in lesser forms. For example, keeping a walkway clear at all times prevents snow from being tracked into a home or office. The advantage of having a clear walkway is enhanced as the risk of slipping and falling is greatly reduced. Driveways that are kept clear of snow and ice provide better traction and less risk of damage to vehicles.

A Typical Installation

A typical installation for a radiant snow removal system is not particularly complicated. Basic heat loops are installed much the same way that they would be in a radiant floor heating system. But a heat exchanger is usually used with a snow-removal system. There are, of course, needs for a heat source, valves, controls, and so forth. The heat source can be the same boiler that provides heated supply water to the interior heating system. There are two main reasons why a heat exchanger should be included in the design plan of a melting system. Many boilers can't cope with extremely low fluid temperatures. This, combined with the fact that most melting systems carry a lot of antifreeze with the heating fluid, is good reason to use a heat exchanger.

Standard boilers simply don't work well with low-temperature water, especially in the heat-up stage. A lot of condensation builds up in a standard boiler that is working with low-temperature water. This is not good. And a standard boiler that is providing heat for other loads may be producing supply water that is too hot for a radiant snow-melting system. For all of these reasons, a heat exchanger is highly desirable in a heat design for snow melting.

Cracking is a concern when heat tubing is installed in a walkway or driveway. When a slab is heated too quickly or to a high temperature around a heat tube, cracking is a possibility. This is known as thermal stress.

Precautions

Precautions must be considered when designing and installing a snow-melting system. Usually snow-melting systems are installed in either concrete or asphalt. Before an installation is made, someone must be sure that the heating plan will not weaken the structural integrity of the slab or pavement where it will be installed. Special provisions may be needed to accommodate heat tubing in a slab or in asphalt. Another concern is that too much heat will be delivered to the heat mass at one time. This can cause cracking in the driveway or walkway. And a heating design must be laid out to ensure that the entire surface of the area to be treated will maintain melting temperatures as needed. This means, basically, keeping heat tubes close together, say about 8 to 10 in. apart, and never more than 12 in. apart. If the tubing is too far apart, cold spots in the slab or asphalt surface will allow snow and ice to build up.

Concrete that is not thick enough will crack. A minimum thickness of concrete over a heating pipe should be not less than 2 in. Now, we're not talking about indoor, thin-slab systems here. We are talking about walkways and driveway areas. An additional 2 in. of concrete should be under the heat tubing. And an allowance for the diameter of the tubing should be made. In other words, a 4-in. slab is not ideal. The slab would be 4 in. thick, plus additional thickness that is equal to the outside diameter of the tubing. So, for example, you might have a slab that is 4½ in. thick to make a suitable installation for tubing that has an outside diameter of ½ in.

Asphalt is more forgiving than concrete when it comes to tubing placement. Since asphalt is more flexible than concrete, tubing can be installed closer to the surface of the finished driving, parking, or walking area. If a finished area will have to handle stress from vehicles, the slab or asphalt will have to be designed to be stronger. This means having an increased thickness in the concrete or asphalt.

Cracking is a concern when heat tubing is installed in a walkway or driveway. When a slab is heated too quickly or to a high temperature around a heat tube, cracking is a possibility. This is known as thermal stress. One method used to minimize this risk is the use of tubing with a small diameter. The tubing material should have a low conductivity, as PEX tubing does. A moderate fluid temperature also helps to prevent cracking. Another way of reducing cracking is starting the system pump without preheating the fluid. This allows the circulating fluid and the slab or asphalt to warm up together, reducing the risk of damage to the heat mass.

An added precaution against cracking is to space heat tubing close together. Installing the tubing on a 6-in. center is acceptable. Any spacing from 6 to 12 in. should be all right. Many contractors favor a spacing of 8 in. This type of spacing minimizes thermal stress. The system fluid should operate at a temperature of no more than 120°F. This temperature should be developed gradually. A warmup period of at least 1 hour should be allowed when the system is coming up to heat.

Coil Design

Coil design is a serious consideration when laying out a melting system. The two options are a grid coil and a serpentine coil. Most contractors favor a serpentine design. Either design can work well, but grid coils are somewhat more difficult to work with. Sometimes the geometry of the area to be headed determines which coil type should be used. Usually the serpentine coil is favored. This is the same type of tubing layout that is most often used in radiant floor heating systems.

Grid coils are made up of straight sections of tubing that run across the heated area. One end of the tubing is connected to a supply header while the other end is connected to a return header. Getting desired performance out of a grid system can be troublesome, but with the right design and installation, a grid coil works fine. Grid systems have low pressure drop over the elements in the grid. In contrast, serpentine coils normally have moderate pressure drop over the length of the coil.

Serpentine coils are sometimes installed with the supply header buried below the surface of the heated area. However, many contractors don't favor this process. Contractors and installers generally prefer to have the header accessible. When the header is buried, there is no access to the header or valves to isolate individual tubing. It is better to have an above-ground header where valves can be used to control each branch of a system. This is especially helpful if a leak develops in one of the branches.

Tubing Selection

The tubing selection for a radiant melting system is a substantial part of a system design. Fluid flow rate is, obviously, a factor in determining tubing size. A limit for a flow rate is affected by the pressure drop performance, the amount of heat to be delivered per coil, the available pump pressure, and the availability of tubing material to tolerate high flow rates. Designers like to use tubing with small diameters to minimize the amount of paving material needed to complete a job. But the downside of smaller tubing is that there are more limits on the total developed length of a coil. However, smaller tubing is easier to bend in short turns, and this is an advantage, since melting loops are usually spaced closely to each other. There are times when tubing with a small diameter may be the only available option, owing to the tight turning radius.

Factors in tubing selection include tubing size, flow rate, coil length, and the heat output per coil. The following are examples of sizing recommendations:

Tubing size (in. OD)	Flow rate (gal/min)	Coil length (ft)	Heat output per coil
3/8	0.5	50	5,000
1/2	0.7	75	7,500
1/2	1.5	150	15,000
3/4	3.5	300	35,000

Not the Same

The installation of a melting system is not the same as that of radiant floor heating systems. Yes, the procedures are very similar, but there are differences. The differences can have a major impact on how well a system works. If you are going to launch an effort to become known as an installer or contractor for melting systems, get to know the ropes early on. Manufacturers are usually very cooperative in providing information to contractors. Take advantage of this. Don't hesitate to contact suppliers and manufacturers for detailed information on products and procedures that are recommended with specific products. This is the best way to stay out of trouble.

What works for one type of material may not be appropriate for a different type of material. You should take the time to get to know your product line very well. If you do this, you should find that melting systems in regions of the country where they can be warranted are a nice addition to a company's financial statement. It's fairly rare to get a request to install a stand-alone melting system, but if a building will be equipped with a hot-water boiler anyway, the addition of a melting system is not such a stretch. Give it some thought and run it past your customers. You might well enjoy the additional work.

CHAPTER 18

Purging Air from Systems

Radiant heating systems don't do well when they contain air. In fact, most water-based heating systems suffer when air is introduced into a system. When new systems are installed, air is a natural problem. Getting the air out before considering the system functional is essential. Existing heating systems also suffer from air entering them. Service mechanics are often called to correct problems that are associated with air being in a hydronic heating system. The problems can be minor annoyances or major obstacles. It's possible for a system to become tainted by air to a point where heat production is greatly reduced. There are ways to reduce and sometimes avoid the problems caused by air being trapped in a closed system.

Radiant floor heating systems may not seem subject to many of the traditional problems associated with air in a heating system. Anyone who has lived with an older hydronic heating system has probably heard gurgling noises in pipes or banging from the pipes. Radiant floor heating systems are installed under floors, but the noise can still be a factor. This is especially true in dry systems and thin-slab systems. The plastic tubing used in modern radiant floor heating systems does not bang, but it can still make plenty of unwanted noise if it becomes polluted with air.

A lot of people in the trade have, at one time or another, faced the phantom of air in a hydronic system. Sometimes finding the cause of an air problem and eliminating it seems nearly impossible. In truth, with the right approach, most problems can be diagnosed and corrected. With a little advance planning, much can be done to reduce the risk of having to deal with air problems after a system is put into use.

All new hydronic heating systems contain air. Air is trapped in pipe and tubing when it is installed. Fresh water used to fill a new system contains oxygen. A well-designed hydronic system will be purged when first started and will, usually within a few days, rid itself of most of the air it contained as a new installation. In theory, once air is out of a system, it should stay out of it. This, however, is not always the case.

Some hydronic heating systems seem plagued by problems associated with trapped air. Is this just bad luck? No, it's far more likely to be the result of a careless installation or design. Contractors who install heating systems that are frequently flawed by air problems may not be able to identify and correct the problem. In most cases, the contractors would not have made the design or installation mistakes if they had known better. Looking for a problem that you will not recognize when you find it is not only frustrating, it is futile. Contractors must take the time to learn the fundamentals of acceptable design and installation methods.

Problems with Air

The problems associated with air in a heating system can be serious. Few people enjoy hearing gurgling noise coming from their walls or floors. When air is present in a hydronic heating system, some noise is to be expected. Certainly, the noise will be a bother, but it should be much more. When pipes are gurgling with air, a lot of potential damage could be occurring within the heating system. Except after initial start-up of a new system, gurgling or banging sounds should not be ignored. Severe problems can manifest when air invades a radiant heating system.

TRADE SECRET It's entirely possible that a complete heating system will surrender all of its heat output to air. An air lock, or binding as it is often called, can occur when a large air pocket blocks a heating pathway.

It's entirely possible that a complete heating system will surrender all of its heat output to air. An air lock, or binding as it is often called, can occur when a large air

TROUBLESHOOTING

Air pockets in a heating system translate into trouble. There are different types of air-associated problems that are common to water-based heating systems. The three main considerations are entrained air bubbles, stationary air pockets, and air that is dissolved in a fluid.

pocket blocks a heating pathway. This generally occurs near a circulating pump or when a low-head circulator cannot lift water above an air pocket. If the problem is at a circulating pump, it indicates that a substantial air mass is present near an impeller in the pump. If the impeller can't clear the air, the entire zone can be denied heat. Total failure of this type is rare, but it can occur.

Systems that are not designed to handle an induction of air can suffer from corrosion when air is present. The chemical reaction of oxygen mixing with fluid in a heating system can lead to rapid rusting. If this type of situation is allowed to exist, the rate of deterioration on the components of a heating system can be greatly accelerated.

Microbubbles created by air in a system can wreak havoc on the heating system. The busings in wet-rotor circulators can be damaged when microbubbles are present. The foamy-type substance created when microbubbles are present might hamper the lubrication of bushings. Additionally, microbubbles can reduce heat transfer, which reduces the effectiveness of a boiler. If there is an air cushion between a boiler and the boiler water, the direct transfer of heat is reduced.

Circulators which are required to work with air in a heating system may not be able to distribute a suitable amount of heat. When air and water meet at a circulator, it can reduce the circulator's ability to transfer mechanical energy to fluid. Pump cavitation is also a threat when air is trapped in a heating system. There is a lot that can go wrong when air enters a heating system or is never removed from the system.

Air Pockets

Air pockets in a heating system translate into trouble. There are different types of air-associated problems that are common to water-based heating systems. The three main considerations are entrained air bubbles, stationary air pockets, and air that is dissolved in a fluid. Some

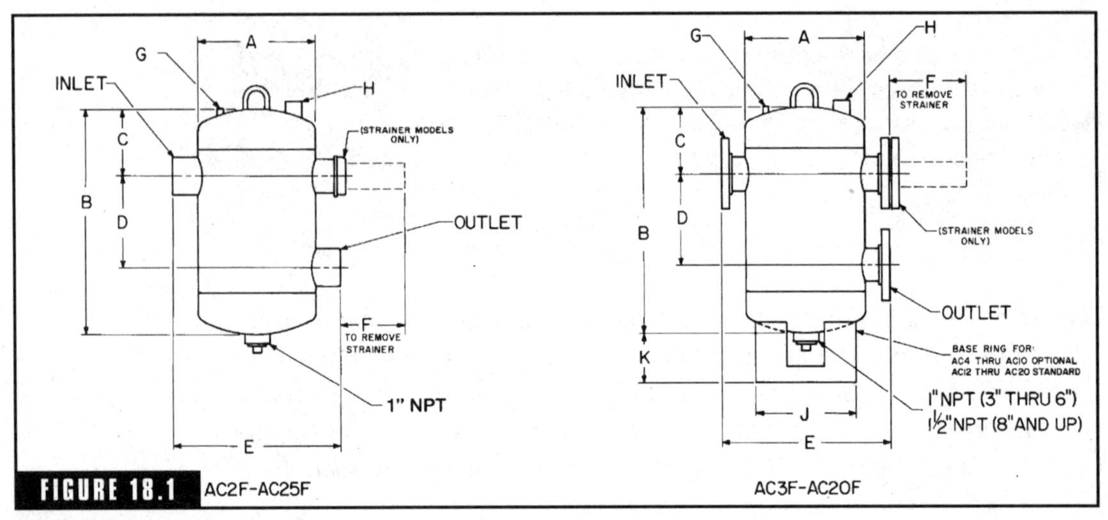

FIGURE 18.1

Air separator. *(Courtesy of Taco.)*

people might assume that they would be dealing with one or the other of these problems, but in fact you might face all three types of problems at once. This is especially true of a new heating system being started up for the first time. Troubleshooting symptoms for each type of air problem differ.

A system that is suffering from entrained air bubbles is able to move the air bubbles through the heating system. This is not always a big problem. If the air can be separated from the heating fluid when it reaches a central deaerating device, the process is effective and not so bad (Fig. 18.1). But if the air cannot be removed, problems will exist. Air bubbles want to travel upward. If they are forced downward, it may not be such a bad thing. Ideally, an operating heating system should be purged of air. If air must exist, it should be forced downward rather than allowed to rise. Many factors come into play when dealing with the movement of air in a heating system (Figs. 18.2 and 18.3).

The size of an air bubble affects its ability to rise in heating fluid. A large bubble will rise faster than a small one. In order to make bubbles move downward, the velocity of the flow of heating fluid must be greater than the velocity of the bubble rise. In addition to the diameter of an air bubble, the bubble's density affects its ability to rise. An air bubble that is entrained in a heating fluid with high viscosity will not rise very quickly if the bubble's density is low. Of all the factors to consider, the size of the bubble is the most likely component to determine its direction of flow.

How much impact does the diameter of a bubble have on its ability to rise? A lot, actually. For example, an air bubble that has a diameter of half the size of a competitive bubble will rise at a rate equal to about one-fourth of the rate of the larger air bubble. The size is cut in half, but the rate of rising is cut to one quarter. Other factors can come into play, but the diameter of an air bubble is the primary determinant of how fast the bubble will rise.

Microbubbles, as the name implies, are very small. But they tend to group together. Microbubbles cause tap water to appear cloudy. Trying to spot a single microbubble is quite difficult, owing to the diminutive size. When dissolved air comes out of solution when heated, it can create microbubbles. Since the microbubbles are tiny, they are difficult to get out of a heating system. The rise velocity of microbubbles is low. Since most air is removed from a heating system with purge valves and vents that depend on air bubbles rising, the low rising ability of microbubbles tends to keep them trapped within a system. Most hydronic systems are not designed or installed in a manner to rid themselves easily of microbubbles. Over time, the microbubbles manifest themselves into larger bubbles and can be removed from the heating fluid. However, this can take days to occur, during which the air can wreak havoc with a heating system.

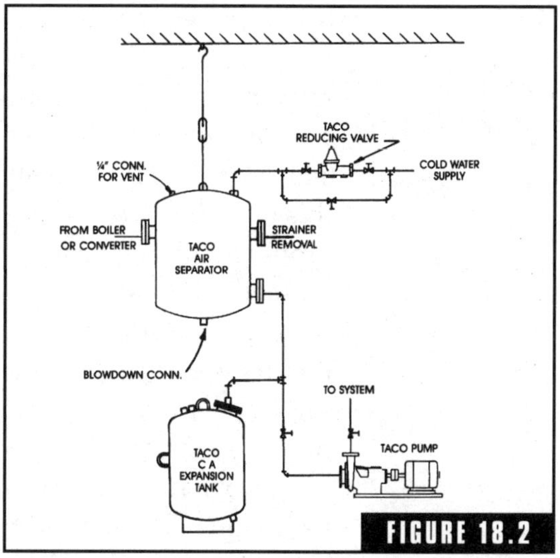

FIGURE 18.2

Recommended method of air control installation. (*Courtesy of Taco.*)

Stationary Air Pockets

Stationary air pockets can form at many points within a heating system. Most people think of the pockets being created at the highest points of heating systems, but

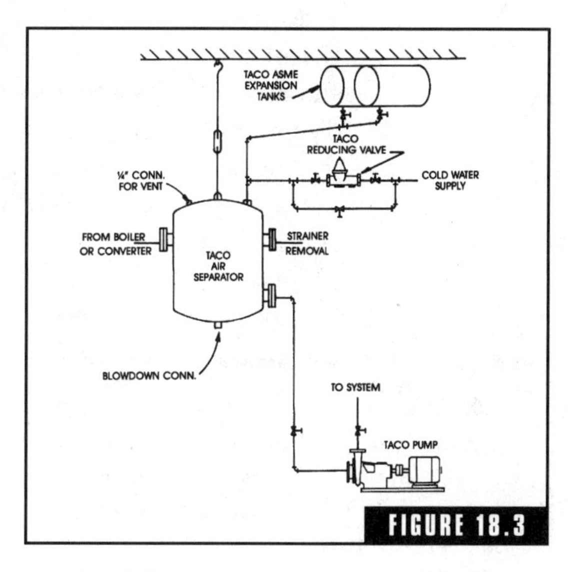

FIGURE 18.3

Conventional method of air control installation. (*Courtesy of Taco.*)

this is not always the case. Air is lighter than water and should therefore rise within the fluid. But this does not mean that the air will migrate to the overall highest elevation. In fact, stationary air pockets are commonly encountered near heat emitters. Horizontal heat piping and tubing is also a likely place to find stationary air pockets, and it is this area that most often affects a radiant floor heating system. If heat piping or tubing is running horizontally and then forced to rise over an obstacle, the area of the piping or tubing that is above the lower sections of horizontal conduits is a potential location for air pockets. Problems associated with this type of air infiltration and tubing or piping installation are common, especially when a heating system is first put into service.

When a new system is purged properly, most stationary air pockets are removed. However, the pockets can form again after a system has been running. This doesn't happen too often in residential applications. Common causes for recurring air pockets include large heat emitters, storage tanks, and large-diameter piping. Systems operating at low flow velocities are more susceptible to recurring stationary air pockets. Purging a system will usually remove the air and keep the system running well until new pockets of air form.

Dissolved Air

Dissolved air in a heating system can be a perplexing problem. The dissolved air is invisible. Water temperature and water pressure affect the amount of air that exists in solution within the heating system. Typically, hotter water contains less dissolved air. As fluid in a heating system cools, it attempts to collect air. The fluid will gather air from any source available. So the point is, hot water is less likely to contain dissolved air.

Just as water temperature affects the ability of the water to maintain air, so does water pressure. When the water pressure is low, less air is maintained in the fluid. If water pressure is high, the likelihood of dissolved air being in the fluid increases. Keeping the fluid in a heating system at a high temperature while operating at low pressure is the best combination for avoiding dissolved air in the system fluid.

MATERIALS

Air vents (Fig. 18.4) are installed in heating systems to provide a means for air removal. There are automatic vents and manual vents.

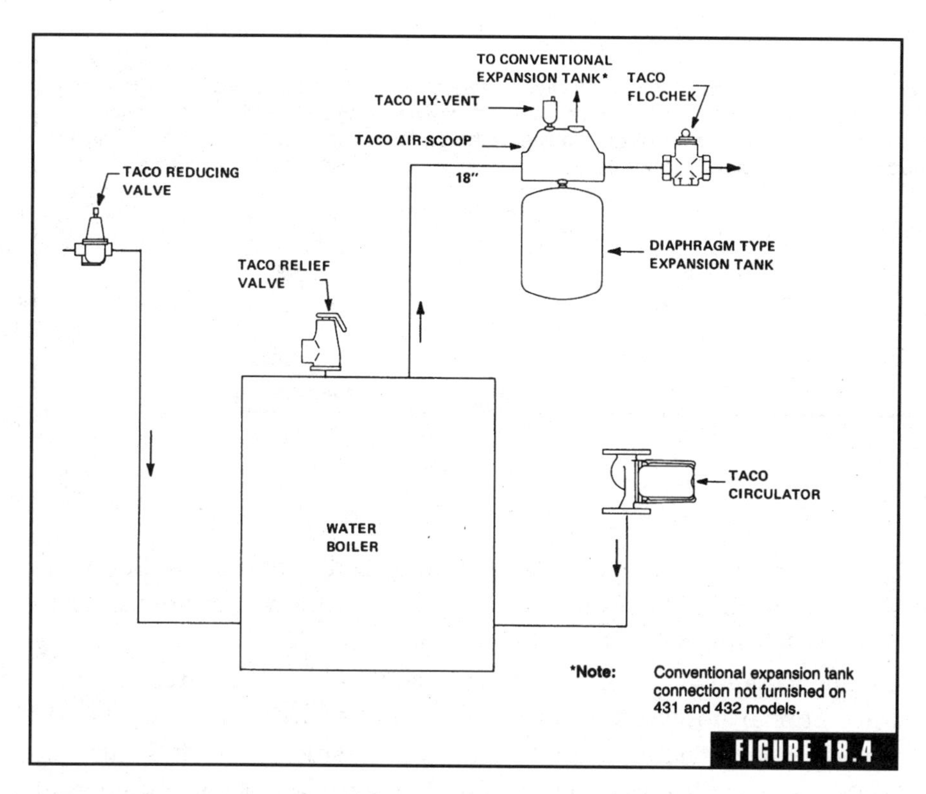

High air vent installed on a typical heating system. (*Courtesy of Taco.*)

Air Vents

Air vents (Fig. 18.4) are installed in heating systems to provide a means for air removal. There are automatic vents and manual vents. In terms of classification, the two types of vents used on small heating systems are high-point vents and central deaerators, with high-point vents being the more common of the two. A high-point vent is installed at a high point on the heating system. In the case of radiant floor heating, the vents are installed at the top of both the supply and return manifolds. When a central deaerator is used, it is installed near the outlet of the boiler. All system water passes through the device, and entrained air is removed by the deaerator (Fig. 18.5).

Manual air vents are both simple to install and inexpensive. They are more of a valve than a vent. Most manual air vents are designed to be threaded into a tapped fitting. Different types of manual vents are

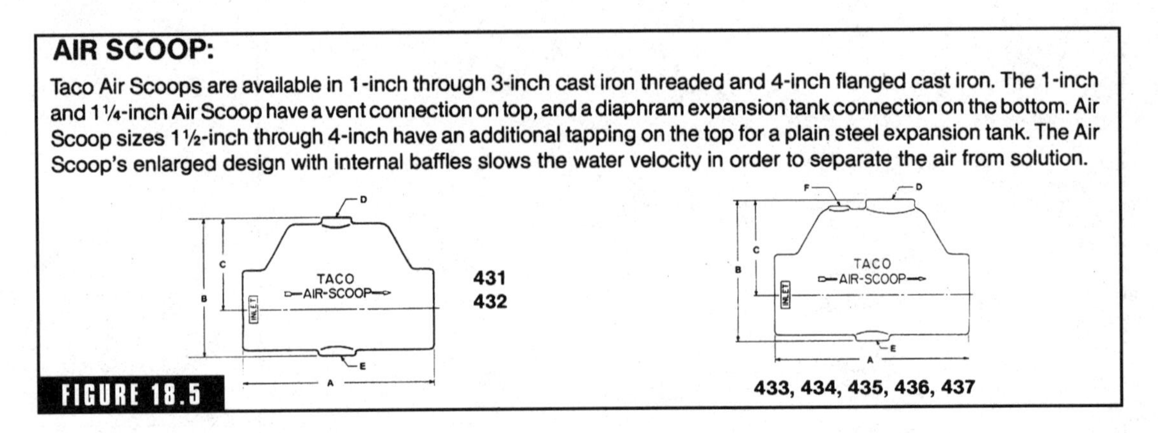

FIGURE 18.5

Air scoop. (*Courtesy of Taco.*)

sold, but most of them are opened and closed with either a flat-blade screwdriver or a square-head key. People in the trade often refer to these little vents as coin vents or bleeder vents. To use the vents, they are opened and air is allowed to escape, then the vents are closed. In conventional hot-water heating systems, the bleeders are frequently placed at each heat emitter. In the case of radiant floor systems, the vents are placed at supply and return manifolds. When purging air from a system, the vent should be left open until a steady stream of heating fluid is produced. Be prepared for the liquid by having a can or small pail available. Once the liquid is flowing without air, the vent can be closed.

Automatic air vents are available for use as a high-point vent. These vents are fine when used in locations where dripping heating fluid will not be a problem. But, owing to the method of operation, simple automatic vents do occasionally allow water or fluid leakage in small quantities. Do not install basic automatic vents in areas where heating fluid will damage floors, floor coverings, or ceilings. When automatic venting is needed and leakage is not acceptable, use float-type air vents.

Air can be removed automatically with float-type air vents. These vents are ideal for installations where access for manual purging is limited. The vents consist of an air chamber, a float, and an air valve. When enough air is present in the air chamber, the float will drop. As the float is dropping, the air valve is opening. Air leaves the vent and water or heating fluid enters the air chamber. Water coming into the air chamber raises the float, which closes the air vent. The result is an automatically vented system with no spillage.

Float-type vents are available in many shapes, sizes, and designs. Depending upon the model chosen, the vent can be mounted horizontally or vertically. One potential problem with float-type vents and automatic vents is that they may allow air into a heating system during the venting process. This happens if the water pressure in the system is less than atmospheric pressure. Improper placement of an expansion tank is one common cause that leads to automatic vents taking in air. To avoid air infiltration, you should make sure that there is a minimum of 5 psi of positive pressure at the top of a heating system at all times. This is done by setting the feedwater valve to maintain the pressure required.

Central deareating devices, also known as purgers and air scoops (Fig. 18.6) are used to control entrained air. The devices are located close to the heat source, and all system fluid passes through the device. A purger can have an internal baffle that will send air bubbles to a separation chamber. Once in the chamber, the air can rise to the top, where a float-type air vent is installed to allow the air to escape. The baffle in the purger also keeps pressure low, which allows more dissolved air to be removed. One important thing to know and remember is that purgers must be installed with the proper flow direction in order to work. There is an arrow on the body of a purger. The point of the arrow must be pointing in the direction of flow.

The flow rate of a system is important when a purger is installed. If the flow rate exceeds 4 feet per second, the purger may not function as

SIZE & DIMENSIONS

<table>
<tr><th rowspan="2">Prod. No.</th><th rowspan="2">Size</th><th colspan="2">A</th><th colspan="2">B</th><th colspan="2">C</th><th rowspan="2">D</th><th rowspan="2">E</th><th rowspan="2">F</th><th colspan="2">Weight</th></tr>
<tr><th>in</th><th>mm</th><th>in</th><th>mm</th><th>in</th><th>mm</th><th>lb.</th><th>Kg.</th></tr>
<tr><td>431</td><td>1"</td><td rowspan="2">6</td><td rowspan="2">152</td><td rowspan="2">4</td><td rowspan="2">102</td><td rowspan="2">2½</td><td rowspan="2">64</td><td rowspan="2">N/A</td><td rowspan="7">½" NPT</td><td rowspan="6">⅛" NPT</td><td rowspan="2">4</td><td rowspan="2">1.8</td></tr>
<tr><td>432</td><td>1¼"</td></tr>
<tr><td>433</td><td>1½"</td><td rowspan="2">8</td><td rowspan="2">203</td><td rowspan="2">6</td><td rowspan="2">152</td><td rowspan="2">4</td><td rowspan="2">102</td><td>¾" NPT</td><td rowspan="2">7</td><td rowspan="2">3.2</td></tr>
<tr><td>434</td><td>2"</td><td rowspan="2">1" NPT</td></tr>
<tr><td>435</td><td>2½"</td><td rowspan="2">10</td><td rowspan="2">254</td><td rowspan="2">8</td><td rowspan="2">203</td><td rowspan="2">5½</td><td rowspan="2">140</td><td>15</td><td>6.8</td></tr>
<tr><td>436</td><td>3"</td><td>1¼ NPT</td><td>14</td><td>6.4</td></tr>
<tr><td>437</td><td>4"</td><td>16 5/16</td><td>414</td><td>11⅝</td><td>295</td><td>7⅛</td><td>181</td><td>1½" NPT</td><td>¼" NPT</td><td>52</td><td>23.6</td></tr>
</table>

FIGURE 18.6

Specifications for an air scoop. (*Courtesy of Taco.*)

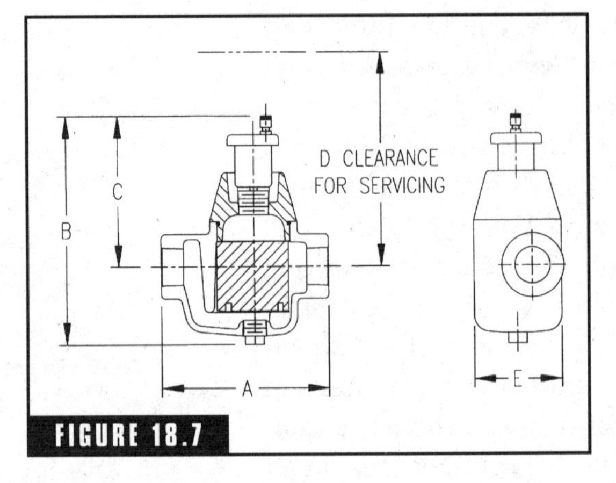

Air separator. (*Courtesy of Taco.*)

well as it should. Faster flow rates can allow small air bubble to remain entrained. To get maximum efficiency, the purger should be installed on the outlet side of the boiler and close to the boiler. The purger should be close to the inlet port of the circulator (Fig. 18.7). Being installed close to the boiler means that water temperature will be high. The purger keeps water pressure low. As you should recall from earlier in this chapter, high temperature and low pressure is the best way to keep air out of a system.

The Purging Process

The purging process for a heating system can be done in different ways. Existing systems will purge themselves automatically when equipped with the proper vents. When manual vents are used, they must be operated as explained earlier to release air from a system. The filling and purging of new systems can be done in different ways. Many installers use the gravity method. This entails opening all high-point vents and allowing the heating system to run until all air is removed through the vents. This method is effective and has been used for decades. However, many installers choose to use a forced-water method to speed up the purging process. Gravity purging can take a considerable amount of time to complete. When forced-water purging is done, the time required is much less.

Forcing water into a new heating system at a high velocity makes it possible to rid the system of air much more quickly than would be possible with gravity purging. The high-velocity water entrains air bubbles and carries them to the purge outlet for quick removal. Usually a boiler drain and a gate valve are added to the return pipe on a boiler to facilitate forced-water purging. The drain should be installed near the return connection of the boiler. An alternative to installing a boiler drain and a gate valve is to install a purge valve (Figs. 18.8 and 18.9), which is a specialty valve that fulfills both requirements. This helps to get unwanted debris, such as tubing shavings, out of the heating system. Once a new heating system is ready to be filled and purged, the

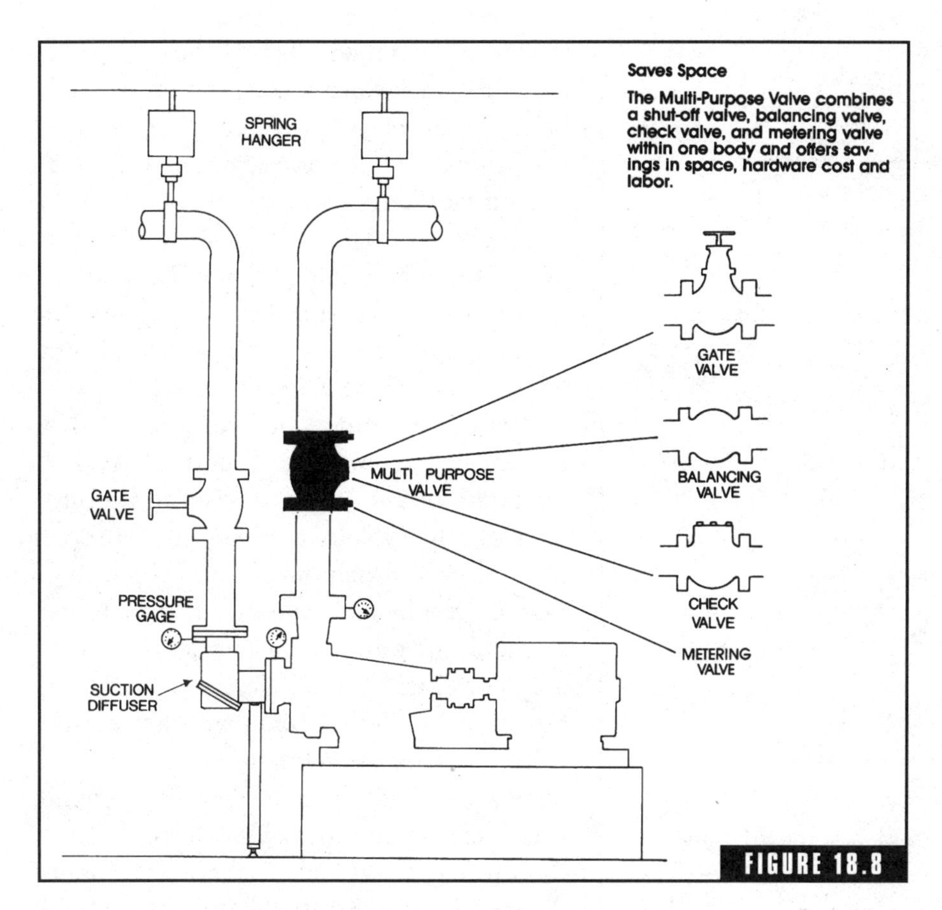

Multipurpose valve. (*Courtesy of Taco.*)

fast fill level at the top of the feedwater valve must be lifted to allow water into the system at a high rate of flow. The drain valve should be opened and normally has a hose attached to it to convey the discharge water to a suitable destination. The gate valve on the return pipe should be closed before water is introduced into the heating system.

After water rushes into the boiler and fills it, the incoming water is driven down the heat tubing in the system. Air that is trapped in the boiler is released through

TRADE SECRET

Getting air out of a hydronic heating system is an important part of a new installation, and it can be vital to keeping an existing system running in a proper manner. Don't rush the process of purging. Spend an adequate amount of time to get as much air out of a system as is reasonably possible.

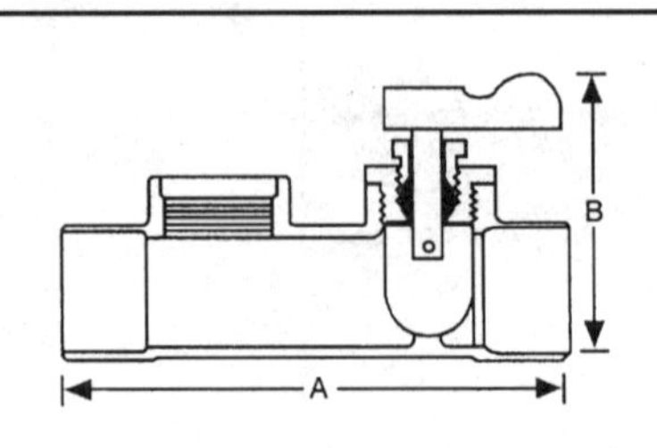

Specifications and Dimensions

No.	Size	Dimensions A	Dimensions B	Material	Drain Tapping	Shipping Weight
PV-0	1/2"	3 1/2"	1 15/16"	Bronze	1/2"	3 1/2 ozs.
PV-1	3/4"	3 7/8"	2"	Bronze	1/2"	8 ozs.
PV-2	1"	4 3/8"	2 1/2"	Bronze	1/2"	12 ozs.
PV-3	1 1/4"	4 3/8"	2 5/8"	Bronze	1/2"	15 ozs.

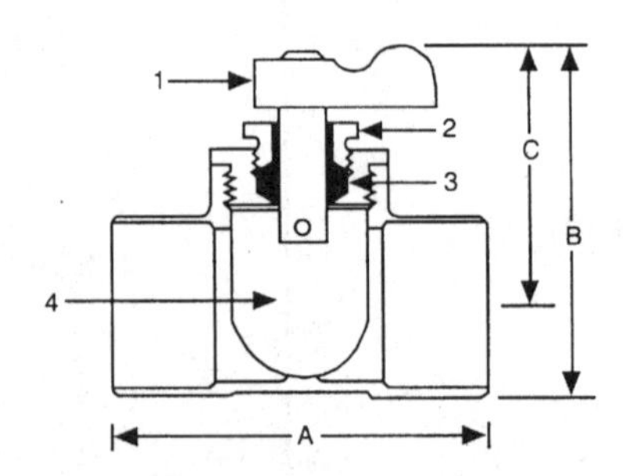

Specifications and Dimensions

No.	Size	Dimensions A	Dimensions B	Dimensions C	Material	Shipping Weight
BR-1	1/2"	1 5/8"	1 7/8"	1 7/16"	Bronze	3 ozs.
BR-2	3/4"	2 1/8"	2 1/8"	1 9/16"	Bronze	5 ozs.
BR-3	1"	2 3/4"	2 7/16"	1 3/4"	Bronze	9 ozs.
BR-4	1 1/4"	2 3/4"	2 1/2"	1 3/4"	Bronze	12 ozs.

FIGURE 18.9

Purge valves. (*Courtesy of Taco.*)

the air vent on the boiler. As the entire system fills, the water is rushing toward the boiler drain or purge valve at the end of the piping line. Air and water exit the system quickly once reaching the boiler drain or purge valve. The water is allowed to run until few visible bubbles of air are left in the discharge stream. After purging the system, the fast-fill lever is released and the drain or purge valve is closed.

Getting air out of a hydronic heating system is an important part of a new installation, and it can be vital to keeping an existing system running in a proper manner. Don't rush the process of purging. Spend an adequate amount of time to get as much air out of a system as is reasonably possible. It's not a hard job, but it can take more time that some contractors and installers care to devote to the process. In the end, you will be better served to spend a few extra minutes while doing the job than to be called back for more problems associated with air that should have been eliminated in the purging.

CHAPTER

19

Solar Heating Systems

Solar heating systems are a great concept, and they can work very effectively. However, the cost of the systems and the regional climatic conditions often outweigh the advantages of heating systems powered by the sun. A number of considerations must be taken into account when thinking of a solar heating system. For example, some subdivisions prohibit the use of solar panels, owing to their unconventional appearance. Cost is a big factor to consider, and so is the amount of time during winter days when a solar system can effectively collect energy. Simply put, solar systems are not for everyone, but they can be very good under the proper conditions.

Many types of solar heating systems are available. When they were first introduced in mass marketing the systems gained a good bit of attention. However, over the past 10 years or so, interest in solar heating systems seems to have fallen off. I suspect part of the reason for this is the high cost of creating complex systems. Another reason for the decreased interest in solar heating may be lower costs for fuel oil or the difficulty of maintaining quality heat with solar heating systems in some parts of the country.

The late 1970s and the 1980s were the solar heating system's spotlight. This is when people were turning to solar systems to overcome highly priced fuel oil that was not guaranteed to remain available. As the oil situation leveled out, so did the interest in solar heating systems. During the high times of active solar heating systems much was learned about design issues, effectiveness, and so forth. A great number of system designs either failed, were too expensive, or presented extreme servicing problems. Many systems were very complex in their design. Through more than a decade of evolution, solar heating systems have had their ups and downs.

The cost of creating a solar heating system has been a problem that still exists. It just isn't cheap to build a working solar heating system. And, with today's lower cost of fuel oil, recovering the high installation cost of a solar heating system takes a long time. This situation influences builders and customers to opt for more traditional heating systems. Through all of the good and bad experience gained with solar heating systems one type of solar system remains favored; it is the flat plate solar collector.

Flat Plate Solar Collectors

Flat plate solar collectors are one remaining type of active solar heating system that continues to have its followers. Throughout most of the United States the flat plate solar collector has proved to be one of the most cost-effective types of solar heating system. Availability of these collectors is also good. When these collectors are used, they can be grouped together as components to make almost any type of collector arrangement. They are most often installed on the roof of a home and face in a southerly direction. However, the collectors can be mounted at ground level and still function.

TRADE SECRET

The positioning of solar collector panels is crucial to the effectiveness of a heating system. Typically the collectors should have a slope angle that is equal to the local latitude, plus an additional 10 to 15°. This type of setting generally maximizes the winter heat output.

The positioning of solar collector panels is crucial to the effectiveness of a heating system. Typically the collectors should have a slope angle that is equal to the local latitude, plus an additional 10 to 15°. This type of setting generally maximizes the winter heat output. Roofs are a preferred mounting place for solar collectors. There are a couple of reasons for this. First, it is generally less expensive to mount the panels on a roof, since independent racks or holders are not needed, as they are with ground-mounted systems. Protection from damage is another reason why roofs are a preferred location. When collector panels are set at ground level, they are more likely to fall victim to vandalism or simple accidents which may break the panels.

Storing solar energy that is collected during the day is a major aspect of a successful solar heating system. Since more heat is needed at night, when no sunlight is available for collection, the heat collected during the day must be stored for use at night. Different methods are used for this purpose. Some homes are designed with mass or heat storage elements that will absorb heat and then radiate it for several hours after the sun is no longer available. Concrete floors covered with thick tiles are one way of creating a storage mass for radiant heat. This, however, is not the same as storing energy to produce more even heat throughout a home.

To accomplish energy storage in larger amounts, water-filled thermal storage tanks are used. The tanks absorb energy during the day and retain it for use during the night. Circulating water through collectors and storing it is a common means of producing heat storage. In some cases, antifreeze solutions are circulated between the collectors and a heat exchanger that is coupled to the storage tank.

Installation Methods

Just as many types of solar heating systems are available, so are there different ways to install various systems. There are four primary methods of installation. One type of system is the direct circulation system. Another type is a draindown system. Or you might find that a gravity drainback system is more appropriate. The fourth type of system is the closed-loop system. To further our discussion on these four main systems, let's look at them individually.

Direct Circulation Systems

A direct circulation system offers many advantages. When a direct circulation system is used, there is no need for a heat exchanger between the solar collector and the storage tank. Since a direct circulation system is moving, it does not require an antifreeze solution. This also saves money. Other elements that are required with a closed-loop system that are eliminated with direct circulation systems include an expansion tank, a pressure-relief valve, and air purging devices. The elimination of so many components naturally makes this type of system less expensive to install. When you eliminate system components, you are not only saving money on the cost of an installation, you are reducing the number of potential problems that might occur after installation.

When a heat exchanger is used between a storage tank and a collector, some heat is lost. This is not the case with a direct circulation system. The amount of heat, in degrees, that may be lost in a closed-loop system can be 10 to 15° more than what is lost with a direct circulation system. This is a significant amount of heat. Heat loss from collectors working with a closed-loop system reduce the effectiveness of a heating system. The fact that direct circulation systems cost less to install, work more efficiently, and reduce the components of a system which may require service makes direct circulation systems a wise choice.

There is one drawback to a direct circulation system. Since these systems don't incorporate the use of antifreeze solutions, they must be protected from freezing temperatures. Any part of the circulation system that is not in heated space must be otherwise protected from freezing when the temperature drops below the freezing point. To do this, pipes in the system are drained of all water prior to freezing conditions. The need for freeze protection is the only real negative aspect of a direct circulation system.

Draindown Systems

Draindown systems work automatically to protect circulation systems from freezing. Electronic controls are installed to facilitate the draining process. When sensors associated with the electronic controls sense temperatures near the freezing point, a command is sent to motorized valves to open them. When the valves open, the system is drained in the sections where drainage is required to prevent freezing.

QUICK»TIP **There is one drawback to a direct circulation system. Since these systems don't incorporate the use of antifreeze solutions, they must be protected from freezing temperatures.**

When collectors and piping are installed with a draindown system, the piping and the collectors must be installed with a minimum grade of 1/4 in. to the foot of downslope for drainage. While a draindown system may seem convenient, it is not without its risk. Any failure of operation within the system can result in frozen pipes and collectors, which will shut down the system and may require costly replacements of damaged parts that were swollen or ruptured during the freezing.

Since draindown systems use fresh water when they refill after a draindown, the systems must be piped as open-loop systems. Since fresh oxygenated water is allowed into the system after a draindown, it is not practical to use cast-iron or steel components with the system. Rust created from the oxygen-filled water will damage such components.

Since draindown systems use fresh water when they refill after a draindown, the systems must be piped as open-loop systems. Since fresh oxygenated water is allowed into the system after a draindown, it is not practical to use cast-iron or steel components with the system. Rust created from the oxygen-filled water will damage such components.

Gravity Drainback Systems

Some systems rely on gravity for drainage. In a gravity system, the pipes and collectors are drained automatically, by natural gravity, when the circulating pump stops. This type of system eliminates the needs for motorized valves and sensors that may fail to function properly. By eliminating the valves and electronics, the cost of a gravity drainback system is less than that of a draindown system. Just as is required with a draindown system, a drainback system must have all piping installed with a minimum of 1/4 in. per foot drop on the pipes for proper drainage.

A drainback system does not introduce new water into the system. Existing water is reused each time the circulator runs. When designing this type of system, you must be sure that the circulating pump is capable of pumping water to the highest point of the system. Gravity drainback systems are fairly simple and quite effective.

The concept of a gravity drainback system requires water to drain down whenever the circulating pump stops. Air enters the system to replace the water that is drained down. The air enters the system through an open tee that is located above the water level in the tank. It's possible to size the tank so that it works as an expansion tank for the collector loop and the distribution system.

Return piping from collectors to a tank should be sized to accept a minimum flow velocity of 2 ft/s. When this is the case, water returning from the collectors can entrain air bubbles and return them to the storage tank. The result is a pipe that becomes something of a siphon. When the siphon is created, it minimizes gurgling noises in the piping and increases the rate of water flow. By increasing the speed with which water is flowing through a collector, the efficiency of the collector is improved.

Gravity drainback systems are liked because of their simplicity. By not relying on electronics and motorized valves, a gravity system is typically more dependable and produces fewer service problems. The ability of this type of system to work almost flawlessly is advantageous in preventing frozen pipes and collectors. Another big advantage that a drainback system has over a draindown system is the fact that new water is not introduced into the system. This means that less expensive components such as cast-iron circulators can be used. A direct circulation system with a gravity drainback system is generally considered the best type of active solar heating system to install.

Closed-Loop Antifreeze Systems

Closed-loop antifreeze systems require an external heat exchanger. The heat exchanger links the collectors to the storage tank. Some of this type of system is exposed to freezing temperatures at times, and it must therefore be protected from freezing. This is done by filling the system with antifreeze. A glycol-based antifreeze solution is used most often. The antifreeze solution is what transports heat from the collectors to the heat exchanger. Water is moved between a storage tank and the heat exchanger with a circulating pump. When solar energy is being collected, two circulating pumps are running. Since all pipes and collectors are filled with antifreeze, they can be installed in unheated areas and do not have to be installed so that they can be drained down effectively, although it is still preferable to install the piping so that it can be drained down.

Planning a Solar Heating System

Some people think of solar collectors as running at high temperatures. In reality, this is exactly the opposite of what is desirable. Ideally collectors should operate at temperatures that are as low as possible.

When collectors operate at low temperatures, they are more efficient. The lower operating temperature allows a system to collect more energy.

It is rare for an active solar heating system to provide complete heating for a home or building. While the solar heating system may be the primary system, it is usually backed up by some other type of system. Without a backup system, it would be very difficult to have regular dependable heat output. Radiant heating systems that are installed in floors, especially concrete floors, usually are the most effective way to distribute heat from an active solar heating system. Since radiant heat in a floor can perform well at lower water temperatures, it is ideal for use with solar collectors. Hydronic fan-coil units that are sized to work with low water temperatures are also a viable means of distributing heat from an active solar heating system.

Backup heating systems used in conjunction with solar heating systems should be sized to work independently in heating the required space without assistance from the solar heating system. However, the auxiliary heating system must not maintain the thermal storage tank at temperatures suitable for use by the distribution system during nonsolar periods. This is a mistake that some people make, so retain the fact in your mind.

When the solar heating system is in use, the storage tanks must be allowed to cool down to room temperature. This is how the tank transfers its heat. Assuming that the tank is located in heated space, the heat from the tank cools off and offsets a portion of the heating load in the structure. Solar collectors will reestablish the higher temperature of the tank and allow the process to repeat itself.

The Cost Factor

The cost factor is probably the largest contributor in deciding when a solar heating system is sensible. Since a full-scale backup heating system of a traditional design should be installed in conjunction with a solar heating system, the cost of installing a solar system is substantially higher than the cost of installing a typical heating system. Customers are paying for a traditional heating system and then paying extra, usually thousands of dollars in extra money, for the benefit of heating with energy from the sun. This added cost is not always practical.

The decision on whether or not to go solar revolves around cost and energy conservation. Certainly solar heating systems do reduce the need for fuel oil or electricity when running a heating system. However, the cost savings may not be enough to justify the installation of a solar heating system. When money is a motivating factor, and it usually is, a customer has to determine how long is too long to wait for a payback from the installation cost of a solar system. With present prices of heating fuels, it can take many years to recover the cost of installing a solar system. And, if the installation is being done in a region where extremely cold temperatures are encountered, the auxiliary heating system may be carrying the majority of the heating load. A lot of thought and research is required to install a solar heating system that will pay for itself in a reasonable period of time.

Depending on an active solar heating system to function properly with some types of heating, such as hot-water baseboard heating units, can be asking too much of the solar system. As noted earlier, radiant heat pipes embedded in concrete floors are the most sensible means of heat distribution when the heat source is the sun. Wall-mounted hot-water convectors require a much higher water temperature to operate well than in-floor radiant heat does.

Before any active solar heating system is installed, experts in the design and function of the system should be consulted. It will be expensive to have a professional design a complex solar heating system, but the money will have been well spent. If such a system is designed by someone with limited solar experience, an expensive system may be installed that simply will not perform adequately.

Solar energy definitely has its advantages, but it also has disadvantages. Each job has to be considered on an individual basis. The topography and location of a building lot may make a solar system impractical. Geographic regions vary in how well they will allow a solar heating system to work. Installing a solar system can add thousands of dollars to the cost of building a home. Maintenance of a solar system is an additional burden that is not required when only a traditional heating system is used. To fully plan a system and to make a wise buying decision, each potential buyer must have expert advice that is directed to a particular case study. If you are considering the installation of a solar heating system, talk with experienced professionals and spend enough time in the planning stages to avoid the displeasure so many people have suffered from poorly planned systems.

CHAPTER 20

Coal-Fired Heating Systems

Coal-fired heating systems are not often used in modern homes. Some people like coal-fired stoves and continue to use them, but coal-fired boilers are rare in modern construction. However, most boilers that will burn wood can also be set up to burn coal. These are called solid-fuel boilers. Coal is a good fuel, but it's dirty to handle and requires a substantial amount of space for storage. If a boiler has to be fed coal by hand on a regular basis, like loading wood into a boiler, the time commitment and effort can be more than an average person might want to deal with. Many older homes and buildings do still have coal-fired heating systems, and there are probably some new homes being built where coal will be responsible for the heating system's operation.

People who shovel coal into their heating system are faced with the routine of making sure that the fire doesn't go out. Of course, if they are using a combination boiler, this is not a major issue. But the opening of the door to the combustion chamber to fuel the fire lets colder air in that cools the fire's temperature. It is also a fact that after a coal fire is loaded, there is a period of smoking that takes place prior to normal combustion conditions, and this reduces efficiency. As with any solid fuel, the fuel tends to burn down to a level where it is not at peak efficiency. All of these are drawbacks to hand loading coal. The alternative is an automatic loader that is most often called a stoker.

MATERIALS

There are different types of coal that can be burned. Three of the main types are anthracite coal, bituminous coal, and semibituminous coal. Of the three, anthracite coal is the one used most often in residential applications.

Coal and wood are both solid fuels, and usually both can be used in the same boiler, but not under the same conditions. Wood has different characteristics from coal, and a boiler that is capable of using either fuel must be set up for one type or the other. In the case of coal, the draft requirements depend on several factors. The size of the coal grate, the size and type of the coal to be burned, the thickness of the fuel bed, and the boiler pass resistance all affect the draft requirements. Perhaps the most significant factor is the degree of resistance offered by the boiler passes. If the gases can't pass through the boiler fast enough, they will back up in the combustion chamber and take away oxygen that is needed to keep the fire burning hot.

Types of Coal

There are different types of coal that can be burned. Three of the main types are anthracite coal, bituminous coal, and semibituminous coal. Of the three, anthracite coal is the one used most often in residential applications. This type of coal burns a cleaner flame and produces a steadier heat than the other types of coal. It also burns longer than the other types of coal. While anthracite coal holds many advantages, it does cost more than other types of coal and does require a higher temperature for combustion. Still, it is the coal of choice in residential applications.

Since coal should be purchased in a size that will work well with the fire grate and heating unit being fueled, it is common for coal to come in different sizes. Anthracite coal, for example, is available in many sizes, some of which are listed below:

- Buckwheat
- Stove
- Pea
- Egg
- Chestnut
- Broken

Buckwheat-size coal comes in four sizes or grades. When using buckwheat coal, you should always maintain an even low fire. A smaller mesh grate or a domestic stoker should be used. Make sure that new coal is exposed to hot coal so that combustion will not be delayed. And keep the heating system warm at all times. If a system is allowed to cool down, the amount of time and coal required to bring the system back up to temperature is excessive.

Egg-sized coal is typically used in large firepots with 24-in. or even larger grates. To obtain maximum performance, this type of coal should be placed in a bed that is at least 16 in. deep. When stove-sized coal it used, it should be used with a grate that is at least 16 in. in size, and the bed of coal should be a minimum of 12 in. deep. Chestnut-size coal is used in firepots that are up to 20 in. in diameter, and the recommended bed depth ranges from 10 to 15 in. Pea-sized coal can be used on a standard grate, and the bed depth should be added to gradually until it reaches the sill of the fire door. A strong draft is required when burning pea-sized coal.

Bituminous Coal

Bituminous coal ignites and burns easily. The flame from this type of coal is long and produces excessive smoke and soot when it is not fired properly. To avoid smoking and soot buildup, a side banking method of coal placement is recommended for bituminous coal. Basically, this means moving hot coals to one side of the grate and placing new coal on the opposite side. Side banking allows for a slower and more even release of volatile gases.

When feeding a fire with bituminous coal, you should add small quantities of coal at a time. This allows for better combustion. You should never fire bituminous coal over the entire fuel bead at a single loading. Doing so will result in a very smoky fire. A stocking bar can be used to break up a fresh charge of coking coal about 20 to 60 min after firing the coal. The stoking bar should not be brought up to the surface of the coal. When the stoking bar breaks up, the fired coal ashes should fall through the grate without any need for being shaken. Because of its tendency for smoking and sooting, bituminous coal is not an ideal choice for residential applications.

Semibituminous Coal

Semibituminous coal burns with a great deal less smoke than bituminous coal. However, semibituminous coal lights with greater difficulty than bituminous coal. A central cone method of loading is recommended for semibituminous coal. The coal is heaped into a pile to form a cone shape. Large lumps of coal fall to the sides, and the fine cones remain in the center. Poking the coal in a cone formation should be done so that the cone is not stirred or knocked over. Regardless of what type of coal is used, the question of how to get it into the firebox is one that must be answered; so let's look at this question now.

Stokers

Many a ton of coal has been shoveled into a fire, but few people today opt for the manual labor associated with feeding a coal fire. Stokers are normally used to deliver coal to a firebox. There are four basic classes of stokers normally used with different types of heating systems. A class 1 stoker delivers between 10 and 100 lb of coal per hour and is the type most often associated with residential applications. These stokers are usually an underfeed type that is designed to burn anthracite, bituminous, semibituminous, lignite coal, and coke. Ash deposits can be removed automatically or manually. Anthracite coal is the type most often used in domestic heating systems. Underfeed stokers feed the coal upward from underneath the boiler. The action of a worm or screw carries the coal back through a retort from which it passes upward as the fuel above is being consumed. Ash from the fire is left on dead plates on either side of the retort.

Class 2 stokers deliver between 100 and 300 lb of coal per hour. Larger heating systems can be served by class 3 stokers at a rate of 300 to 1200 lb per hour or class 4 stokers that will produce over 1200 lb of coal per hour. Some systems are designed with stokers that work with hoppers. The stokers work fine, but the hopper must be filled routinely with coal. If a stoker works with a bin instead of a hopper, the coal doesn't have to be loaded by hand. When a coal delivery is made to the bin, that is the end of the handling process, the stoker does the rest.

Start-Up Procedures

Start-up procedures for coal-fired heating systems vary. There are, however, some general procedures which usually apply. Room thermostats should be set above room temperature so that they will call for heat. Both the coal feed and air settings must be set to their proper rate.

When the line switch is in the on position, the stoker can start. If the stoker has a hopper, fill it with coal and set the over-fire air door about one-quarter to one-half open. Once the stoker has filled the retort with coal, the fire must be started. This is often done by putting paper, kindling wood, and a little coal on top of the retort pile and lighting it.

The draft for a coal-fired heating unit should be as little as possible without causing smoking from the fire door. As a fuel bed builds up, an adjustment of the air damper may be needed. The goal is to get a fire that is burning yellow with hardly any smoke. Air control is very important in maintaining a consistent burn rate. Automatic air controls don't normally require any manual adjustment. Motors and transmissions on stokers should be equipped with overload protection. Reset buttons should be present on both the motor and transmission that can be pushed to restart the equipment after the cause of an overload is eliminated.

The general maintenance for a coal-fired heating system is not too intense. Motors should usually be lubricated before the heating season and at least twice during the heating season. Any high grade of motor oil should be fine, but always read and comply with manufacturer recommendations. Transmissions normally require about 1 pt of good engine oil for lubrication. Some people change transmission oil once every 2 years, and others change it annually.

Troubleshooting

Troubleshooting coal-fired systems is unlike working with an oil-fired boiler. If there is a problem where noise is the complaint, you should check to see if any pulleys or belts are loose. If they are, either tighten or replace them. Noise may come from dry motor bearings, but oiling the bearings will correct this problem. Worn gears and gears that need to be oiled can make noise. Once you check the gears, you should either oil or replace them, subject to their condition.

There are usually three reasons why a motor will not start. If there are hard clinkers over or on the retort, a motor may not start. To remedy this problem you must remove the clinkers. If something gets caught in the feed screw of a stoker, it can cause the motor to not start. Removing whatever is blocking the feed screw will solve this problem. Sometimes the end of the feed screw becomes worn to a point where it

When a stoker won't shut off, it may mean that the controls are out of adjustment or that there is a dirty fire, no fire, or a dirty heating unit. If the fire is dirty, it must be rebuilt to burn clean. A fire that is out must be restarted. Clean the heating unit if it is dirty, and correct control adjustments if needed.

causes coal to be packed in a retort. This can keep a motor from starting. Removing the coal from the retort and replacing the worn part will solve the problem.

When a stoker won't shut off, it may mean that the controls are out of adjustment or that there is a dirty fire, no fire, or a dirty heating unit. If the fire is dirty, it must be rebuilt to burn clean. A fire that is out must be restarted. Clean the heating unit if it is dirty, and correct control adjustments if needed.

If a heating system becomes filled with unburned coal, you should check for clinkers that may be clogging the retort. Remove any clinkers that are in the way. Sometimes the coal feed is set too high and causes an overstock of coal. When this happens, adjust the feed setting. A lack of air in the firebox can cause a buildup of coal. Open the manual damper first. If the damper being open doesn't solve the problem, check the air ports to see if they are clogged and clear them if they are. Also check the wind box; it may be full of siftings that should be removed.

When you are faced with a stoker that will not run, you should look for a problem with the limit control. It may have shut the system down because of high temperatures. Let the limit control cool off and then see if the stoker will run. Another possibility for a stoker not running is that a boiler has shut down because of low water. Check the boiler to see that it has an adequate supply of water. A less likely cause of the problem is a gear case that has been submerged in water. If this has happened, you should drain and flush the gear case immediately, with the stoker turned off, and then refill the gear case with oil.

If smoke is backing up into a hopper, the hopper may be low on coal and may need to be refilled. There could be a clinker blocking the retort that must be removed, or there could be a clog in the smoke back connection that must be cleared. It's also possible that a fire is burning down in the retort. If this is the case, check the air supply to see that the fire is not getting too much air. You should also check to see that the coal feed is not too low.

When there is no fire in the combustion chamber, you should check to see that enough coal is being delivered to burn properly. Remove any clinkers that are blocking the retort and check to see if the switch to the unit may have been turned off accidentally. Check the fuse box or circuit breaker box to make sure that the failure is not due to a tripped circuit or blown fuse. The last likely cause will be a failure in the electric controls. Refer to the control specifications to confirm that they are functioning properly.

Coal is not a common fuel for residential heating purposes any longer, but coal-fired heating systems are still in use. Very few new homes are built with coal-fired heating systems. It is unlikely that the average heating company serving residential customers will have much contact with coal, but if you do, you now have a pretty good idea of how the systems work and what to look for when they are not operating properly.

CHAPTER 21

Wood-Fired Heating Systems

Wood has served as a heating fuel from generation to generation. Whether placed in a campfire, a fireplace, a wood stove, or a boiler, wood is a dependable source of heat. And it is a renewable resource. A great number of people rely on wood for some or all of their home heating needs. Wood stoves are very popular in rural areas, and wood-fired boilers can be found in houses all across cold climates. When a home is equipped with an efficient wood-burning heating unit, the cost of fuel can be quite low. Some people cut their own wood and others buy it. There are, however, drawbacks to burning wood.

Anyone who has never used wood as a primary heat source may not realize what a major time commitment the process is. During different phases of my life, I have heated my homes entirely with wood stoves. I enjoy the smell of wood burning, and the heat from a wood stove is unlike that of any other type of heater. When I was younger, cutting, splitting, stacking, and hauling wood was no big deal. In fact, it was good exercise and a great stress reliever. But there is no mistaking the fact that burning wood is a lot of work, and the dust created from wood stoves is a bother.

A friend of mine has a wood-fired boiler. He cuts his own wood and heats his home exclusively with the wood-fired boiler and hot-water baseboard heating elements. Like a wood stove, a wood-fired boiler requires a lot of effort to keep it running. The physical requirements of moving and loading wood can be too much for some people. Time, however, is one of the biggest problems most people encounter with wood-burning heaters. Unless someone is home most of the time during the heating season, getting a wood-burning heater to keep a house warm is difficult.

Airtight wood stoves are capable of burning for several hours without reloading with new wood. Wood-fired boilers have large fireboxes so that a lot of wood can be loaded at one time, and they too will burn for hours. Some wood-fired boilers are made as combination boilers. This means that they can burn more than one type of fuel. For example, a boiler that will burn wood may also be capable of burning fuel oil. Almost any boiler that will burn wood can also be set up to burn coal. Having a combination boiler that will burn wood or oil is a very good idea.

The benefits of combination boilers include being able to leave the boiler unattended for days at a time, when it will be running from the burning of fuel oil. If a home is equipped with only a wood-fired boiler, leaving the home during freezing temperatures for an extended stay somewhere else is not practical, as the plumbing in the home will freeze. A combination boiler eliminates this problem.

Wood heat became very popular in the mid-1970s. This is about the same time that solar power was becoming popular. The oil crisis and high prices of fuel oil had people looking for alternative heating options. Both solar and wood gained popularity, but wood maintained its following more successfully. Solar heating became less fashionable over time, but wood heating remains popular. During the growth process of wood heat, a number of new boilers and wood stoves were designed. Today's modern wood-fired boilers are the result of many years of research.

Many Choices

People in the market for a wood-burning boiler (Fig. 21.1) have many choices available to them. Some wood-fired boilers rely on the natural draft created by a standard chimney. Others incorporate the use of a small blower to force air into the combustion chamber. Boilers that are

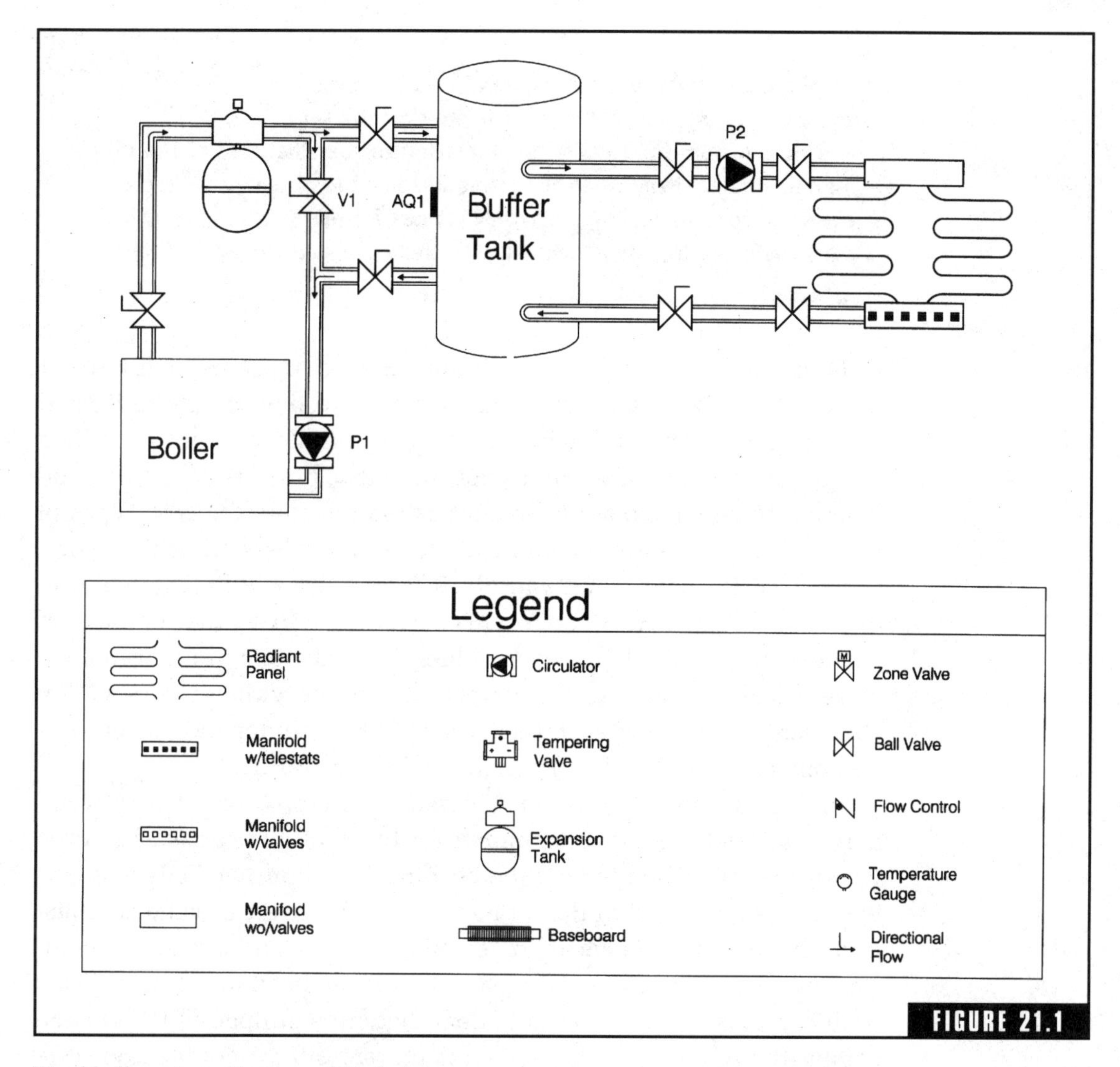

Wood boiler setup. (*Courtesy of Wirsbo.*)

called solid-fuel boilers are designed to burn only wood or coal. Most solid-fuel boilers can be set up to burn either wood or coal. These boilers partially control the heat output of a fire with the use of a thermostatically controlled air damper. Combustion chambers for solid-fuel boilers are quite large to accommodate large loads of wood. There is an ash cleanout drawer located at the bottom of these boilers that must be maintained periodically.

TROUBLE-SHOOTING

Problems can arise when a combination oil burner and wood burner is used. It's not uncommon for the soot and fly ash from the wood-fired operation to coat the head of the oil burner. This happens when the flame head of the oil burner is located in the same combustion chamber that is used for burning wood. Under such circumstances, it is necessary to clean the flame head frequently.

Other choices include combination boilers that can use solid fuel or fuel oil. These boilers are more expensive but well worth the additional cost. Having a combination boiler makes it possible to burn wood when it's convenient and oil when it's not suitable to burn wood. Some combination boilers have a single combustion chamber for both types of fuel. This type of setup is not ideal. When a single combustion chamber is used for both wood and oil, it is not possible for the boiler to operate as efficiently. A compromise is required to set the combustion chamber up for the two types of fuel. To make combination boilers more efficient, some manufacturers build boilers with two combustion chambers. This allows each fuel type to have an independent combustion chamber and results in a higher efficiency rating.

Problems can arise when a combination oil burner and wood burner is used. It's not uncommon for the soot and fly ash from the wood-fired operation to coat the head of the oil burner. This happens when the flame head of the oil burner is located in the same combustion chamber that is used for burning wood. Under such circumstances, it is necessary to clean the flame head frequently. This problem does not occur with boilers that are equipped with separate combustion chambers. Some manufacturers offer a double-door system on their single-chamber combination boilers. When this is the case, a user chooses which door to close, depending upon the type of fuel being used. With this type of design, the oil burner's flame head is not in the combustion chamber when wood is burning, so there is far less soot and fly ash to contend with. The disadvantage to the system is that the boiler has to be in one mode or the other and cannot have the oil fuel kick in automatically if the wood fire goes out.

When wood-fired boilers are used, they must basically burn out the wood that is in them. Unlike other fuel types which can be turned on and off, wood cannot be turned off. It is possible to reduce the intensity of a wood flame by closing an air damper, but this is not something that should be done too often or for long periods of time. When a damper is closed, the fire in the combustion chamber burns more slowly and colder, creating creosote. This means that experience is required to regulate a wood-fired boiler so that it will not burn too hot or too cold.

Efficiency

The efficiency of a wood-fired boiler is difficult to predict. Heat output from a wood-fired boiler is highly dependent on the quality of the firewood being burned. If the wood is unseasoned, has a high moisture content, or is a type of wood that gives off low Btus, the boiler cannot produce as well as it would with seasoned hardwood as a fuel. Efficiency estimates on wood-fired boilers range from a low of about 50 to a high of about 80 percent. However, you must keep in mind that any efficiency rating with a wood-fired boiler is subject to the quality of the wood being burned.

In order to have high efficiency from a wood-fired boiler, the boiler must reach a high combustion temperature. For example, a wood-fired boiler that is equipped with a forced draft might hit a combustion temperature of 2000°F, which is plenty for good efficiency. At this temperature, the pyrolytic gases released from the firewood ignite and burn hot. This reduces the buildup of creosote in the chimney. However, if the combustion temperature doesn't rise to an excess of approximately 1100°F, the pyrolytic gases will not ignite and will coat the chimney with a creosote buildup.

Transfer Storage Tanks

Wood-fired boilers lose efficiency when they are forced to burn out at low temperatures and creosote is created. One way to overcome this problem is the installation of a transfer storage tank. Instead of cutting a boiler back to a low burning temperature, the boiler is allowed to burn at a high temperature with the unwanted heat being transferred to the storage tank. Then, when the heat is wanted from the storage tank, it can be delivered through a hydronic distribution system.

Another advantage to installing a storage tank is that the firing time for the boiler is reduced to only a few hours a day. This makes it easier for people who cannot be around their home for hours at a time to keep the boiler burning wood.

If a heat-storage tank is going to be used, the boiler must have a much higher heat output rating than would be needed without the storage tank. The reason for this is that the boiler has to produce 24 hours worth of heat in only a few hours. It is generally recommended to use a forced-draft type of boiler when a storage tank is installed. This is because the forced-draft boiler can produce a higher firing rate. Don't expect a system like this to be cheap. The larger boiler and the insulated storage tank run up the cost of installation.

Another Option

Another option when a wood-fired boiler is desired but another fuel type is also wanted is to install two different systems. To obtain optimum performance with minimum problems, both a wood-fired boiler and an oil-fired boiler should be installed. This, of course, is expensive and is worthwhile only when wood is available at a reasonable cost and will be used steadily for heating purposes. There are different ways to pipe the two boilers to work together. One way is to set one boiler up as a primary boiler and have the other boiler piped as a secondary heat source. It's also possible to pipe the two boilers in a parallel or a series fashion.

When boilers are installed in a series piping design, with the wood boiler being the primary boiler, the wood boiler burns first. When the wood-fired boiler drops to a colder temperature, a temperature control brings on the conventional boiler to supplement and eventually take over for the wood-fired boiler. While this type of hookup is simple to install, it does not offer the most efficiency.

> **QUICK»TIP** **If a heat-storage tank is going to be used, the boiler must have a much higher heat output rating than would be needed without the storage tank. The reason for this is that the boiler has to produce 24 hours worth of heat in only a few hours.**

Using a parallel piping system allows each boiler to have its own circulator and check valve. This improves efficiency. The small cost of the extra circulator and check valve is returned soon in higher efficiency. Controls installed for the boilers allow the wood-fired boiler to run until it reaches a colder temperature, and then the conventional boiler kicks in. The controls should be set to avoid having both boilers running at the same time. If they are both running simultaneously, the flow rate on the system will be increased and may become noisy.

Choosing between a combination boiler and two separate boilers is a personal decision. Cost is clearly a factor. The highest efficiency is achieved with the use of two independent boilers, but the cost may be more than a customer is willing to pay. All factors must be taken into consideration when choosing a wood-fired system. For some people, wood-fired boilers are an excellent choice, but they are not for everyone.

CHAPTER 22

Troubleshooting Gas-Fired Boilers

Troubleshooting gas-fired boilers is a bit different from the same job done with oil-fired boilers. There are significant differences in the ways in which the two types of boilers function. Gas-fired boilers are very common in many parts of the country. The boilers may be fueled by either natural gas or manufactured gas. Fuel oil burns but will not explode under normal conditions. Gas, on the other hand, can explode. This in itself makes the troubleshooting process somewhat different. A leak in a gas line can prove fatal, either from inhaling the fumes or from being blown up. This doesn't happen with oil-fired systems.

The type of fuel used is not the only difference between oil-fired and gas-fired boilers. Since the fuels are different, the firing mechanisms are also different. This requires different troubleshooting techniques for the two types of boilers. One of the first considerations when troubleshooting a gas-fired system is safety. If excessive gas fumes have collected in the boiler area, the least little spark could ignite disaster. Gas is not a fuel to be taken lightly.

Standard Safety Measures

There are certain standard safety measures to take into consideration whenever you are working with a gas-fired heating system. The first

> **QUICK»TIP** Always shut off the gas supply to any device you are planning to work on. Cut the gas valve off and wait several minutes before proceeding with your work. If there are multiple cutoff valves, close all of them. Wait at least 5 min to give residual gas time to vacate the system.

rule is to pay attention to any odors. If you smell any appreciable amount of gas or other strange odors, ventilate the area before working in it. Avoid electrical devices, if possible. Open windows when you can. If gas accumulates heavily, it can explode. The quicker you can rid the area of fumes, the safer you will be.

Always shut off the gas supply to any device you are planning to work on. Cut the gas valve off and wait several minutes before proceeding with your work. If there are multiple cutoff valves, close all of them. Wait at least 5 min to give residual gas time to vacate the system. You may already know, but if you don't, be advised that LP gas is heavier than air and will not dissipate upwardly in a natural fashion. LP gas lingers below air, gathering in low areas and creating a possibility for an explosion risk. In minute quantities, this is not a problem, but it is a fact that any service technician should be aware of.

If you will be working with or near electrical wiring, you should turn off the power supply. This may mean removing a fuse, but in most homes it is as simple as flipping a circuit breaker. There may also be an individual disconnect box near the heating unit. If this is the case, you can turn off the power by throwing the switch on the disconnect box.

If you will be working with electrical matters, don't attempt to jump or short the valve coil terminals on a 24-V control. Doing this can short out the valve coil or burn out the heat anticipator in the thermostat. Also, never connect millivoltage controls to line voltage or to a transformer. If you do, you may be burning out the valve operator or the thermostat anticipator.

> **QUICK»TIP** If you will be working with or near electrical wiring, you should turn off the power supply. This may mean removing a fuse, but in most homes it is as simple as flipping a circuit breaker. There may also be an individual disconnect box near the heating unit. If this is the case, you can turn off the power by throwing the switch on the disconnect box.

If you have any reason to suspect a gas leak, check all connections where leaks may be possible. Don't use any type of flame to test for leaks. I've known plumbers who tested joints on natural-gas piping with their torch flames, but don't do this, and certainly, never do it with LP gas. Natural gas usually just burns in a leak situation, but LP gas is likely to explode. I say again, don't search for leaks with any type of flame. Soapy water is the right substance to use when looking for leaks in gas piping. Obviously, to find a gas leak, the gas supply must be turned on.

When in search of a leak, ventilate the area by opening windows and other ventilating openings. With the gas on, apply a generous quantity of soapy water to all gas connections. When the water is applied to a connection where a leak exists, the gas escaping will create bubbles in the soapy solution. Take note of the leak and continue testing the rest of the gas connections. Once you have checked all connections, you can turn off the gas supply and fix the leaks. When fixing leaks, don't take shortcuts. Treat each leak with respect. Loosen all leaking joints, apply thread sealant as required, and tighten the joints. Sometimes just tightening a fitting is enough. Regardless of what you have to do to correct a leak, always test the connection carefully after it is repaired. Test the joint just the way you did to find the leak in the first place.

Never bend the pilot tubing at the control once the compression nut has been tightened. If the tubing is bent after the compression nut is tightened, a leak can result. Compression nuts don't take well to vibration and manipulation. Any substantial movement of a compression fitting after it is tightened can create a leak. Compression fittings are fine when they are installed and treated properly.

The Troubleshooting Process

The troubleshooting process for gas-fired boilers doesn't involve a lot of steps (Figs. 22.1 and 22.2). Only about nine types of failures are typically encountered with gas-fired boilers. While there are not a lot of potential problems, there are several possible solutions. But using a methodical plan during the troubleshooting process makes the job easier. Experience is the best teacher, in many ways, but a good checklist of what to do and in what order to do it is the next best thing. And that's what I'm about to give you. So, with the safety issues covered, let's move into the cause-and-repair section to give you some guidance in finding and fixing problems with gas-fired boilers.

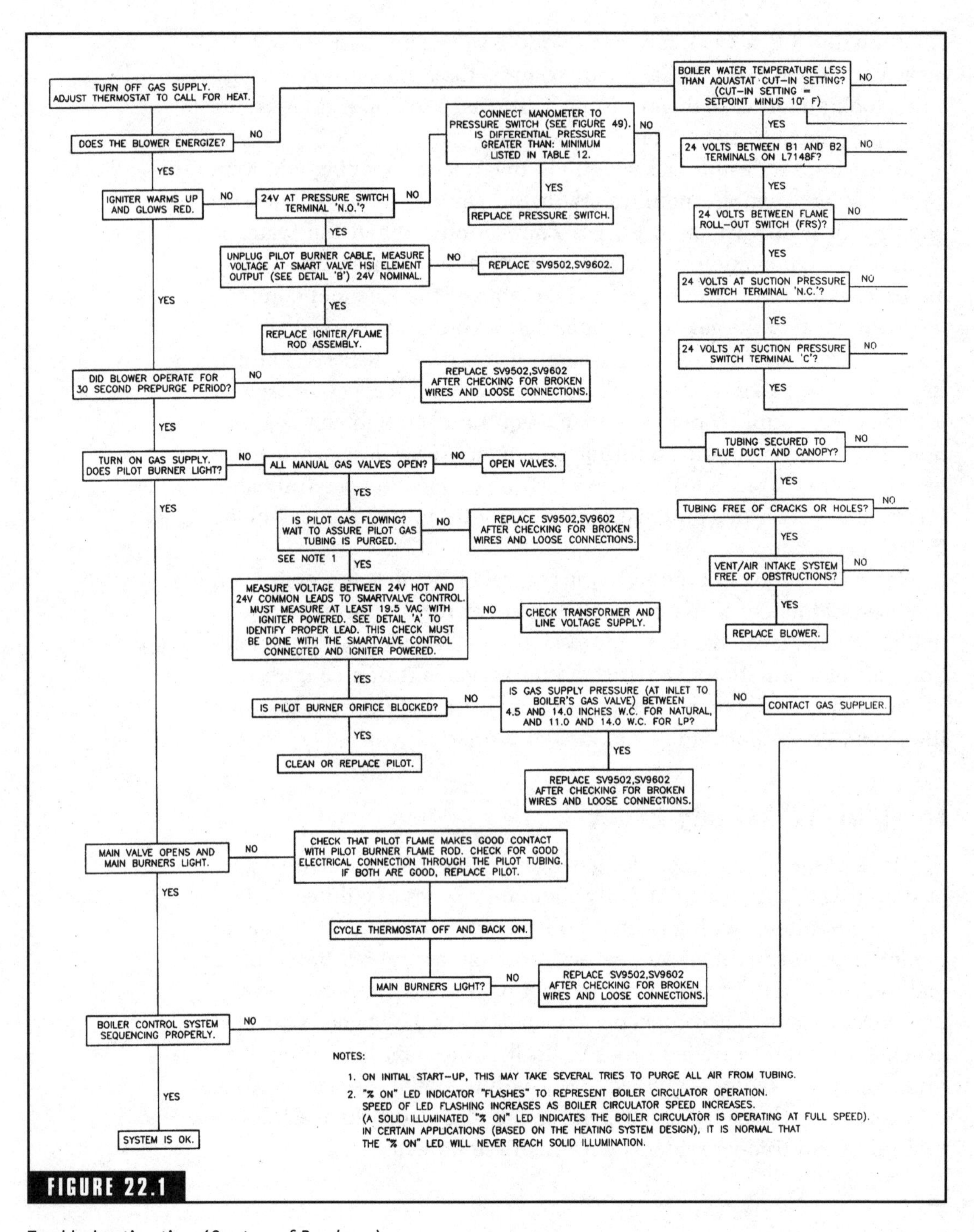

FIGURE 22.1

Troubleshooting tips. (*Courtesy of Burnham.*)

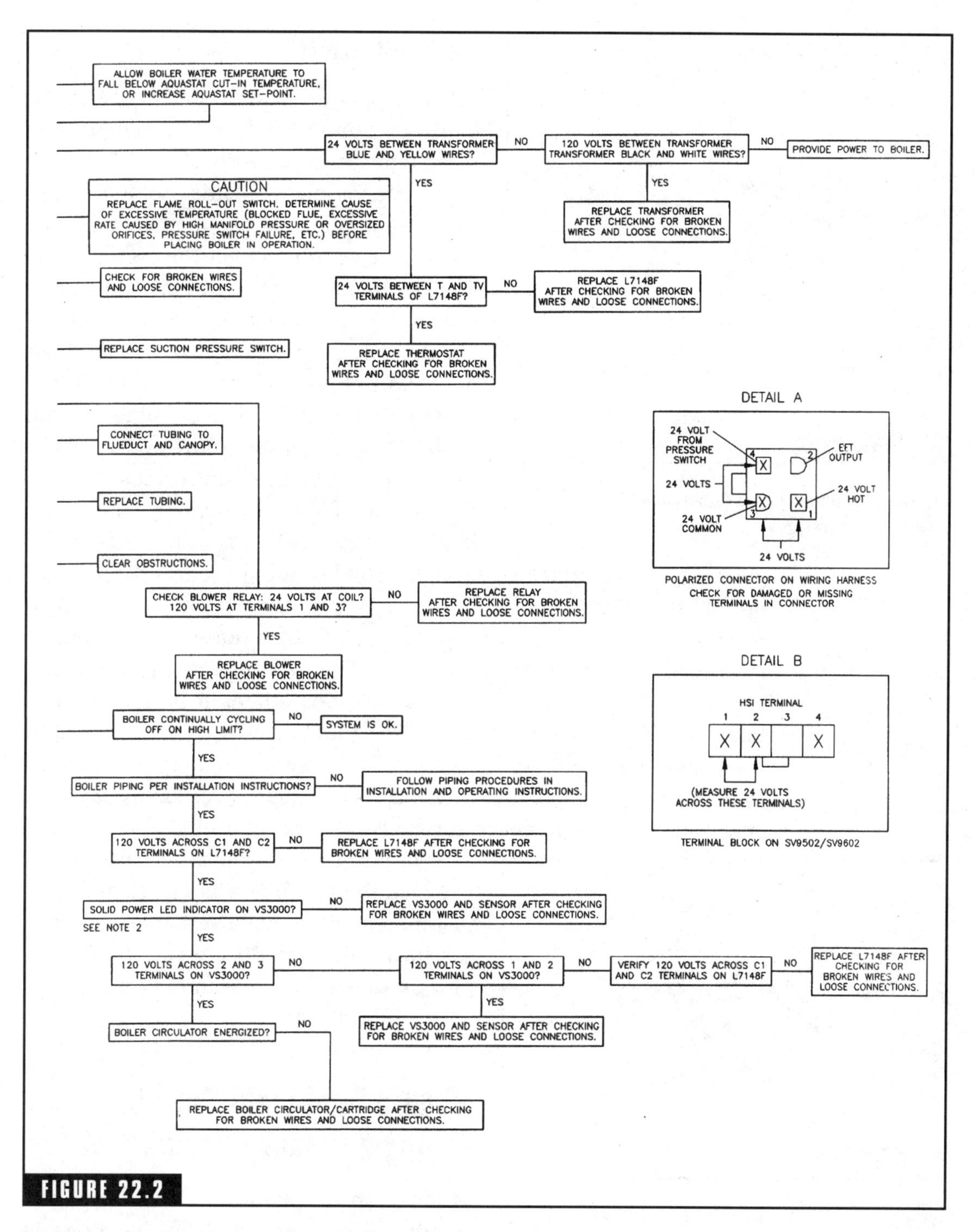

FIGURE 22.2

Troubleshooting tips. (*Courtesy of Burnham.*)

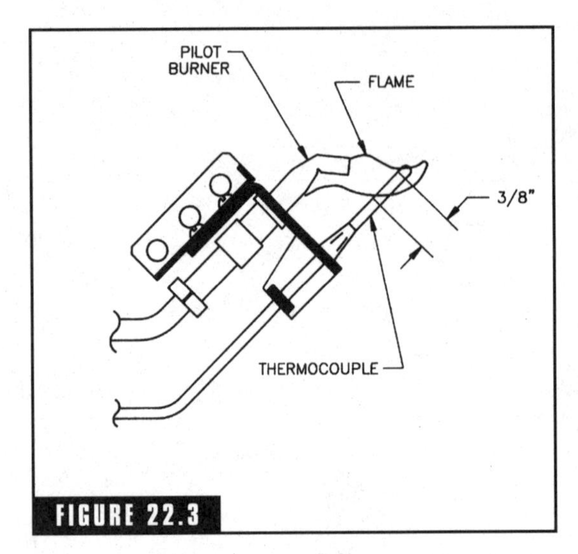

FIGURE 22.3

Typical pilot flame with a Robertshaw 7CL-6.
(Courtesy of Burnham.)

No Pilot Light

When a gas-fired boiler fails to operate, it could be due to the lack of a functioning pilot light (Figs. 22.3 to 22.6). There are four basic reasons why this problem might exist. First, check to see if there might be air in the gas line. Open the connections and purge any air that may be trapped in the tubing or piping. A second thing to check for is the gas pressure. Is it too high or too low? Either situation can result in a failing pilot light. It's also possible that the problem is a blocked pilot orifice. Do a visual inspection and confirm that the orifice is clear of any obstructions. Obviously, if gas cannot make its way through the opening, it cannot fuel a pilot light. The last thing to check is the position of the flame runner. If it is not positioned properly, it can be responsible for the lack of a pilot light. One of these four reasons is most likely the cause of your problem. This, of course, is assuming that the gas to the pilot light is turned on. Sometimes it is the simplest things in life that are the most difficult to recognize. Always check to make sure that you have a good gas flow before you go too far in your troubleshooting steps. Gas valves can be turned off for many reasons, and someone may forget to turn the valve back on. It's also possible that children will somehow turn off a valve and not realize what they are doing. Just remember to make sure that there is a gas flow before you begin the more complex aspects of troubleshooting a problem with a pilot light.

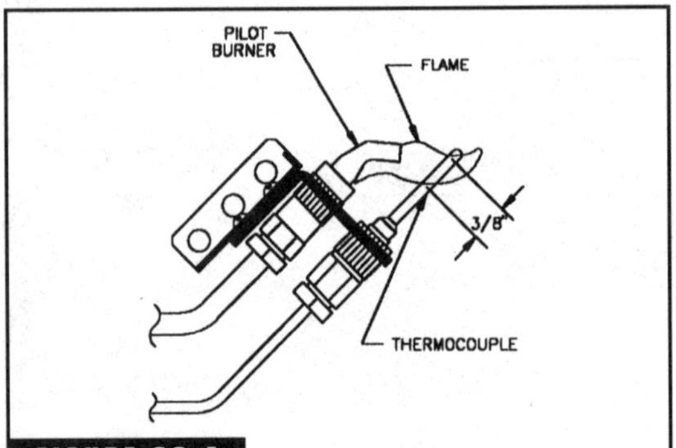

FIGURE 22.4

Typical pilot flame with a Honeywell Q327.
(Courtesy of Burnham.)

When a Pilot Light Goes Off During a Standby Period

When a pilot light goes off during a standby period, it might be that there is some type of obstruction in the tubing leading to the pilot light. Look to see if there are any kinks in the tubing. If not, remove the section of tubing and confirm that it is not blocked. This can be done just by blowing air through it. If there is a kink or obstruction, either replace the tubing or remove the obstruction. Low gas pressure can also trigger a pilot light to go off during standby. Check the gas pressure and confirm that it is within acceptable levels for the equipment that you are working on.

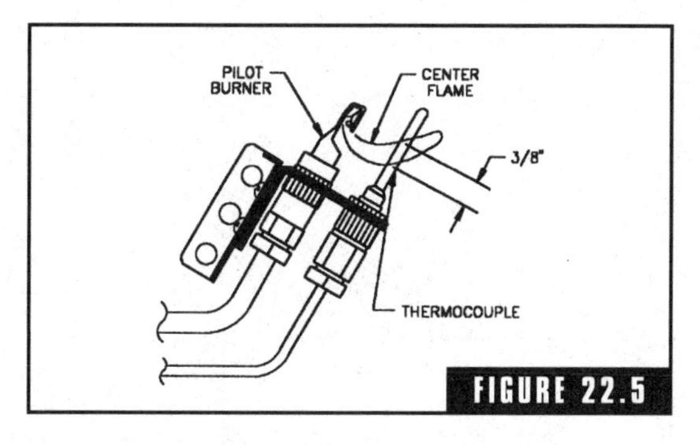

Typical pilot flame with a Honeywell Q350. (*Courtesy of Burnham.*)

You already know that a blocked pilot orifice can prevent a pilot light from burning. But did you know that a blocked orifice can also be responsible for a pilot light that goes off during standby? It can, so check the orifice if your problem persists. If there is a loose thermocouple on 100 percent shut off, a pilot light can go off during a standby period. It's also possible that the thermocouple is defective. When a pilot safety is not working properly, it can cause a pilot light to go off during standby. Another consideration to check is the draft condition. If there is a poor draft, a pilot light may not stay lit. A final possibility is a draft tube that is set into or flush with the inner wall of the combustion chamber. These same seven reasons for failure can apply to a safety switch that needs frequent resetting.

If a Pilot Light Goes Off When the Motor Starts

If a pilot light goes off when the motor of a unit starts, there are only three probable causes. The first one to look into is the possibility of some type of restriction or obstruction in the tubing to the pilot light. You need to loosen the connection on the tubing and confirm gas flow.

TROUBLE-SHOOTING

When a pilot light goes off during a standby period, it might be that there is some type of obstruction in the tubing leading to the pilot light. Look to see if there are any kinks in the tubing. If not, remove the section of tubing and confirm that it is not blocked.

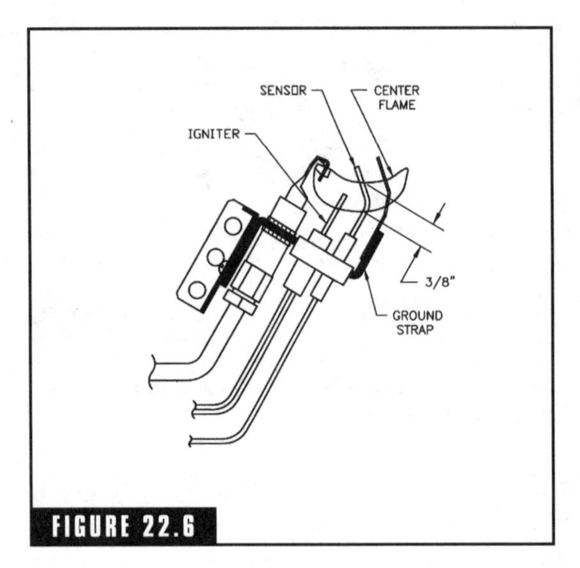

Typical pilot flame with a Honeywell Q3480.
(Courtesy of Burnham.)

If the tubing is blocked or kinked and cannot be repaired, replace the tubing. Once again, gas pressure could be at fault. If the pressure is too high or too low, the pilot light function can be affected. Confirm that the gas pressure is within suitable limits for the equipment that you are working with. A final consideration is the possibility that there may be a substantial pressure drop in the gas piping when the main gas valve opens. Putting the equipment through cycles until you can observe what happens as the main gas valve opens might be necessary.

Motor Won't Run

The first thing to check when you encounter a boiler with a motor that will not run is the electric circuit. If there is a local disconnect box, check it first to see that it has not been turned off. Next go to the electrical panel for the home or building and check all fuses or circuit breakers. If all appears to be in order, use your electrical meter to confirm that power is reaching the motor. Also make sure that the wiring is connected to the controls properly. Before you get too involved, move around the building and check the thermostats. Make sure that they are turned up to a sufficient temperature to be calling for heat. A mechanic can feel quite foolish after spending considerable time performing major troubleshooting steps and then finding out that the thermostats were turned down for some reason. Assuming that the thermostats are calling for heat, you must check to see that the thermostats and limit controls are not defective or calibrated improperly.

TROUBLE-SHOOTING

The first thing to check when you encounter a boiler with a motor that will not run is the electric circuit. If there is a local disconnect box, check it first to see that it has not been turned off. Next go to the electrical panel for the home or building and check all fuses or circuit breakers.

Sometimes the bearings in a motor will seize up, owing to a lack of lubrication. See if this might be the case on your job. Try turning the equipment. If it will not turn freely, you have probably identified your problem. In a worst-case scenario, you are dealing with a motor that has burned out, which will, of course, require replacement.

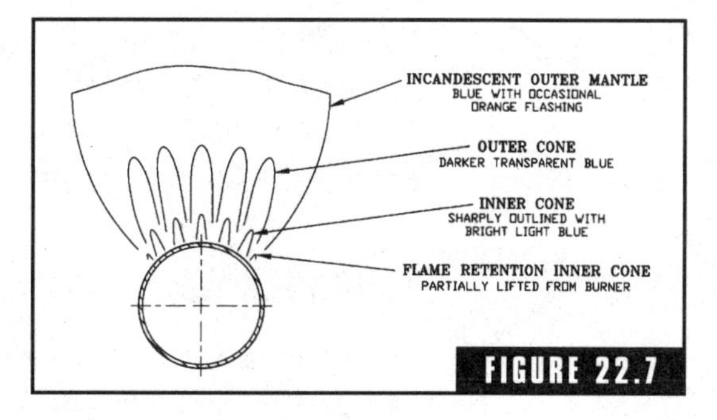

Main burner flame with 50-mm burner.
(*Courtesy of Burnham.*)

It Runs but Doesn't Heat

If you have a boiler that runs but doesn't heat, check to see if the pilot light is burning. If the pilot is out, relight it. You may discover that the root of the problem lies within a defective pilot safety control. This could be as simple as resetting the safety, or it could require replacement of the control. Another possible cause for the problem of a motor that runs with no flame is that the thermocouple might not be generating enough voltage for the system. The causes are limited, so the troubleshooting should go quickly. Gas pressure has a large effect on a gas-fired system. Check the pressure to make sure that it is within recommended working parameters. If it is, and you still have the problem, check to see if the motor is running more slowly than it should be. This could be the cause of the motor's running without a flame (Figs. 22.7 and 22.8).

A Short Flame

A short flame in a gas-fired boiler is not good. It might be caused by any one of about five possible problems. The place to start is with the pressure regulator. If it is set too low, the flame will be short. Also check the air shutter. An air shutter that is open too wide can cause a short flame to burn in the equipment. Any major pressure drop in the gas supply line could also be responsible for the short flame. A defective regulator is always a possibility when

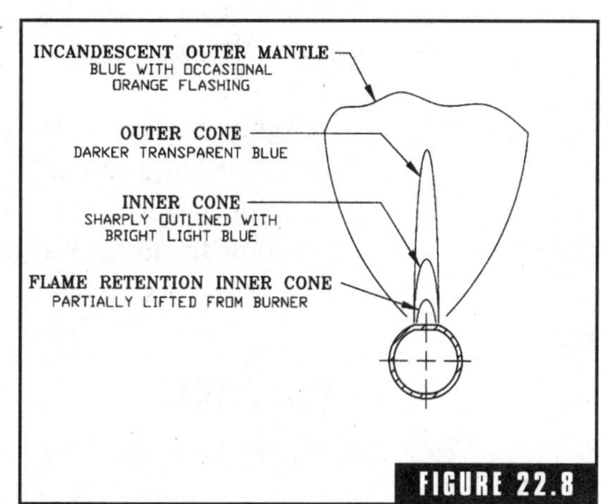

Main burner flame with 1-in. burner.
(*Courtesy of Burnham.*)

you are looking for the cause of a short flame. And a vent that is plugged in the regulator can also cause the problem of a short flame. Some trial-and-error techniques will be needed, but one of the causes listed above is almost certain to be the reason for a short flame.

Long Flames

Long flames in gas-fired equipment are usually limited to one of three causes. The first thing to check for is an air shutter that is not far enough open. When it is open too far, the flame will burn short, but if not open far enough, the flame will be long. The air shutter must be balanced to produce a flame of a desired length. If there are obstructions in air openings or at the blower wheel, the result can be a long flame. Too much input to the flame source will also produce a long flame. Since the causes of long flames are few in number and easy to check for, this problem is not very time-consuming when it comes to troubleshooting it.

A Gas Leak at the Regulator Vent

A gas leak at the regulator vent is a sign of a damaged diaphragm. The diaphragm most likely has a hole in it. In any case, the diaphragm damage must be replaced. Gas cannot be allowed to leak from any part of a system. In such a case, all attention should be focused on the regulator vent, since that is where the problem is.

Won't Close

Sometimes the main gas valve on a unit will not close when the blower stops. This is a problem that is usually caused by a defective valve or some type of obstruction on the valve seat. You can inspect the seat for damage or buildup, but be prepared to replace the entire valve. It is unsafe to have a gas valve that does not close properly when it should.

Be Careful

When you are working with gas, whether natural or manufactured, you must be careful. The fuel is volatile. It is said that familiarity breeds contempt. Don't allow this to be the case in your work. If you make a mistake with an oil-fired boiler, you might only get oil on a floor or incur some odors. A mistake with a gas system could result in a major explosion. Pay attention to what you are doing and stay focused on your task. Make sure anyone who might enter your work area is aware of what you are doing and what your needs are. For example, the last thing you need is for a casual observer to come in and fire up a cigarette lighter when you are bleeding off some gas. Stay sharp and stay safe.

CHAPTER 23

Troubleshooting Oil-Fired Boilers

Oil-fired boilers are used in many parts of the country. The boilers are usually very dependable. But, like any type of equipment, boilers can suffer from mechanical problems. If a boiler fails to run during the cold of winter, plumbing and heating systems in a home can freeze. Much damage can occur if a boiler fails and no one is aware of the problem. Granted, this doesn't happen often, but it does happen. When the heat goes off in a house or building, it is usually noticed by the occupants of the building. However, if a house is being marketed for sale and no one's living in it, a boiler failure can be disastrous. Even when there is someone to notice a boiler failure, the risk of damage to plumbing and heating systems exists. Whether property damage is imminent or not, the comfort level of residents in buildings with a failed heating system is at stake. Knowing what to look for when a boiler fails and how to correct the problem quickly is a big asset for anyone in the heating business.

Many items associated with a boiler can fail. However, most of the problems can be traced with relative ease. Still, pinpointing a problem quickly can be tricky. The job is easier if you use a logical approach in troubleshooting. Anyone with a basic understanding of boilers can eventually find the cause of failures by using a trial-and-error method.

TROUBLE-SHOOTING

A systematic approach is best when troubleshooting, but there are some times when trial and error is the only way to determine the exact cause of a problem. The trick is to use a systematic method of trial and error.

This type of troubleshooting takes time. A systematic approach is faster. Knowing what to look for and where to look for it is half the battle. Once a problem is identified, correcting it is usually not very difficult.

Some problems encountered with oil burners can be potentially dangerous. Smoke or fumes can fill a home quickly when a boiler becomes defective. The smell and residue left behind can be hard to live with. In some cases, the risk of fire may exist. If a safety control malfunctions, big trouble could be right around the corner. The types of complaint common with oil burners range from noisy operation to no heat. And there's a lot of possible trouble in between. Getting to the heart of a problem quickly is something of an art. In the case of boilers, the service technician is the artist.

Noise

Noise is not an uncommon complaint with oil-fired boilers. Since many boilers are located under or near living space, excess noise can be a real problem. A boiler that is properly tuned should not produce a lot of noise. When a noise is noticed, it is a sign that something is wrong with some aspect of the boiler. The root cause of the problem can be one of any number of things. The type of noise heard can be an indicator of what to look for. There are three typical types of noise associated with oil burners. Boilers sometimes give off pulsing sounds. Thumping noises are also common. Rumbling is another type of noise that you might experience with an oil burner. Once you have a noise to deal with, what do you do? The nozzle in the burner is always a good place to start your investigation.

A systematic approach is best when troubleshooting, but there are some times when trial and error is the only way to determine the exact cause of a problem. The trick is to use a systematic method of trial and error. In the case of noise, you should systematically turn to the nozzle in the burner. Then you have to gamble with some trial-and-error work. Start by replacing the existing nozzle with one that has a wider spray angle. Try the unit and see if it is still producing excess noise. With a little luck, you will have solved the problem. If not, you have to take a few more steps.

When a nozzle with a wider spray angle will not quiet the noise made by a burner, try a nozzle that has a smaller opening. Then installation of a new nozzle that is one size smaller than the one previously installed could be all it will take to quiet the boiler. It is necessary to make adjustments and test after each one to see if the problem is corrected. Another course of action, especially if the oil burner is producing pulsating noises, is to install a delayed-opening solenoid on the nozzle line.

There is another consideration. If you are getting a noisy fire and the nozzle replacements don't fix the trouble, check to see where the oil supply tank is located. Assuming that nozzle replacement and a delayed-opening solenoid has not satisfied you, the problem might be cold oil. When an oil tank is installed outside, the oil can get cold enough to create a noisy burner. To remedy this, pump the fuel oil through a nozzle that is the next size smaller at a pressure of about 125 psi. One of these four courses of action should put the burner into good operating condition.

A Smoky Joe

If you run across a boiler that smokes like a barbeque grill, there could be serious problems at hand. The problem could be as simple as a defective nozzle to something as major as a damaged combustion chamber. If the problem turns out to be the combustion chamber, your customer is going to be very unhappy, owing to the high cost required to get the heating system back up and running. Don't just assume that the problem is in the combustion chamber or in the burner tube. Start with the simple fixes.

When you are called to troubleshoot a smoking boiler, check first for dirt. If the air-handling parts of the oil burner are dirty, smoke output can occur. The dirt prohibits the burner from functioning properly. A good cleaning may be all that it will take to solve the problem. Assuming that dirt is not the problem, turn your attention to the nozzle in the burner. If it's the wrong size, the nozzle could be the cause of the smoke. The worst scenario for your customer is a cracked combustion chamber.

The air-handling parts to check for dirt include the fan blades, air intake, and air vanes in the combustion head. When these components are clean, they allow the system to work much more efficiently. Nozzle problems are not difficult to fix. In the case of a smoking boiler, the replacement of an existing nozzle with a smaller one can be all it takes to stop the smoking. Another tactic with the nozzle is to install one that has a narrower spray angle. If the cause of the smoke is a cracked combustion chamber, it's likely that the boiler will need to be replaced. It may be possible to repair the combustion chamber, but if repairs are made, you must make sure that no smoke or gas is escaping the combustion chamber.

It Stinks

There are times when boilers just simply stink. Fuel odors can occur when oil burners are used. Many people are very sensitive to the smell of oil, and the odor can be difficult to get rid of once it is present. There are three basic reasons why oil burners produce odors. Chimney obstructions are one of the first things to consider when you are faced with a boiler that is stinking up a mechanical room. If the chimney checks out, the next logical step is to check the ignition time on the burner. If it is delaying, that can account for the buildup of excess fumes. The third possibility is that too much air is going through the boiler.

Since chimney problems are the first consideration when odors are present, let's start with troubleshooting them. The draft in a chimney or flue over a fire from a boiler should not be less than 0.02 to 0.04. If you do a draft test and find the draft to be below suitable levels, you will have to do a physical inspection of the chimney or flue. Maybe a bird has started building a nest in the flue. It's possible that leaves or other obstructions are blocking the draft. There is a good chance that odors being produced from oil burners will be associated with a poor draft.

Once you know that the chimney or flue is clear, look for delayed ignition. This is a sure cause of odor buildup. Many factors may contribute to delayed ignition. A simple test may prove that the electrode setting is not correct. If there are insulator cracks, delayed ignition can result. Dirt, such as soot or oil, can foul an electrode and cause it to fire improperly. When a pump is set with an incorrect pressure setting, you can experience delayed ignition. A bad spray pattern for a nozzle can also be responsible for delayed ignition. If a nozzle becomes clogged, it can cause slow ignition. Another potential cause of delayed ignition is an air shutter that is opened too far.

Electrode settings and nozzles are the main issues to consider. Sometimes replacing an existing nozzle with one that has a solid spray pattern will correct the problem. More likely than not, a simple nozzle replacement will solve the problem. If this is not the case, resetting the electrode settings can solve the problem. Check manufacturer's suggestions for nozzle angles, flow rates, and electrode settings. If you have a burner that is using a hollow spray pattern and firing 2.00 gal/hour (gph), try replacing it with a nozzle that produces a solid spray pattern.

In addition to the problems that we have just discussed, there are many other types of problems that arise from time to time with oil burners. To make this information as accessible as possible and as easy to use as we can, let's group the remainder of the troubleshooting steps together under the headings which are most appropriate.

Noisy Operation

You should consider the possibility that there may be a bad coupling alignment in the system. To correct this, you must loosen the fuel-unit mounting screws slightly and shift the unit in different positions until the noise is eliminated. If the noise stops, tighten the mounting screws and pat yourself on the back.

If air gets into an inlet line, noise can occur during operation. To eliminate this possibility, check all connections for leaks. All connections should be made with good flare fittings. It may be worth checking the flaring of all tubing. If you find a leak, either tighten the connection or replace it with a new one.

Have you ever run into a tank hum on a two-pipe system with an inside tank? If you haven't, you may. This type of problem is fixed by installing a return-line hum eliminator in the return line. As you can see, there are not a lot of causes for noisy boilers, so this is one aspect of service and repair that shouldn't take long to troubleshoot.

No Oil

It's common knowledge that oil burners need oil to function. As basic as this is, sometimes the possibility of a lack of oil is ignored. When you've got a burner that will not fire, check the supply tank to see that it contains an adequate supply of oil. Don't trust the oil gauge; it may be stuck. Use a dip stick to probe the tank and to determine the fuel level. If the tank has oil in it, check the fuel filter. It may be clogged. Remove and inspect all fuel filters and strainers. Clean them if they appear to be blocked.

Nozzles frequently become clogged. Obviously, an obstructed nozzle is going to produce less oil for the burner to fire on. This is easy enough to fix. Just replace the nozzle with a new one. If you've followed all these steps and still have a problem, look for a leak in the intake line. If you find one, try tightening all fittings where you suspect the leak. Check the unused intake port plug to make sure that it is tight and not leaking. Air can also invade a system around a faulty filter cover or gasket, so check these components to make sure that they are in good shape and are not leaking. When you are checking the intake line for leaks, make sure that there are no kinks in the line which may be obstructing it.

Occasionally a two-pipe system will become air-bound. If this happens, you must check to see about the bypass plug. Insert one if necessary and make sure that the return line is below the oil level in the supply tank. When you have an air-bound single-pipe system, you should loosen the port plug or easy flow valve and bleed oil for about 15 seconds after all foam is gone from the bleed hose. Then check all connections and plugs to make sure that they are tight. Getting the air out is not hard, and it can be all it takes to get a system back up and running.

There are a couple of other things to check if you are having trouble getting oil to the burner. For example, check to see that there are no slipping or broken couplings. If there are, they must be tightened or replaced. It may sound weird, but you may find a situation where the rotation of a motor and fuel unit is not the same as indicated by the arrow on the pad at the top of the unit. Should this happen, install the fuel unit so that the rotation is correct. The last thing to consider is a frozen pump shaft. If you find this to be the problem, the unit must be sent out for approved servicing or factory repair. Before you install a replacement unit, check for water and dirt in the oil tank.

Leaks

Leaks happen in oil systems. The vibration of equipment can create small leaks. There are many reasons why a system may suffer from leakage, but the leaks should not be left unattended. Sometimes removing a plug, applying a new coat of thread sealant, and reinstalling the plug will take care of a leak. When leaks show up at nozzle plugs or near pressure-adjustment screws, the washer or O-ring at the connection may need to be replaced. Time can take its toll on both washer and O-rings. The covers on housings sometimes leak. When this happens, you may get by with just tightening the cover screws, but be prepared to replace a defective gasket. If a seal is leaking oil, the fuel unit must be replaced.

Blown seals can occur in both single-pipe and two-pipe systems. A blown seal in a single-pipe system may mean that a bypass plug has been left out. Check it and see. You may have to install one. You may find that you will have to replace the whole fuel unit. Two-pipe systems with blown seals could be affected by kinked tubing or obstructions in the return tubing. If this is not the case, plan on replacing the fuel unit.

Bad Oil Pressure

Bad or low oil pressure might be as simple as a defective gauge. It's worth checking. If the gauge proves to be defective, replace it and you are home free. Assuming that the gauge is reading true, look for a nozzle where the capacity is greater than the fuel unit capacity. Should you find this to be the case, replace the fuel unit with one that will offer an adequate capacity.

Pressure Pulsation

Pressure pulsation may be caused by a partially clogged strainer or filter. When this is the problem, all you have to do is clean the strainer and the filter element. Air leaks are another possible cause for pulsating pressure. If you are experiencing an air leak in an intake line, you should tighten all fittings where leaks may be present. Another place to look for air leaks is the cover of the unit. If you suspect leaks on a cover, tighten the cover screws first and then check for leaks. If leaks persist, remove the cover and install a new cover gasket. Then, of course, replace the cover and tighten the cover screws.

Unwanted Cutoffs

Unwanted cutoffs can be caused by several factors. If you have a burner that is cutting off when it shouldn't be, start by testing the nozzle port with a pressure gauge. Put the gauge in the nozzle port of the fuel unit and run the system for about 1 min. Then shut the burner down. If the pressure drops from normal operating pressure and stabilizes, the fuel unit is doing fine. This means that air is the culprit of your troubles. But if the pressure drops to zero, the fuel unit has given up the ghost and needs to be replaced.

Filter leaks can cause unwanted cutoffs. Check the face of the cover and the cover gasket to see if there are any leaks. If there are, replace the gasket and secure the cover tightly. Another thing to check is the cover strainer. This can be fixed by tightening the cover screws of the strainer. Sometimes an air pocket builds up between the cutoff valve and the nozzle. If this happens, run the oil burner while stopping and starting it until all smoke and afterfire disappear.

An unwanted cutoff could be the result of an air leak in the intake line. Check all fittings for possible leaks. Tighten any suspicious connections. Check the unused intake port and return plug to make sure that it's tight. If you are still having trouble, look for a partially clogged nozzle strainer. Clean obstructed strainers or replace the nozzle. If there is a leak at the nozzle adapter, change the nozzle and the adapter.

Well, there you have it. You are now ready to set forth and troubleshoot oil burners. Oh, well, maybe not quite; it helps to have on-the-job experience, but the approaches given here will get you going in the right direction on a fast track. If you follow the systematic procedures discussed in this chapter, you should have less lost time and a more productive work day, not to mention fewer headaches in the field.

CHAPTER 24

Selling Radiant Heating Systems

Selling radiant heating systems may be easier now than it has ever been. The trend toward radiant floor heating is strong, and many consumers are vocal about how good the systems are. Even with this to work with, a lot of heating contractors have not tapped into this lucrative market. Why is this? There are many reasons. A common excuse is that the installation procedures are something that some installers are not familiar with. Some contractors simply prefer to do their work the way they have always done it, regardless of changes in technology and consumer demand. A long list of excuses could be built, but they would be just that, excuses. In reality, radiant floor heating systems offer a great deal to both contractors and consumers. It's time that contractors informed their customers of their options as they pertain to radiant floor heating systems. Short of doing this, contractors are, in effect, cutting their customers out of what may well be the best heating system for their needs.

It's common for contractors to be nervous about changes in their industries. Seasoned contractors sometimes feel threatened by new ways of doing their trade. Compare radiant floor heating systems to computer systems for a moment. Not so many years ago, many contractors grimaced at the thought of computerizing their businesses.

Today contractors who are not computerized are at a definite disadvantage. You could build a case that contractors who think that the only way to use hot-water heat is with baseboard heat emitters are just as far behind the times. In truth, radiant floor heating is a wave of the future, even though the heating systems have been around for more years than the contractors who oppose them.

Customers often trust the contractors with whom they work to advise them on the right heating system for a job. If contractors show potential customers only hot-water baseboard heating systems, the odds are good that the customers will opt for the baseboard systems. Certainly there are plenty of occasions when a baseboard system does make the most sense. But there are also more than enough opportunities that warrant the use of in-floor heating systems.

When radiant floor heating was considered to be effective only when embedded in concrete, the systems were more limited in their acceptable usage. With modern improvements that allow dry systems, this is no longer the case. Consumers need to be educated in the advantages offered by radiant floor heating. Unfortunately, a large number of heating contractors don't know as much as they should about radiant floor systems. If a contractor is not aware of the features and benefits available with radiant floor systems, it's very difficult for the contractor to sell a customer on such a system. Since you have read the previous chapters, you now have a good general understanding of what radiant floor systems can do, and this gives you a marketing edge.

The various contracting fields are crowded with competition. It doesn't matter if you are a heating contractor, a plumber, or an electrician, there is sure to be more than enough competition to go up against. To prevail over your competitors regularly, you need an edge. Radiant floor heating systems may very well be that edge. Any advantage that you can gain over the competition is valuable. The first step, which you have already taken, is to stay abreast of current trends and improvements in the heating industry.

QUICK»TIP **When radiant floor heating was considered to be effective only when embedded in concrete, the systems were more limited in their acceptable usage. With modern improvements that allow dry systems, this is no longer the case.**

Buyers of new homes are more likely to request information on radiant floor heating than homeowners who are planning to remodel their homes. A large number of homeowners have no idea that they can combine radiant floor heating with their existing baseboard heating systems. If they knew what could be done and what the installation might do for them, more remodeling customers would probably choose radiant floor heating. It is your job to make customers aware of all the options available to them.

> **QUICK»TIP** **How are you going to expand your business? One solid path involves radiant floor heating.**

How are you going to expand your business? One solid path involves radiant floor heating. Installing this type of heating system is not difficult. If you are an experienced heat installer, learning the ropes for radiant floor systems will not take long. Everything that you take the time to learn can be passed on to your potential customers. The marketing angles are nearly boundless. If you work in a small community, you might be the only heating contractor who offers radiant floor heating systems. It may not be wise to specialize in these systems, but it certainly doesn't hurt to be known as the local pro when it comes to radiant floor systems.

No Concrete Needed

One of the first mental images that a contractor selling radiant floor heating systems must overcome is the impression that such systems will not function properly unless they are covered in concrete. A large group of people firmly believe that radiant heat tubing depends on concrete flooring to deliver heat. This simply isn't true. Dry systems work very well when installed properly. To get this point across, you should arm yourself with literature from manufacturers, magazine articles, and books like this one to show your prospective customers. Educating the consumer is the first step in expanding your business to include more radiant floor heating systems.

Most dry systems used today are below-deck systems. This simply means that they are installed on the underside of plywood subflooring. Above-deck dry systems are installed on the top of subflooring and then covered with a layer of underlayment. Both types of systems function well, but above-deck systems cost more to install. This is due to the added labor and materials involved. Because of cost and simplicity, below-deck systems are much more popular than above-deck systems.

When heat-transfer plates are used, with either above-deck or below-deck systems, the efficiency of the heating system is improved. Transfer plates are not a requirement, but they do allow a system to perform better. Yes, the plates add to the cost of an installation, but the added expense is well worth the cost. Increased comfort and reduced operating costs are both factors that favor the use of heat-transfer plates.

Concrete is an excellent thermal mass for radiant floor heating, but the added weight and cost of concrete can make a wet system difficult to justify. In some cases, installing a wet system is not practical. Certainly, if a concrete slab is a planned part of construction, it is an excellent place to install radiant floor heating. But don't assume that radiant floor heating should be ruled out if concrete is not to be used. The concept of installing dry systems is relatively new, but the process has proved itself year after year and is fine to recommend to your customers.

Can't Combine

Some customers, and even some heating contractors, feel that they can't combine radiant floor heating systems with hot-water baseboard heating systems. This is not true. It is a fact that radiant heating systems require a lower temperature for the heating fluid supplied to the heating system, and this is why so many people assume that one boiler cannot serve both types of systems at the same time. In fact, a single boiler can be used for both types of heat. Adaptations made at the boiler allow control of fluid temperatures so that both types of heating system can run well together. This can be a hard sell to stubborn customers, but it is easy enough to document. Again, gather up your literature from manufacturers, magazine articles, and books to have on hand for customers who want to see the proof in print.

TRADE SECRET **Some customers, and even some heating contractors, feel that they can't combine radiant floor heating systems with hot-water baseboard heating systems. This is not true.**

Contractors sometimes assume that the only time when baseboard systems should be combined with radiant floor systems is when adding onto a job or remodeling a building. This is the most likely occasion to combine the two type of systems, but it is not the only situation when a combined system makes sense. New construction can present conditions where a combined system is warranted. Don't rule out the use of different types of hot-water heating systems being served by a single boiler.

Won't Last

Some customers are resistant to radiant floor heating systems because they feel that the PEX tubing won't last. There is a fear that the tubing will fail and cause leaks that are hard to fix and potentially damaging. For the most part, this is not a viable concern. PEX tubing will generally last a lot longer than copper tubing. Since most heating systems are installed with no concealed joints, there is no reason why leaks should develop. Of course, a poor installation might result in a leak. For example, if a piece of PEX tubing is installed close to the point of a nail or near metal bridging between joists, some chafing might occur that would result in a leak. However, installed properly, there is no reason to believe that PEX tubing will not outlast most other components of a home or other type of building.

The bad press that PB tubing got in recent years has been difficult for some contractors to overcome. But, PEX is not the same as PB; so the two should not be compared as apples to apples. If you want to make a visual impression on your prospective customers, take along some PEX tubing on your sales call. Give the customer a nail and let the customer attempt to puncture the tubing. Stand on the tubing. Jump up and down on it. Let the customer abuse the material to see just how resistive it is to damage. This type of demonstration can go a long way in removing any fears that customers may have. Also, stress that your installation will not have joints in concealed locations. The combination of this approach should make most customers comfortable with the use of PEX tubing as a component in their new heating system.

Lower Operating Cost

Since radiant floor heating systems operate at a lower fluid temperature than baseboard heating systems, the result is often a lower operating cost. Most consumers will be happy to hear this, so don't leave it out of your sales pitch. Point out how a boiler doesn't have to work as hard to maintain the setpoint temperature for a radiant floor system.

Average consumers will not have trouble understanding how a lower water temperature translates into lower operating costs. It doesn't hurt to carry some examples of operating costs for various types of heating systems when you are making your sales calls. If you can lay out a spreadsheet or a graph to give a visual picture of cost savings, your customers might be more receptive to your cost data. The lower operating cost can be a selling point to offset higher installation costs, if needed.

Space-Saving Features

When you are selling radiant floor heating systems, you should play up the space-saving features of the system. Since radiant floor systems are contained in a floor, they do not restrict furniture placement. This can be a major sales factor, especially in smaller homes, where floor space is at a premium. If you want to work this angle aggressively, show your prospective customers how much clearance is needed between furniture and baseboard heating units for optimum performance. Point out how forced-air ducts would have to be accommodated with specific furniture placement. Of all the various types of heating systems, a radiant floor heating system is the least invasive when it comes to furniture placement.

In addition to taking up less space, you can tell your customers how radiant floor systems are not the dust magnets that baseboard units are. Anyone who has lived with hot-water baseboard heat knows how dusty the heating units can get. Time saved in not having to clean the heat emitters can be put to better use. Not only will radiant floor heat not collect dust, it will not send it airborne in the way that baseboard heat and forced-air ducts will. Anyone with allergies will appreciate this fact. There are a great many benefits to radiant floor heating systems, and you should show as many of them as you can to your potential customers.

QUICK»TIP **When you are selling radiant floor heating systems, you should play up the space-saving features of the system. Since radiant floor systems are contained in a floor, they do not restrict furniture placement.**

Maintenance

Maintenance is not a big factor in forced-air heat or hot-water baseboard heat, but both of these types of heating systems require more maintenance than radiant floor heating systems do. I'm not talking about the boiler or furnace, but the heat emitters themselves. The ducts used to transport forced hot air should be vacuumed periodically. The grilles and registers used with ducts can get banged up and require touch-up painting. Covers used on baseboard heating units get dinged and require touch-up paint. Baseboard covers sometimes rust and become ugly. None of these problems occur with in-floor heat tubing. The tubing is out of sight and out of mind.

Comfort

Comfort associated with radiant floor heating is unsurpassed. There's nothing quite like a warm floor under bare feet, and this is a feeling that is not achieved with other types of heating systems. Since radiant floor heat rises evenly throughout living space, there are no cold zones. Heating systems that are installed on the perimeters of living space cannot possibly provide the even heating that radiant floor heating does. Since heat rises, heat from most heat emitters starts above floor level and goes up. With radiant floor heat, the heat originates in the floor and then rises. This provides a bit more heat, and it keeps the floor and the air close to the floor warmer than other types of heat emitters do.

Comfort should be a serious concern to customers. After all, heat is installed to create a comfort zone. If comfort is as important as it should be, radiant floor heating definitely should be considered as a heating system. How much is too much to spend for comfort? Every customer that you encounter may have a different answer for this question. Not everyone can afford maximum comfort. However, the lower operating cost of radiant floor heat helps to offset potentially greater installation cost. And the installation cost may turn out to be about the same as some other forms of heating systems. You should explore all aspects of costs with your customers when selling a heating system. There is more to consider than just the installation cost. If a customer wants a cheap heat installation, electric baseboard heat is about as cheap as it gets. But when it comes to running the electric baseboard heat in cold climates, the money saved on installation is soon lost in operating costs. Make sure that you factor all elements of cost into the equation when selling a heating system.

It's New

Some people resist radiant floor heating since they believe it's new. In reality, radiant floor heating has been in use for far longer than most other types of heating systems. Granted, it is relatively new when compared to other types of modern heating systems. But new is not always bad. Still, there are people who are leery of anything that is not traditional in their minds. If a customer doesn't want radiant heat because none of the neighbors have it, you could be up against a brick wall when it comes to selling a radiant floor system. While most of the reasoning here is not rational, some of it is. When a house is appraised for real estate value, it might suffer from having a different type of heating system.

Assume that someone is having a new home built in a subdivision where hot-water baseboard is the only type of heating system being used. In a case like this, installing radiant floor heating could hurt the value of the home. Maybe it will increase the value, but the fact that the house is not consistent with comparable homes in the neighborhood could have a negative effect on the home's appraised value. You can sell until you turn blue in the face with this type of situation and wind up losing a job by being too pushy. Sometimes it is better to go with the flow and sell a baseboard system, even though a radiant floor system might be a wiser choice in terms of comfort and use. We don't live in a world where a practical approach is always best, so you have to know when to give in on what you are selling and close a sale for what the customer wants, rather than what you would like to install.

Information

Quality information is all that is needed to sell radiant floor heating systems. Consumers are taking a more active interest in researching their purchases. Long gone are the days when customers would take a contractor's word for what was best. Today customers do their own research to determine where their best value can be achieved. If you make it easy for customers to do this, by providing your prospective customers with plenty of detailed information, you are more likely to make sales. Customers often sell themselves if they are given enough data to evaluate. This can backfire, however. There is a saying in sales that goes like this: "A confused mind always says no." Don't flood your customers with information. Make as much information as they want available, but distill it into manageable doses.

Generally speaking, average consumers can become confused quickly by technical documents. A summary sheet from a manufacturer that makes perfect sense to a contractor may read like a foreign language to a consumer. Keep this in mind when you are putting together sales packages for potential customers. Go over the information and condense it for your customers. Highlight specific areas that pertain to a certain job. If necessary, write notes on the data sheets to transform the technical jargon into reader-friendly terms. Make it easy for your customers to understand what they are looking at. This takes some extra time, but more sales are likely to occur when you put the effort into giving your customers information that they can understand.

All of the charts, tables, graphs, and spreadsheets in the world are not as valuable as a solid relationship between customer and contractor. Developing an honest, open dialogue has proved to be the best sales tool that I've ever discovered, but this takes time. A contractor who rushes though a sales meeting is not going to create the trust needed for a solid sale. There have been many times when I've spent two hours or more with a single customer while making a sale. This is a lot of time, but once the sale is made it generally sticks. Time is a precious commodity, but it is also an essential element in forming a relationship. Spending a lot of time with customers means that you may not be able to meet with as many potential customers, but if you have a stronger closing ratio on your sales, it's justified. After all, business profit is not earned by the number of customers you sit down with but rather by the number of customers you sell. If you remember this, you should do well in sales.

GLOSSARY

above-deck dry system Radiant floor heating system that is installed on or above subflooring, usually with the use of sleepers, coiled tubing, and heat-transfer plates.

air binding A condition that occurs when a circulator cannot dislodge an air pocket at a high point in a heating system.

air purger A device that separates air from the fluid of a heating system and ejects the air through a vent.

air vent A device that expels air from a heating system by either manual manipulation or automatic function.

aquastat A device that measures water temperature in a heating system and opens or closes electrical contacts as needed for the water temperature to maintain its setpoint rate.

auxiliary heating loads Demands on a heating system that are in addition to general heating of living or work space, such as a heating zone that is used to heat a garage or swimming pool.

below-deck dry system Radiant floor heating system that is installed below subflooring, usually with the use of coiled tubing, heat-transfer plates, and staples.

bend supports Devices used to guide flexible heat tubing into turning positions while avoiding kinking of the tubing.

boiler drain A valve that can be installed on piping that will allow the connection of standard garden-hose threads for draining purposes.

cavitation Vapor pockets forming when the pressure on a liquid drops below its vapor pressure create the condition called cavitation.

circulator A device, also called a pump, that creates fluid motion in a hydronic system by adding head energy to the fluid.

closed-loop system A heating system that allows no air to enter it.

deaerated fluid Fluid that has had most of the existing air removed from it.

dissolved air Air mixed with water at a molecular level, which cannot be seen.

domestic water Potable water, or water that is safe for human consumption.

dry system A radiant floor heating system that is not covered by a poured material, such as concrete.

entrained air Air that is carried in bubble form along with the flowing fluid in a heating system.

flow rate The rate of flow, calculated in gallons per minute, of water at a temperature of 60°, needed to create a pressure drop of 1 psi across a piping component.

flow velocity A measurement of fluid movement, usually measured at a rate of feet traveled per second.

gaseous cavitation Cavitation that is caused by air bubbles entrained in heating fluid, rather than being caused by vapor pockets.

heat emitter A term used to describe a device that delivers heat from a circulating system of heated water.

heat transfer plate An aluminum plate that fits over floor heating tubing and provides lateral heat conduction away from tubing.

high point vents Air vents that are placed at high points on a heating system to allow for the removal of air.

impeller A disk with curved vanes that rotates and that is housed in a centrifugal pump used to add head to fluid.

manifold A piping device that provides a common starting or ending point for multiple piping zones.

manifold station Multiple manifolds used to provide supply and return water to various zones of a heating system.

microbubbles Tiny bubbles of air in water that tend to make the water appear cloudy.

open-loop system A heating system that is not closed to the atmosphere or where fresh water is added to the heating system periodically.

pump head The head added to fluid by a pump that is operating at a set rate of flow.

purger valve A specialty valve that replaces the need for a boiler drain and a gate valve when forced-water purging is being done with a heating system.

series pumps Multiple pumps installed end to end and having the same direction of flow.

setpoint temperature A defined temperature that is maintained at all times, when possible, and that is set on some type of control device, such as a thermostat.

slab resistance Thermal resistance given by the combination of the wall of heating tubing and the concrete surrounding the tubing in a radiant floor heating system.

stationary air pockets Air that becomes trapped at high points in a heating system.

tankless coil A device used to create domestic hot water when the coil is installed in a boiler.

thin slab A slab floor that is made up of either lightweight concrete or gypsum-based underlayment that is not more than 1.5 in. thick and in which heat tubing for radiant floor heating systems is contained.

total head The total mechanical energy content of a fluid at some point in a piping system.

velocity head Mechanical energy possessed by a fluid due to its motion.

viscosity A way of explaining the natural tendency of a fluid to resist flow.

working pressure The maximum pressure at which a component of a heating system can function safely.

zone valve A device used to control the flow of hot water from a boiler through individual heating zones.

Pressure Losses and Flow Charts

PRESSURE LOSS PER FOOT
3/8" PEX
100% Water

WIRSBO	Head (Feet of Water) Per Foot of Pipe					
FLOW GALLONS PER MINUTE	80°F	100°F	120°F	140°F	160°F	180°F
.1	.0029	.0028	.0027	.0026	.0025	.0024
.2	.0091	.0080	.0082	.0079	.0077	.0076
.3	.0191	.0181	.0173	.0167	.0162	.0159
.4	.0362	.0309	.0295	.0285	.0278	.0272
.5	.0472	.0457	.0427	.0412	.0402	.0394
.6	.0659	.0624	.0596	.0576	.0561	.0550
.7	.0879	.0832	.0795	.0768	.0749	.0734
.8	.1097	.1053	.0993	.0959	.0936	.0917
.9	.1360	.1289	.1232	.1190	.1161	.1138
1.0	.1657	.1570	.1501	.1451	.1415	.1387
1.1	.1942	.1858	.1759	.1700	.1659	.1625
1.2	.2277	.2158	.2063	.1994	.1946	.1906
1.3	.2646	.2508	.2398	.2318	.2062	.2217
1.4	.2993	.2860	.2714	.2623	.2560	.2508
1.5	.3396	.3220	.3080	.2977	.2905	.2847
1.6	.3834	.3635	.3477	.3362	.3281	.3215
1.7	.4242	.4048	.3848	.3720	.3630	.3558
1.8	.4710	.4466	.4273	.4131	.4032	.3951
1.9	.5214	.4945	.4731	.4575	.4464	.4375
2.0	.5699	.5387	.5154	.4984	.4864	.4767

PRESSURE LOSS PER FOOT
1/2" PEX
100% Water

WIRSBO	Head (Feet of Water) Per Foot of Pipe					
FLOW GALLONS PER MINUTE	80°F	100°F	120°F	140°F	160°F	180°F
.1	.0006	.0006	.0006	.0005	.0005	.0005
.2	.0021	.0020	.0019	.0019	.0018	.0018
.3	.0044	.0042	.0040	.0038	.0037	.0037
.4	.0073	.0069	.0066	.0064	.0062	.0061
.5	.0109	.0103	.0099	.0095	.0093	.0091
.6	.0151	.0143	.0137	.0132	.0129	.0126
.7	.0199	.0189	.0180	.0174	.0170	.0166
.8	.0253	.0239	.0229	.0221	.0215	.0211
.9	.0316	.0299	.0286	.0276	.0269	.0264
1.0	.0381	.0360	.0344	.0333	.0325	.0318
1.1	.0451	.0427	.0408	.0394	.0385	.0377
1.2	.0526	.0499	.0477	.0461	.0449	.0440
1.3	.0607	.0575	.0550	.0531	.0518	.0508
1.4	.0693	.0657	.0628	.0607	.0592	.0580
1.5	.0784	.0743	.0710	.0686	.0669	.0656
1.6	.0879	.0833	.0797	.0770	.0751	.0736
1.7	.0980	.0929	.0888	.0858	.0837	.0821
1.8	.1091	.1034	.0989	.0956	.0933	.0914
1.9	.1201	.1139	.1089	.1053	.1027	.1007
2.0	.1317	.1248	.1194	.1154	.1126	.1103
2.1	.1436	.1361	.1301	.1257	.1226	.1201
2.2	.1561	.1479	.1414	.1367	.1333	.1306
2.3	.1691	.1602	.1532	.1480	.1444	.1415
2.4	.1825	.1729	.1653	.1598	.1559	.1527
2.5	.1963	.1861	.1779	.1720	.1677	.1644

FIGURE A1.1

Pressure loss per foot of 3/8-in. and 1/2-in. PEX tubing, pure water. (*Courtesy of Wirsbo.*)

PRESSURE LOSS PER FOOT
5/8" PEX
100% Water

WIRSBO	Head (Feet of Water) Per Foot of Pipe					
FLOW GALLONS PER MINUTE	80°F	100°F	120°F	140°F	160°F	180°F
.1	.0002	.0002	.0002	.0002	.0002	.0002
.2	.0008	.0008	.0007	.0007	.0007	.0007
.3	.0017	.0016	.0015	.0014	.0014	.0014
.4	.0028	.0026	.0025	.0024	.0024	.0023
.5	.0041	.0039	.0037	.0036	.0035	.0034
.6	.0057	.0054	.0052	.0050	.0049	.0048
.7	.0075	.0071	.0068	.0066	.0064	.0063
.8	.0095	.0090	.0086	.0083	.0081	.0080
.9	.0118	.0111	.0106	.0103	.0100	.0098
1.0	.0142	.0135	.0129	.0124	.0121	.0119
1.1	.0168	.0160	.0152	.0147	.0144	.0141
1.2	.0197	.0186	.0178	.0172	.0168	.0164
1.3	.0227	.0215	.0206	.0199	.0194	.0190
1.4	.0259	.0246	.0235	.0227	.0221	.0217
1.5	.0293	.0278	.0266	.0257	.0250	.0245
1.6	.0329	.0312	.0298	.0288	.0281	.0276
1.7	.0367	.0348	.0333	.0321	.0314	.0307
1.8	.0407	.0385	.0368	.0356	.0347	.0340
1.9	.0448	.0425	.0406	.0392	.0383	.0375
2.0	.0491	.0462	.0445	.0430	.0420	.0411
2.1	.0534	.0505	.0483	.0467	.0455	.0446
2.2	.0580	.0549	.0525	.0507	.0495	.0485
2.3	.0628	.0595	.0569	.0549	.0536	.0525
2.4	.0678	.0642	.0614	.0593	.0579	.0567
2.5	.0729	.0691	.0661	.0638	.0623	.0610
2.6	.0782	.0741	.0709	.0685	.0668	.0654
2.7	.0837	.0793	.0758	.0733	.0715	.0700
2.8	.0894	.0847	.0810	.0782	.0763	.0748
2.9	.0952	.0902	.0862	.0833	.0813	.0796
3.0	.1011	.0959	.0917	.0886	.0864	.0847

PRESSURE LOSS PER FOOT
3/4" PEX
100% Water

WIRSBO	Head (Feet of Water) Per Foot of Pipe					
FLOW GALLONS PER MINUTE	80°F	100°F	120°F	140°F	160°F	180°F
.5	.0021	.0020	.0019	.0019	.0018	.0018
.6	.0029	.0028	.0027	.0026	.0025	.0024
.7	.0039	.0037	.0035	.0034	.0033	.0032
.8	.0049	.0047	.0044	.0043	.0042	.0041
.9	.0061	.0057	.0055	.0053	.0052	.0051
1.0	.0073	.0069	.0066	.0064	.0062	.0061
1.1	.0087	.0082	.0079	.0076	.0074	.0072
1.2	.0102	.0096	.0092	.0089	.0086	.0085
1.3	.0117	.0111	.0106	.0102	.0100	.0098
1.4	.0134	.0127	.0121	.0117	.0114	.0111
1.5	.0151	.0143	.0137	.0132	.0129	.0126
1.6	.0170	.0161	.0154	.0148	.0145	.0142
1.7	.0189	.0179	.0171	.0165	.0161	.0158
1.8	.0210	.0199	.0190	.0183	.0179	.0175
1.9	.0231	.0219	.0209	.0202	.0197	.0193
2.0	.0253	.0240	.0229	.0221	.0216	.0211
2.1	.0276	.0262	.0250	.0242	.0236	.0231
2.2	.0300	.0284	.0272	.0263	.0256	.0251
2.3	.0325	.0308	.0294	.0284	.0277	.0272
2.4	.0351	.0332	.0318	.0307	.0299	.0293
2.5	.0378	.0358	.0342	.0330	.0322	.0316
2.6	.0405	.0384	.0367	.0354	.0346	.0339
2.7	.0433	.0411	.0392	.0379	.0370	.0362
2.8	.0463	.0438	.0419	.0405	.0395	.0387
2.9	.0493	.0467	.0446	.0431	.0421	.0412
3.0	.0524	.0496	.0474	.0458	.0447	.0438
3.2	.0604	.0573	.0548	.0529	.0516	.0506
3.5	.0690	.0654	.0626	.0605	.0590	.0578
3.7	.0781	.0741	.0708	.0684	.0668	.0654
4.0	.0877	.0832	.0795	.0769		.0735

FIGURE A1.2

Pressure loss per foot of 5/8-in. and 3/4-in. PEX tubing, pure water. (*Courtesy of Wirsbo.*)

PRESSURE LOSS PER FOOT 3/8" PEX 30% Glycol / Water Mixture

WIRSBO	Head (Feet of Water) Per Foot of Pipe					
FLOW GALLONS PER MINUTE	80°F	100°F	120°F	140°F	160°F	180°F
.1	.0034	.0032	.0029	.0028	.0025	.0026
.2	.0116	.0109	.0101	.0096	.0087	.0088
.3	.0238	.0225	.0208	.0198	.0180	.0182
.4	.0397	.0376	.0347	.0332	.0301	.0304
.5	.0591	.0560	.0517	.0494	.0448	.0452
.6	.0818	.0775	.0716	.0684	.0621	.0627
.7	.1077	.1020	.0942	.0901	.0818	.0825
.8	.1367	.1295	.1196	.1144	.1039	.1048
.9	.1687	.1598	.1477	.1412	.1283	.1294
1.0	.2036	.1929	.1783	.1705	.1549	.1563
1.1	.2414	.2287	.2114	.2022	.1838	.1854
1.2	.2820	.2672	.2470	.2363	.2148	.2167
1.3	.3253	.3083	.2851	.2727	.2479	.2502
1.4	.3714	.3519	.3255	.3114	.2832	.2857
1.5	.4201	.3982	.3683	.3524	.3205	.3234
1.6	.4715	.4469	.4135	.3956	.3599	.3631
1.7	.5255	.4981	.4609	.4410	.4012	.4048
1.8	.5820	.5518	.5106	.4885	.4446	.4485
1.9	.6411	.6078	.5625	.5383	.4899	.4942
2.0	.7027	.6663	.6167	.5901	.5372	.5419

PRESSURE LOSS PER FOOT 1/2" PEX 30% Glycol / Water Mixture

WIRSBO	Head (Feet of Water) Per Foot of Pipe					
FLOW GALLONS PER MINUTE	80°F	100°F	120°F	140°F	160°F	180°F
.1	.0008	.0007	.0007	.0007	.0006	.0006
.2	.0027	.0025	.0023	.0022	.0020	.0020
.3	.0055	.0052	.0048	.0046	.0042	.0042
.4	.0092	.0087	.0081	.0077	.0070	.0070
.5	.0137	.0130	.0120	.0115	.0104	.0105
.6	.0190	.0180	.0166	.0159	.0144	.0145
.7	.0250	.0237	.0219	.0209	.0189	.0191
.8	.0317	.0300	.0277	.0265	.0241	.0243
.9	.0391	.0371	.0342	.0327	.0297	.0300
1.0	.0472	.0447	.0413	.0395	.0359	.0362
1.1	.0560	.0530	.0490	.0468	.0425	.0429
1.2	.0654	.0619	.0572	.0547	.0497	.0502
1.3	.0754	.0715	.0660	.0631	.0574	.0579
1.4	.0861	.0816	.0754	.0721	.0655	.0661
1.5	.0974	.0923	.0853	.0816	.0742	.0748
1.6	.1093	.1036	.0957	.0916	.0832	.0840
1.7	.1218	.1154	.1067	.1021	.0928	.0936
1.8	.1349	.1278	.1182	.1131	.1028	.1037
1.9	.1486	.1408	.1302	.1246	.1133	.1143
2.0	.1628	.1543	.1428	.1366	.1242	.1253
2.1	.1777	.1684	.1558	.1490	.1356	.1368
2.2	.1931	.1830	.1693	.1620	.1474	.1487
2.3	.2091	.1982	.1834	.1754	.1596	.1610
2.4	.2256	.2139	.1979	.1893	.1723	.1738
2.5	.2427	.2301	.2129	.2037	.1854	.1870

FIGURE A1.3

Pressure loss per foot of 3/8-in. and 1/2-in. PEX tubing, 30 percent glycol. (*Courtesy of Wirsbo.*)

PRESSURE LOSS PER FOOT 5/8" PEX 30% Glycol / Water Mixture

WIRSBO	Head (Feet of Water) Per Foot of Pipe					
FLOW GALLONS PER MINUTE	80°F	100°F	120°F	140°F	160°F	180°F
.1	.0003	.0003	.0003	.0002	.0002	.0002
.2	.0010	.0010	.0009	.0008	.0008	.0008
.3	.0021	.0020	.0018	.0017	.0016	.0016
.4	.0034	.0033	.0030	.0029	.0026	.0026
.5	.0051	.0049	.0045	.0043	.0039	.0039
.6	.0071	.0067	.0062	.0059	.0054	.0054
.7	.0093	.0088	.0081	.0078	.0071	.0071
.8	.0118	.0112	.0103	.0099	.0090	.0090
.9	.0146	.0138	.0127	.0122	.0111	.0111
1.0	.0176	.0167	.0154	.0147	.0134	.0135
1.1	.0208	.0198	.0182	.0174	.0159	.0160
1.2	.0243	.0231	.0213	.0204	.0186	.0186
1.3	.0281	.0267	.0246	.0235	.0215	.0215
1.4	.0320	.0304	.0280	.0268	.0245	.0246
1.5	.0362	.0344	.0317	.0303	.0277	.0278
1.6	.0407	.0386	.0356	.0340	.0311	.0312
1.7	.0453	.0431	.0397	.0379	.0347	.0348
1.8	.0502	.0477	.0440	.0420	.0385	.0385
1.9	.0553	.0525	.0484	.0463	.0424	.0425
2.0	.0606	.0576	.0531	.0508	.0465	.0466
2.1	.0661	.0628	.0579	.0554	.0507	.0508
2.2	.0718	.0683	.0629	.0602	.0551	.0552
2.3	.0778	.0739	.0682	.0652	.0597	.0598
2.4	.0839	.0797	.0735	.0704	.0644	.0646
2.5	.0902	.0858	.0791	.0757	.0693	.0695
2.6	.0968	.0920	.0849	.0812	.0744	.0745
2.7	.1036	.0985	.0908	.0869	.0796	.0797
2.8	.1105	.1151	.0969	.0927	.0850	.0851
2.9	.1177	.1119	.1032	.0988	.0905	.0907
3.0	.1250	.1189	.1097	.1049	.0962	.0963

PRESSURE LOSS PER FOOT 3/4" PEX 30% Glycol / Water Mixture

WIRSBO	Head (Feet of Water) Per Foot of Pipe					
FLOW GALLONS PER MINUTE	80°F	100°F	120°F	140°F	160°F	180°F
.5	.0027	.0025	.0023	.0022	.0021	.0020
.6	.0037	.0035	.0032	.0031	.0029	.0028
.7	.0048	.0046	.0042	.0040	.0038	.0037
.8	.0061	.0058	.0053	.0051	.0048	.0047
.9	.0076	.0071	.0066	.0063	.0060	.0058
1.0	.0091	.0086	.0080	.0076	.0072	.0070
1.1	.0108	.0102	.0094	.0090	.0086	.0083
1.2	.0126	.0119	.0110	.0105	.0100	.0097
1.3	.0145	.0138	.0127	.0122	.0115	.0111
1.4	.0166	.0157	.0145	.0139	.0132	.0127
1.5	.0188	.0178	.0164	.0157	.0149	.0144
1.6	.0211	.0200	.0184	.0176	.0167	.0162
1.7	.0235	.0222	.0206	.0197	.0186	.0180
1.8	.0260	.0246	.0228	.0218	.0207	.0200
1.9	.0286	.0271	.0251	.0240	.0228	.0220
2.0	.0314	.0297	.0275	.0263	.0249	.0241
2.1	.0342	.0324	.0300	.0287	.0272	.0263
2.2	.0372	.0353	.0326	.0312	.0296	.0286
2.3	.0403	.0382	.0353	.0338	.0320	.0310
2.4	.0435	.0412	.0381	.0364	.0346	.0334
2.5	.0468	.0443	.0410	.0392	.0372	.0360
2.6	.0501	.0475	.0440	.0420	.0399	.0386
2.7	.0536	.0508	.0470	.0450	.0427	.0413
2.8	.0573	.0543	.0502	.0480	.0456	.0441
2.9	.0610	.0578	.0534	.0511	.0485	.0469
3.0	.0648	.0614	.0568	.0543	.0516	.0499
3.2	.0747	.0708	.0655	.0627	.0596	.0576
3.5	.0853	.0809	.0749	.0716	.0680	.0658
3.7	.0965	.0915	.0847	.0811	.0770	.0744
4.0	.1083	.1027	.0951	.0910	.0864	.0836

FIGURE A1.4

Pressure loss per foot of 5/8-in. and 3/4-in. PEX tubing, 30 percent glycol. (*Courtesy of Wirsbo.*)

PRESSURE LOSS PER FOOT
3/8" PEX
40% Glycol / Water Mixture

WIRSBO	Head (Feet of Water) Per Foot of Pipe					
FLOW GALLONS PER MINUTE	80°F	100°F	120°F	140°F	160°F	180°F
.1	.0037	.0035	.0032	.0030	.0028	.0027
.2	.0127	.0119	.0111	.0104	.0097	.0092
.3	.0261	.0244	.0228	.0213	.0200	.0190
.4	.0436	.0407	.0380	.0357	.0334	.0317
.5	.0649	.0606	.0566	.0531	.0497	.0473
.6	.0898	.0839	.0784	.0735	.0689	.0655
.7	.1182	.1104	.1032	.0968	.0907	.0863
.8	.1499	.1401	.1310	.1229	.1152	.1096
.9	.1849	.1729	.1616	.1517	.1422	.1353
1.0	.2232	.2086	.1951	.1831	.1717	.1633
1.1	.2645	.2473	.2313	.2171	.2036	.1937
1.2	.3089	.2889	.2702	.2537	.2379	.2264
1.3	.3564	.3333	.3118	.2927	.2746	.2613
1.4	.4068	.3805	.3560	.3342	.3136	.2984
1.5	.4601	.4304	.4027	.3781	.3548	.3377
1.6	.5163	.4830	.4520	.4245	.3983	.3791
1.7	.5754	.5383	.5038	.4731	.4440	.4226
1.8	.6372	.5962	.5580	.5241	.4919	.4683
1.9	.7019	.6567	.6147	.5774	.5420	.5160
2.0	.7692	.7198	.6738	.6330	.5942	.5657

PRESSURE LOSS PER FOOT
1/2" PEX
40% Glycol / Water Mixture

WIRSBO	Head (Feet of Water) Per Foot of Pipe					
FLOW GALLONS PER MINUTE	80°F	100°F	120°F	140°F	160°F	180°F
.1	.0009	.0008	.0008	.0007	.0007	.0006
.2	.0030	.0028	.0026	.0024	.0023	.0021
.3	.0061	.0057	.0053	.0049	.0046	.0044
.4	.0101	.0095	.0088	.0082	.0077	.0074
.5	.0151	.0141	.0131	.0123	.0115	.0110
.6	.0208	.0195	.0182	.0170	.0160	.0152
.7	.0274	.0256	.0239	.0224	.0210	.0200
.8	.0348	.0325	.0304	.0284	.0267	.0254
.9	.0429	.0401	.0375	.0350	.0330	.0313
1.0	.0518	.0484	.0452	.0423	.0398	.0378
1.1	.0614	.0574	.0536	.0501	.0472	.0449
1.2	.0717	.0670	.0626	.0585	.0551	.0524
1.3	.0827	.0773	.0723	.0676	.0636	.0605
1.4	.0944	.0882	.0825	.0771	.0726	.0691
1.5	.1067	.0998	.0933	.0873	.0822	.0782
1.6	.1198	.1120	.1047	.0979	.0922	.0877
1.7	.1334	.1248	.1167	.1091	.1028	.0978
1.8	.1478	.1382	.1293	.1209	.1139	.1084
1.9	.1627	.1522	.1424	.1332	.1254	.1194
2.0	.1784	.1668	.1561	.1460	.1375	.1309
2.1	.1946	.1820	.1703	.1593	.1501	.1428
2.2	.2114	.1978	.1851	.1731	.1631	.1553
2.3	.2289	.2142	.2004	.1875	.1767	.1681
2.4	.2470	.2311	.2163	.2023	.1907	.1815
2.5	.2657	.2486	.2327	.2177	.2051	.1953

FIGURE A1.5

Pressure loss per foot of 3/8-in. and 1/2-in. PEX tubing, 40 percent glycol. (*Courtesy of Wirsbo.*)

PRESSURE LOSS PER FOOT
5/8" PEX
40% Glycol / Water Mixture

WIRSBO	Head (Feet of Water) Per Foot of Pipe					
FLOW GALLONS PER MINUTE	80°F	100°F	120°F	140°F	160°F	180°F
.1	.0003	.0003	.0003	.0003	.0002	.0002
.2	.0011	.0010	.0010	.0009	.0008	.0008
.3	.0023	.0021	.0020	.0018	.0017	.0016
.4	.0038	.0035	.0033	.0031	.0029	.0027
.5	.0056	.0052	.0049	.0046	.0043	.0041
.6	.0078	.0073	.0068	.0063	.0059	.0056
.7	.0102	.0095	.0089	.0083	.0078	.0074
.8	.0130	.0121	.0113	.0106	.0099	.0094
.9	.0160	.0149	.0140	.0130	.0123	.0117
1.0	.0193	.0180	.0168	.0157	.0148	.0141
1.1	.0229	.0214	.0200	.0186	.0175	.0167
1.2	.0267	.0249	.0233	.0218	.0205	.0195
1.3	.0308	.0288	.0269	.0251	.0237	.0225
1.4	.0351	.0328	.0307	.0287	.0270	.0257
1.5	.0397	.0371	.0347	.0324	.0305	.0291
1.6	.0446	.0417	.0390	.0364	.0343	.0326
1.7	.0497	.0464	.0434	.0406	.0382	.0364
1.8	.0550	.0514	.0481	.0450	.0423	.0403
1.9	.0606	.0566	.0530	.0495	.0466	.0444
2.0	.0664	.0621	.0580	.0543	.0511	.0486
2.1	.0724	.0677	.0633	.0592	.0558	.0531
2.2	.0787	.0736	.0688	.0644	.0606	.0577
2.3	.0852	.0797	.0745	.0697	.0656	.0625
2.4	.0919	.0860	.0804	.0752	.0708	.0674
2.5	.0988	.0925	.0865	.0809	.0762	.0725
2.6	.1060	.0992	.0928	.0868	.0818	.0778
2.7	.1134	.1061	.0993	.0929	.0875	.0833
2.8	.1210	.1132	.1059	.0991	.0934	.0889
2.9	.1288	.1205	.1128	.1055	.0994	.0947
3.0	.1369	.1281	.1199	.1121	.1057	.1006

PRESSURE LOSS PER FOOT
3/4" PEX
40% Glycol / Water Mixture

WIRSBO	Head (Feet of Water) Per Foot of Pipe					
FLOW GALLONS PER MINUTE	80°F	100°F	120°F	140°F	160°F	180°F
.5	.0029	.0027	.0025	.0024	.0022	.0021
.6	.0040	.0038	.0035	.0033	.0031	.0029
.7	.0053	.0049	.0046	.0043	.0041	.0039
.8	.0067	.0063	.0059	.0055	.0051	.0049
.9	.0083	.0077	.0072	.0068	.0064	.0060
1.0	.0100	.0093	.0087	.0082	.0077	.0073
1.1	.0118	.0111	.0103	.0097	.0091	.0086
1.2	.0138	.0129	.0121	.0113	.0106	.0101
1.3	.0160	.0149	.0139	.0130	.0123	.0117
1.4	.0182	.0170	.0159	.0149	.0140	.0133
1.5	.0206	.0192	.0180	.0168	.0158	.0150
1.6	.0231	.0216	.0202	.0189	.0178	.0169
1.7	.0257	.0241	.0225	.0210	.0198	.0188
1.8	.0285	.0266	.0249	.0233	.0219	.0209
1.9	.0314	.0293	.0274	.0257	.0242	.0230
2.0	.0344	.0322	.0301	.0281	.0265	.0252
2.1	.0375	.0351	.0328	.0307	.0289	.0275
2.2	.0408	.0381	.0357	.0333	.0314	.0299
2.3	.0441	.0413	.0386	.0361	.0340	.0323
2.4	.0476	.0445	.0417	.0390	.0367	.0349
2.5	.0512	.0479	.0448	.0419	.0395	.0376
2.6	.0549	.0514	.0481	.0450	.0423	.0403
2.7	.0588	.0550	.0514	.0481	.0453	.0431
2.8	.0627	.0587	.0549	.0513	.0484	.0460
2.9	.0668	.0624	.0584	.0547	.0515	.0490
3.0	.0709	.0663	.0621	.0581	.0547	.0521
3.2	.0818	.0766	.0716	.0670	.0631	.0601
3.5	.0934	.0874	.0818	.0765	.0721	.0687
3.7	.1057	.0989	.0926	.0866	.0816	.0777
4.0	.1186	.1110	.1039	.0972	.0916	.0872

FIGURE A1.6

Pressure loss per foot of 5/8-in. and 3/4-in. PEX tubing, 40 percent glycol. (*Courtesy of Wirsbo.*)

PRESSURE LOSS PER FOOT 3/8" PEX 50% Glycol / Water Mixture

WIRSBO	Head (Feet of Water) Per Foot of Pipe					
FLOW GALLONS PER MINUTE	80°F	100°F	120°F	140°F	160°F	180°F
.1	.0041	.0037	.0034	.0031	.0029	.0028
.2	.0140	.0125	.0115	.0108	.0100	.0098
.3	.0289	.0257	.0237	.0221	.0206	.0201
.4	.0481	.0429	.0395	.0370	.0345	.0336
.5	.0716	.0639	.0588	.0550	.0514	.0501
.6	.0990	.0884	.0813	.0762	.0711	.0694
.7	.1303	.1163	.1071	.1003	.0936	.0913
.8	.1652	.1476	.1359	.1273	.1189	.1160
.9	.2038	.1821	.1677	.1571	.1467	.1432
1.0	.2459	.2197	.2024	.1897	.1772	.1728
1.1	.2914	.2605	.2400	.2249	.2101	.2050
1.2	.3403	.3042	.2803	.2627	.2455	.2395
1.3	.3925	.3509	.3234	.3032	.2833	.2764
1.4	.4480	.4006	.3692	.3461	.3235	.3156
1.5	.5067	.4531	.4177	.3916	.3660	.3572
1.6	.5685	.5085	.4688	.4395	.4109	.4009
1.7	.6334	.5666	.5224	.4899	.4580	.4469
1.8	.7014	.6276	.5787	.5427	.5074	.4951
1.9	.7725	.6912	.6374	.5978	.5590	.5455
2.0	.8465	.7576	.6987	.6553	.6128	.5981

PRESSURE LOSS PER FOOT 1/2" PEX 50% Glycol / Water Mixture

WIRSBO	Head (Feet of Water) Per Foot of Pipe					
FLOW GALLONS PER MINUTE	80°F	100°F	120°F	140°F	160°F	180°F
.1	.0010	.0008	.0008	.0007	.0007	.0007
.2	.0033	.0029	.0027	.0025	.0023	.0023
.3	.0067	.0060	.0055	.0051	.0048	.0047
.4	.0112	.0100	.0092	.0086	.0080	.0078
.5	.0166	.0148	.0136	.0128	.0119	.0116
.6	.0230	.0205	.0189	.0177	.0165	.0161
.7	.0303	.0270	.0249	.0233	.0217	.0212
.8	.0384	.0343	.0315	.0295	.0276	.0269
.9	.0473	.0423	.0389	.0364	.0340	.0332
1.0	.0571	.0510	.0469	.0440	.0411	.0401
1.1	.0677	.0604	.0557	.0521	.0487	.0475
1.2	.0790	.0706	.0650	.0609	.0569	.0555
1.3	.0911	.0814	.0750	.0703	.0656	.0640
1.4	.1040	.0929	.0856	.0802	.0749	.0731
1.5	.1176	.1051	.0968	.0907	.0848	.0827
1.6	.1319	.1179	.1087	.1018	.0951	.0928
1.7	.1470	.1314	.1211	.1135	.1061	.1035
1.8	.1628	.1455	.1341	.1257	.1175	.1146
1.9	.1792	.1603	.1477	.1385	.1294	.1263
2.0	.1964	.1756	.1619	.1518	.1419	.1384
2.1	.2143	.1916	.1766	.1656	.1548	.1511
2.2	.2328	.2082	.1920	.1800	.1683	.1642
2.3	.2520	.2254	.2079	.1949	.1822	.1778
2.4	.2719	.2432	.2243	.2103	.1966	.1919
2.5	.2924	.2617	.2413	.2263	.2116	.2065

FIGURE A1.7

Pressure loss per foot of 3/8-in. and 1/2-in. PEX tubing, 50 percent glycol. *(Courtesy of Wirsbo.)*

PRESSURE LOSS PER FOOT 5/8" PEX 50% Glycol / Water Mixture

WIRSBO	Head (Feet of Water) Per Foot of Pipe					
FLOW GALLONS PER MINUTE	80°F	100°F	120°F	140°F	160°F	180°F
.1	.0004	.0003	.0003	.0003	.0003	.0002
.2	.0012	.0011	.0010	.0009	.0009	.0008
.3	.0025	.0022	.0020	.0019	.0018	.0017
.4	.0042	.0037	.0034	.0032	.0030	.0029
.5	.0062	.0055	.0051	.0048	.0044	.0043
.6	.0086	.0076	.0070	.0066	.0061	.0060
.7	.0113	.0101	.0093	.0087	.0081	.0079
.8	.0143	.0128	.0117	.0110	.0103	.0100
.9	.0176	.0157	.0145	.0136	.0127	.0123
1.0	.0213	.0190	.0175	.0164	.0153	.0149
1.1	.0252	.0225	.0207	.0194	.0181	.0177
1.2	.0294	.0263	.0242	.0227	.0212	.0206
1.3	.0339	.0303	.0279	.0261	.0244	.0238
1.4	.0387	.0346	.0319	.0298	.0279	.0272
1.5	.0438	.0391	.0360	.0338	.0315	.0308
1.6	.0491	.0439	.0404	.0379	.0354	.0345
1.7	.0547	.0489	.0450	.0422	.0394	.0385
1.8	.0606	.0542	.0499	.0468	.0437	.0426
1.9	.0667	.0596	.0549	.0515	.0481	.0469
2.0	.0731	.0654	.0602	.0564	.0527	.0515
2.1	.0798	.0713	.0657	.0616	.0575	.0561
2.2	.0867	.0775	.0714	.0669	.0625	.0610
2.3	.0938	.0839	.0773	.0725	.0677	.0661
2.4	.1012	.0905	.0834	.0782	.0731	.0713
2.5	.1088	.0973	.0897	.0841	.0786	.0767
2.6	.1167	.1044	.0962	.0902	.0844	.0823
2.7	.1249	.1117	.1030	.0965	.0902	.0881
2.8	.1332	.1192	.1099	.1030	.0963	.0940
2.9	.1418	.1269	.1170	.1097	.1026	.1001
3.0	.1507	.1348	.1243	.1166	.1090	.1064

PRESSURE LOSS PER FOOT 3/4" PEX 50% Glycol / Water Mixture

WIRSBO	Head (Feet of Water) Per Foot of Pipe					
FLOW GALLONS PER MINUTE	80°F	100°F	120°F	140°F	160°F	180°F
.5	.0032	.0029	.0026	.0025	.0023	.0022
.6	.0044	.0040	.0036	.0034	.0032	.0031
.7	.0058	.0052	.0048	.0045	.0043	.0041
.8	.0074	.0066	.0061	.0057	.0054	.0052
.9	.0091	.0082	.0075	.0070	.0067	.0064
1.0	.0110	.0098	.0091	.0085	.0081	.0077
1.1	.0131	.0117	.0107	.0101	.0095	.0092
1.2	.0153	.0136	.0125	.0117	.0112	.0107
1.3	.0176	.0157	.0145	.0135	.0129	.0123
1.4	.0201	.0179	.0165	.0155	.0147	.0141
1.5	.0227	.0203	.0187	.0175	.0166	.0159
1.6	.0255	.0227	.0209	.0196	.0186	.0179
1.7	.0284	.0253	.0233	.0219	.0208	.0199
1.8	.0314	.0281	.0259	.0242	.0230	.0221
1.9	.0346	.0309	.0285	.0267	.0253	.0243
2.0	.0379	.0339	.0312	.0292	.0278	.0266
2.1	.0413	.0370	.0340	.0319	.0303	.0291
2.2	.0449	.0401	.0370	.0347	.0329	.0316
2.3	.0486	.0435	.0401	.0375	.0357	.0342
2.4	.0525	.0469	.0432	.0405	.0385	.0369
2.5	.0564	.0504	.0465	.0436	.0414	.0397
2.6	.0605	.0541	.0499	.0467	.0444	.0426
2.7	.0647	.0579	.0533	.0500	.0475	.0456
2.8	.0691	.0618	.0569	.0534	.0507	.0487
2.9	.0735	.0657	.0606	.0568	.0540	.0518
3.0	.0781	.0699	.0644	.0604	.0574	.0551
3.2	.0901	.0806	.0743	.0697	.0662	.0636
3.5	.1028	.0920	.0848	.0796	.0756	.0726
3.7	.1163	.1041	.0960	.0900	.0856	.0821
4.0	.1305	.1168	.1077	.1010	.0961	.0922

FIGURE A1.8

Pressure loss per foot of 5/8-in. and 3/4-in. PEX tubing, 50 percent glycol. (*Courtesy of Wirsbo.*)

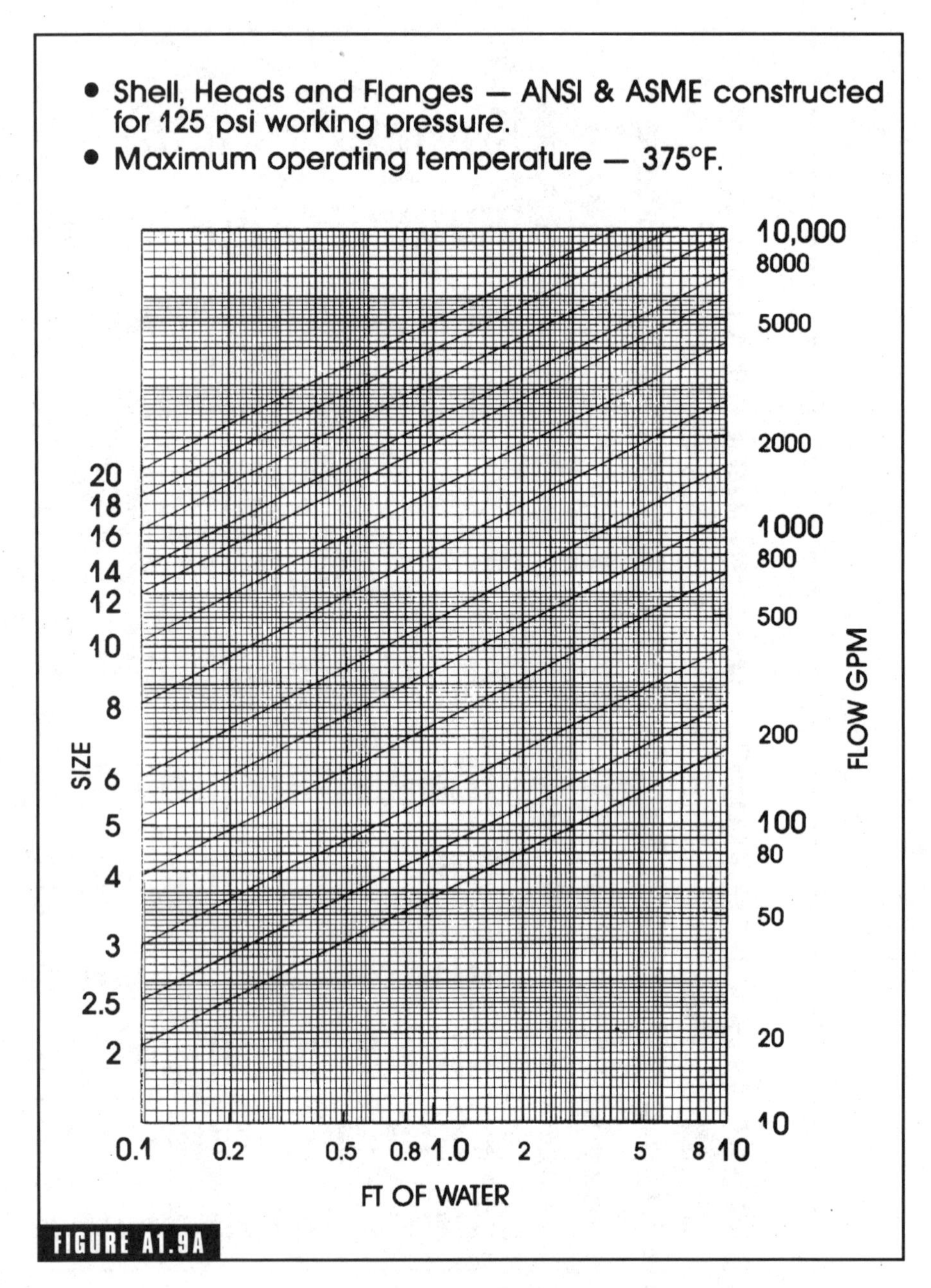

FIGURE A1.9A

Pressure drop for an air separator without strainer. (*Courtesy of Taco.*)

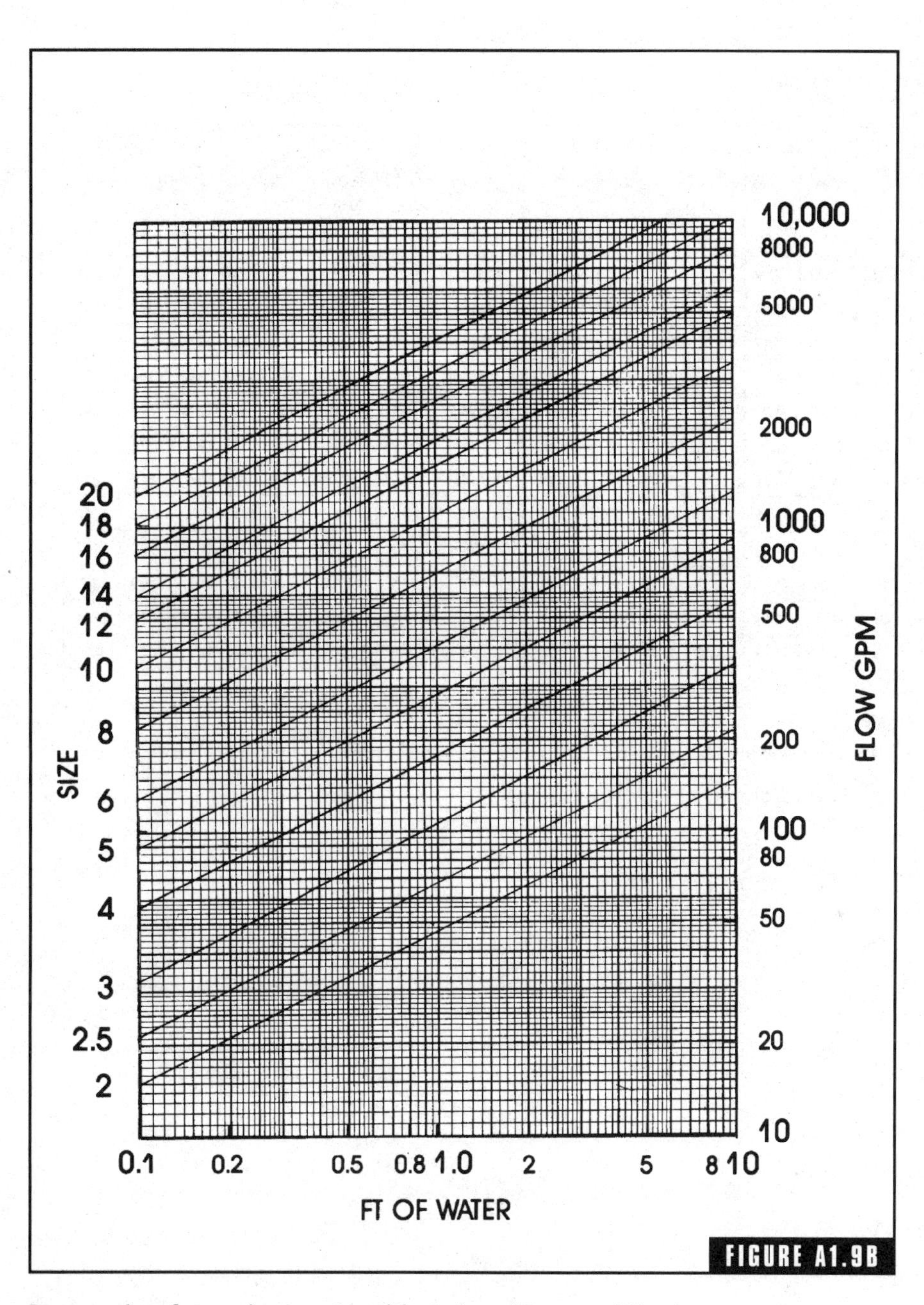

FIGURE A1.9B

Pressure drop for an air separator with strainer. (*Courtesy of Taco.*)

AIR CONTROL PRESSURE DROP WITHOUT STRAINER

AIR CONTROL PRESSURE DROP WITH STRAINER

PIPE SIZE inches	PRODUCT NO.		A	B Max	C	D	E	F	G	H	J	K	MAX FLOW (gpm)	STRAINER FREE AREA (sq. in.)	LESS STRAINER C_v FACTOR	WITH STRAINER C_v FACTOR	APPROX. SHIPPING WT.	
	less strainer	with strainer															less strainer	with strainer
2	AC2	AC2F	8.625	18.00	6.00	5.38	12.00	13	¾"NPT	¾"NPT			80	22	86	72	32	37
2½	AC25	AC25F	10.750	20.00	7.00	5.88	16.00	16	¾"NPT	¾"NPT			130	34	122	102	72	78
3	AC3	AC3F	12.750	24.25	6.88	10.50	18.00	18	¾"NPT	1¼"NPT			190	51	190	162	92	120
4	AC4	AC4F	16	29.13	8.19	12.75	25.25	19	¾"NPT	1¼"NPT		6¾†	330	80	325	272	140	176
5	AC5	AC5F	16	31.25	8.75	13.75	25.25	22	¾"NPT	1¼"NPT		6¾†	550	112	510	422	180	228
6	AC6	AC6F	20	36.75	11.00	14.75	29.25	26	¾"NPT	1¼"NPT		6½†	900	180	750	618	240	290
8	AC8	AC8F	20	41.38	12.00	17.38	29.75	28	¾"NPT	1¼"NPT		6½†	1500	246	1260	1060	322	416
10	AC10	AC10F	24	49.50	14.69	20.12	34.75	32	¾"NPT	1¼"NPT		6¾†	2600	392	2000	1670	545	670
12	AC12	AC12F	30	56.94	16.85	23.25	42.00	37	¾"NPT	1½"NPT	22	7½	3400	548	2900	2400	860	1060
14	AC14	AC14F	36	65.00	19.88	25.25	48.75	43	¾"NPT	1½"NPT	24	7½	4700	732	3500	2850	980	1170
16	AC16	AC16F	36	71.50	21.75	28.00	49.75	44	¾"NPT	1½"NPT	24	7½	6000	845	4600	3800	1200	1300
18	AC18	AC18F	42	74.81	22.59	29.63	55.75	51	¾"NPT	1½"NPT	30	7½	8000	1125	5900	4900	1648	1764
20	AC20	AC20F	48	82.81	25.28	32.25	62.25	58	¾"NPT	1½"NPT	36	7½	10,000	1435	7400	6200	2600	3200

†Optional

FIGUREA1.9C

Dimensions for an air separator. (*Courtesy of Taco.*)

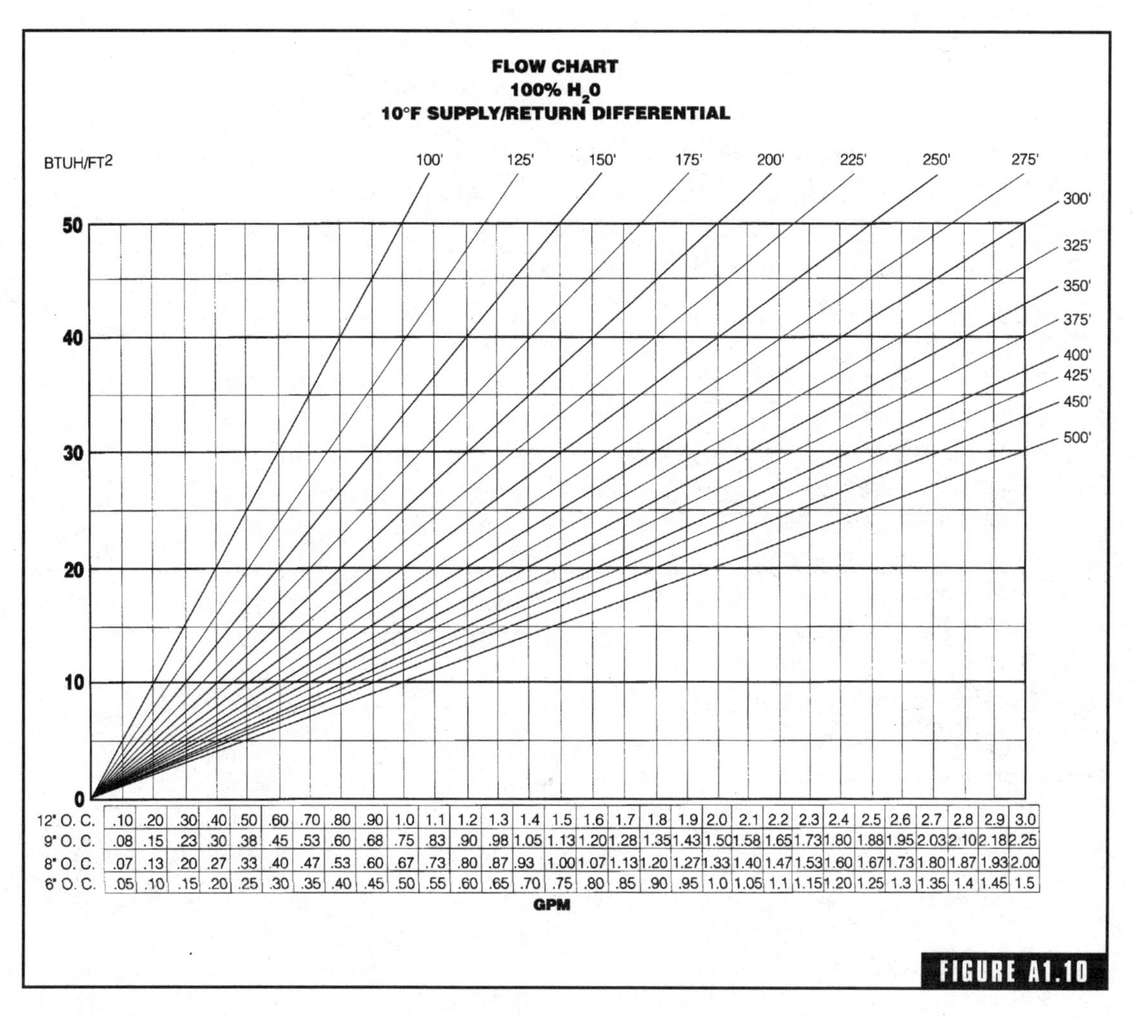

FIGURE A1.10

Flow chart for 100 percent water with 10° differential. *(Courtesy of Wirsbo.)*

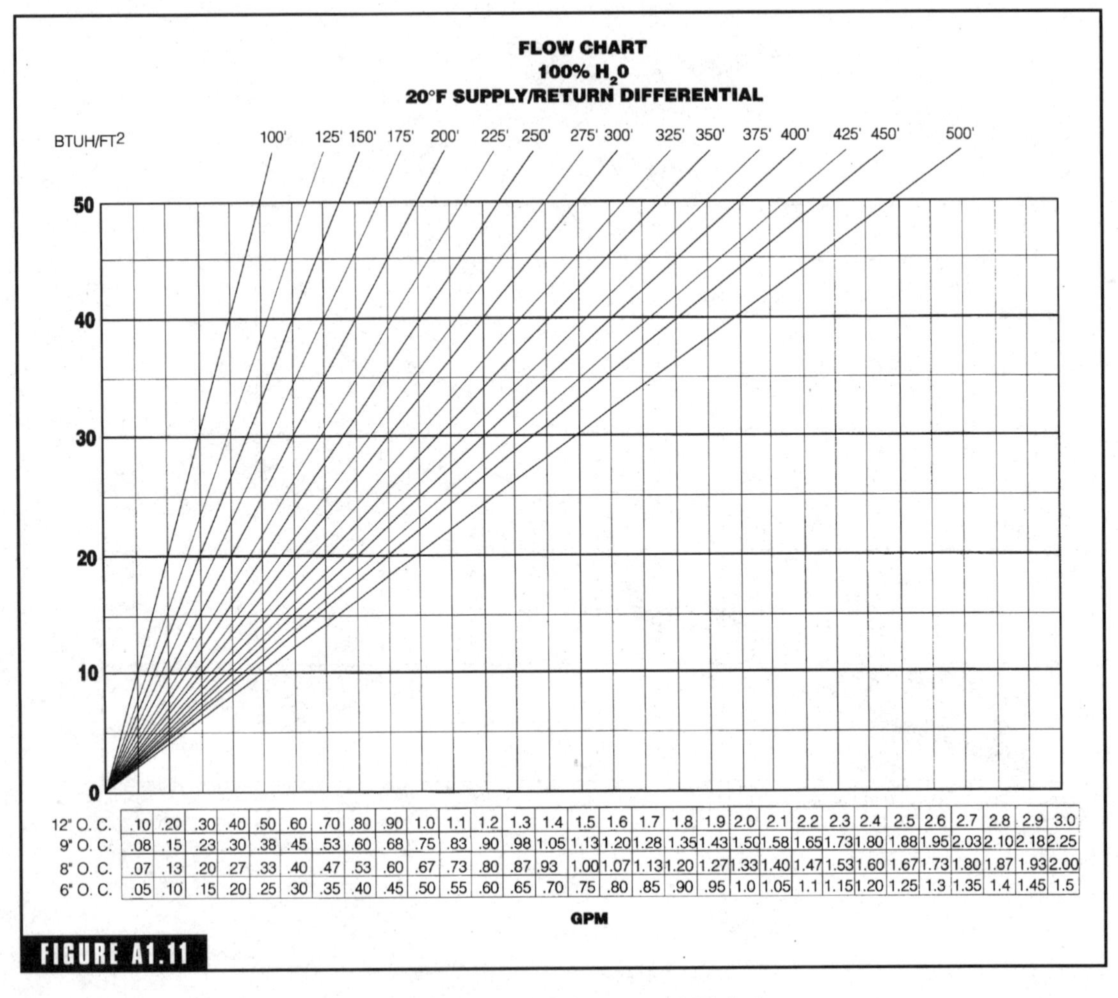

12" O. C.	.10	.20	.30	.40	.50	.60	.70	.80	.90	1.0	1.1	1.2	1.3	1.4	1.5
9" O. C.	.08	.15	.23	.30	.38	.45	.53	.60	.68	.75	.83	.90	.98	1.05	1.13
8" O. C.	.07	.13	.20	.27	.33	.40	.47	.53	.60	.67	.73	.80	.87	.93	1.00
6" O. C.	.05	.10	.15	.20	.25	.30	.35	.40	.45	.50	.55	.60	.65	.70	.75

12" O. C.	1.6	1.7	1.8	1.9	2.0	2.1	2.2	2.3	2.4	2.5	2.6	2.7	2.8	2.9	3.0
9" O. C.	1.20	1.28	1.35	1.43	1.50	1.58	1.65	1.73	1.80	1.88	1.95	2.03	2.10	2.18	2.25
8" O. C.	1.07	1.13	1.20	1.27	1.33	1.40	1.47	1.53	1.60	1.67	1.73	1.80	1.87	1.93	2.00
6" O. C.	.80	.85	.90	.95	1.0	1.05	1.1	1.15	1.20	1.25	1.3	1.35	1.4	1.45	1.5

GPM

FIGURE A1.11

Flow chart for 100 percent water with 20° differential. (*Courtesy of Wirsbo.*)

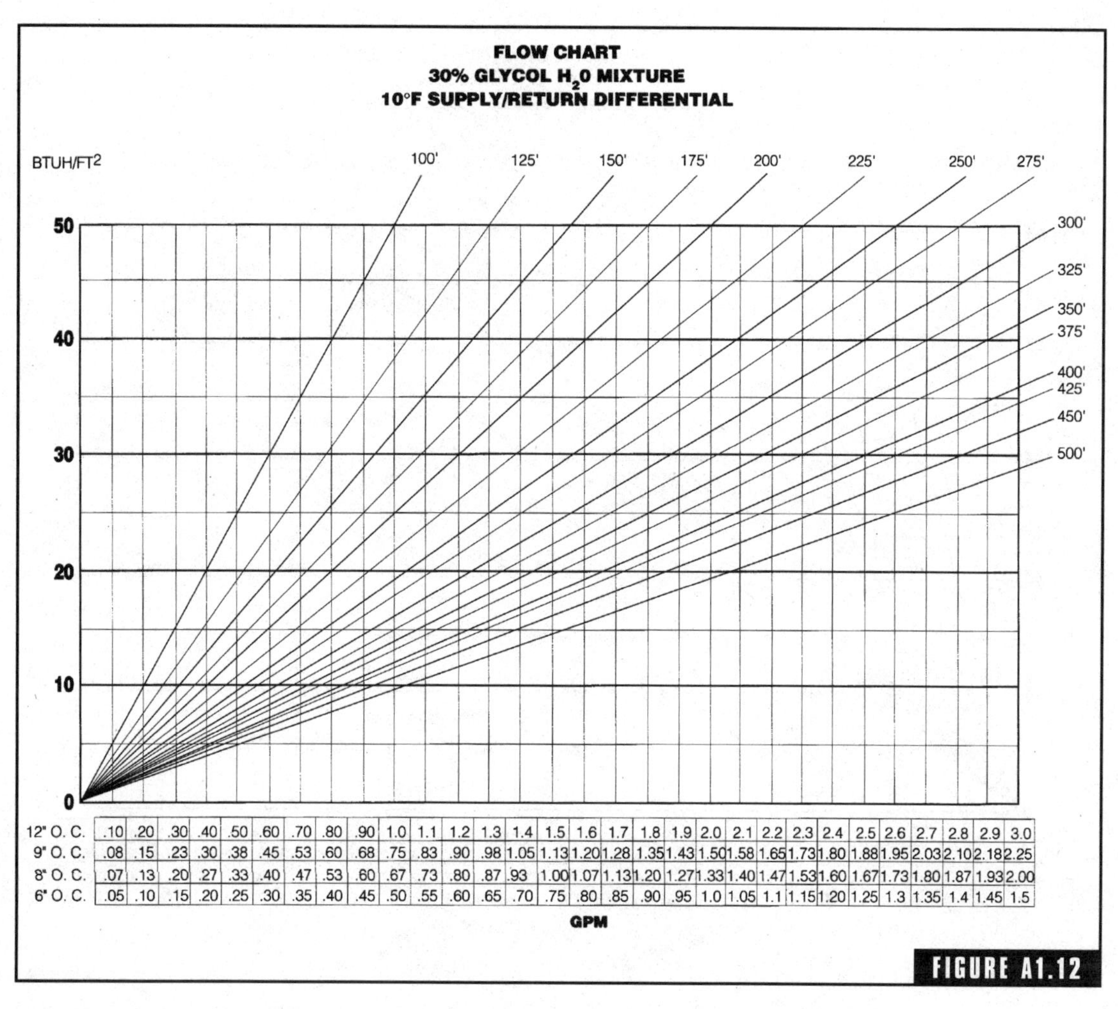

12" O. C.	.10	.20	.30	.40	.50	.60	.70	.80	.90	1.0	1.1	1.2	1.3	1.4	1.5
9" O. C.	.08	.15	.23	.30	.38	.45	.53	.60	.68	.75	.83	.90	.98	1.05	1.13
8" O. C.	.07	.13	.20	.27	.33	.40	.47	.53	.60	.67	.73	.80	.87	.93	1.00
6" O. C.	.05	.10	.15	.20	.25	.30	.35	.40	.45	.50	.55	.60	.65	.70	.75

12" O. C.	1.6	1.7	1.8	1.9	2.0	2.1	2.2	2.3	2.4	2.5	2.6	2.7	2.8	2.9	3.0
9" O. C.	1.20	1.28	1.35	1.43	1.50	1.58	1.65	1.73	1.80	1.88	1.95	2.03	2.10	2.18	2.25
8" O. C.	1.07	1.13	1.20	1.27	1.33	1.40	1.47	1.53	1.60	1.67	1.73	1.80	1.87	1.93	2.00
6" O. C.	.80	.85	.90	.95	1.0	1.05	1.1	1.15	1.20	1.25	1.3	1.35	1.4	1.45	1.5

Flow chart for 30 percent glycol at 10° differential. (*Courtesy of Wirsbo.*)

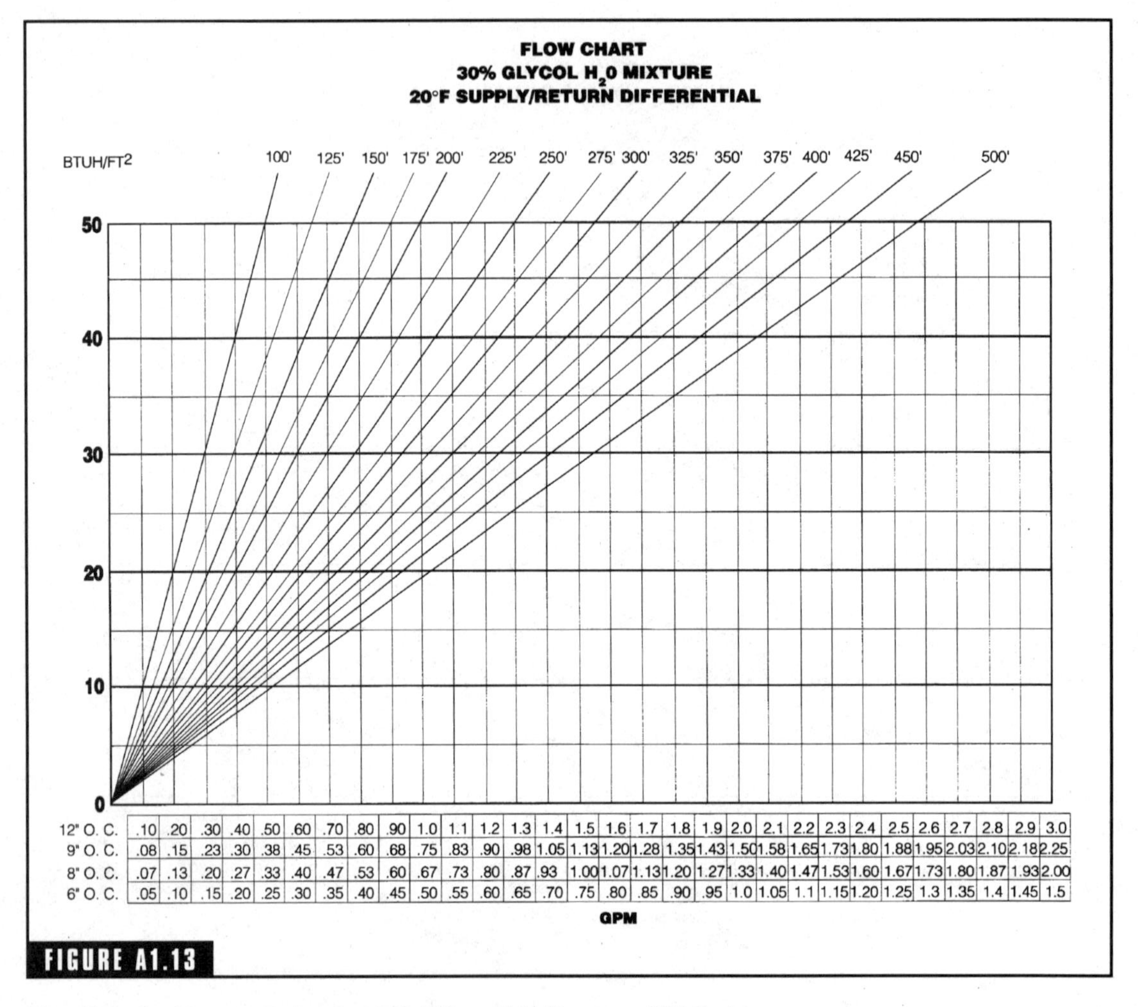

FIGURE A1.13

Flow chart for 30 percent glycol at 20° differential. (*Courtesy of Wirsbo.*)

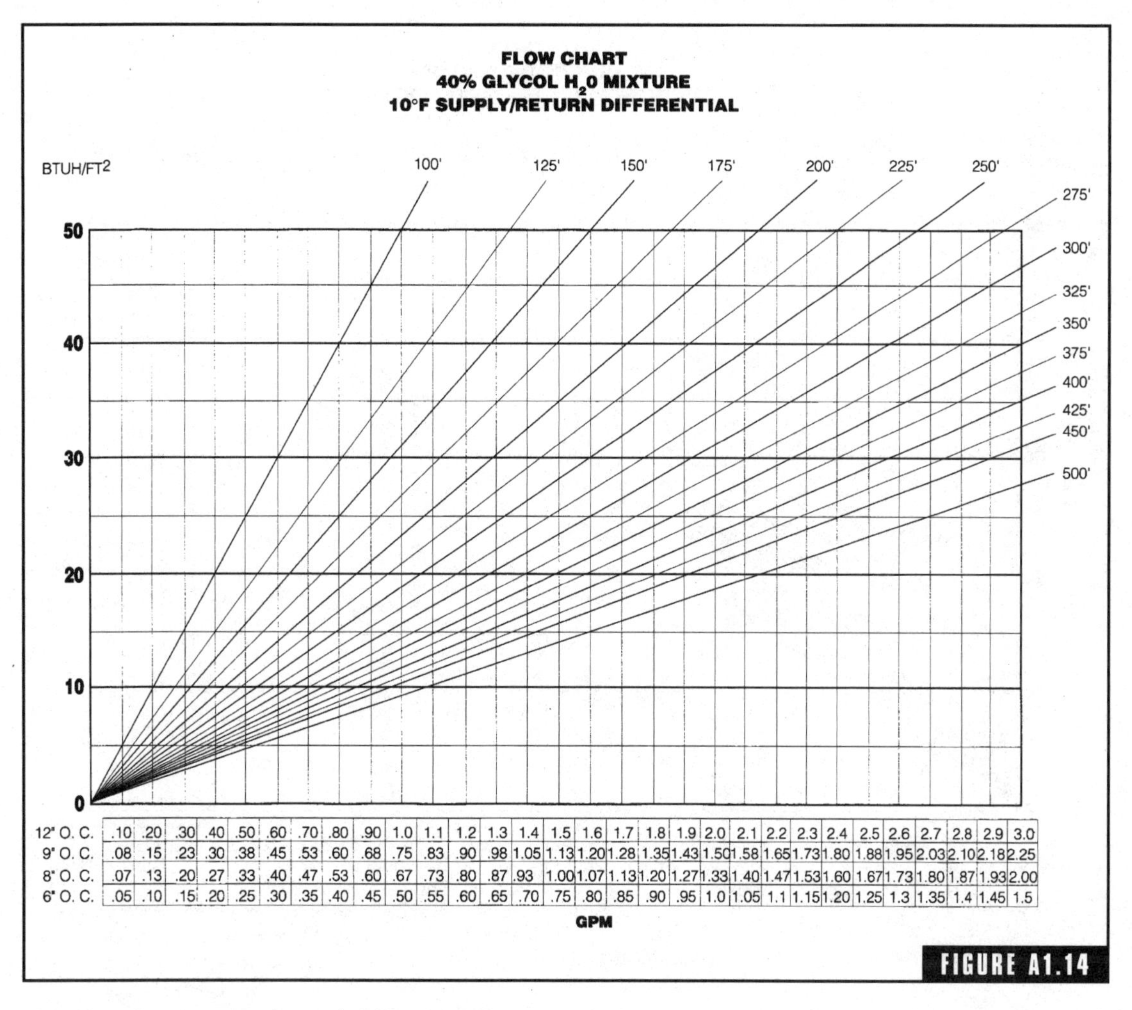

FIGURE A1.14

Flow chart for 40% glycol at 10° differential. (*Courtesy of Wirsbo.*)

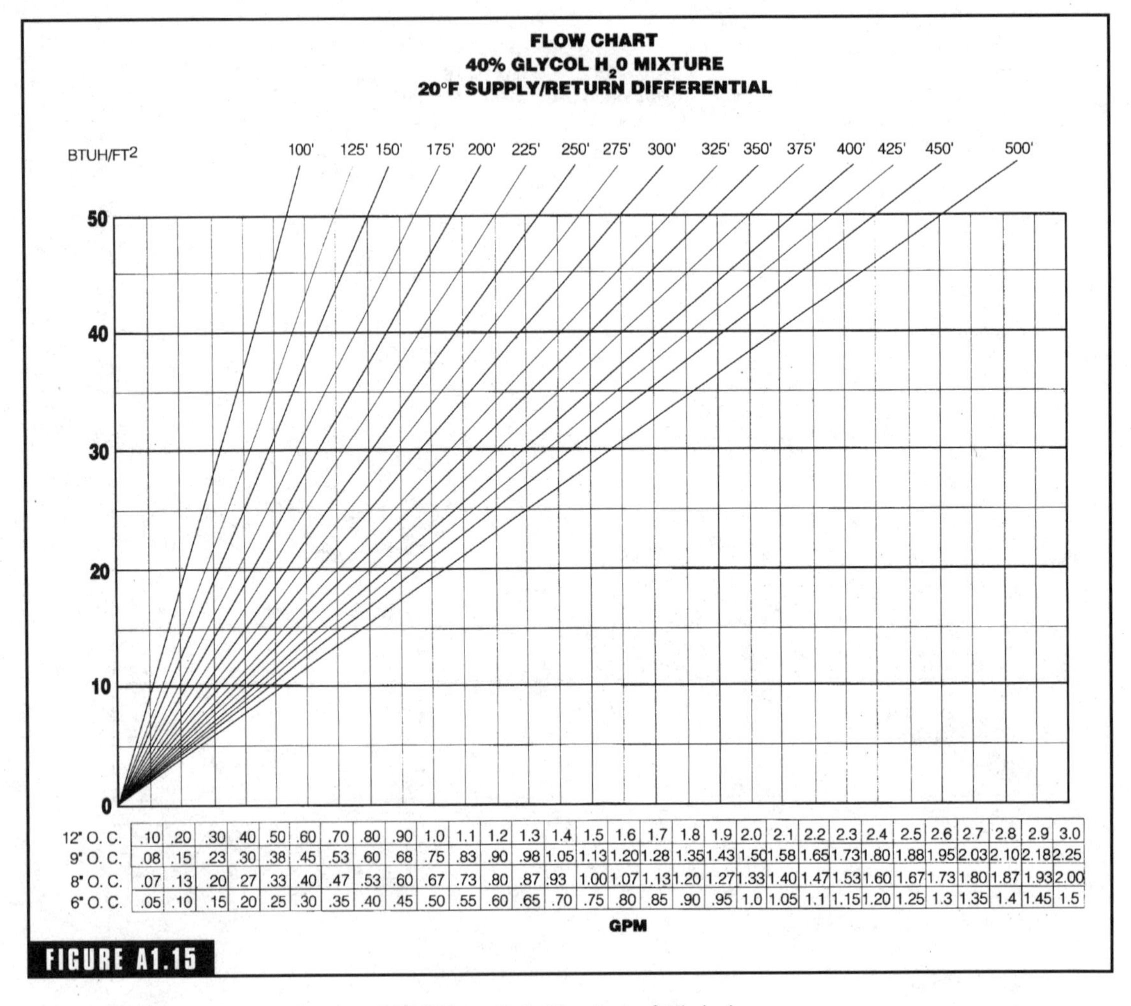

12" O. C.	.10	.20	.30	.40	.50	.60	.70	.80	.90	1.0	1.1	1.2	1.3	1.4	1.5
9" O. C.	.08	.15	.23	.30	.38	.45	.53	.60	.68	.75	.83	.90	.98	1.05	1.13
8" O. C.	.07	.13	.20	.27	.33	.40	.47	.53	.60	.67	.73	.80	.87	.93	1.00
6" O. C.	.05	.10	.15	.20	.25	.30	.35	.40	.45	.50	.55	.60	.65	.70	.75

12" O. C.	1.6	1.7	1.8	1.9	2.0	2.1	2.2	2.3	2.4	2.5	2.6	2.7	2.8	2.9	3.0
9" O. C.	1.20	1.28	1.35	1.43	1.50	1.58	1.65	1.73	1.80	1.88	1.95	2.03	2.10	2.18	2.25
8" O. C.	1.07	1.13	1.20	1.27	1.33	1.40	1.47	1.53	1.60	1.67	1.73	1.80	1.87	1.93	2.00
6" O. C.	.80	.85	.90	.95	1.0	1.05	1.1	1.15	1.20	1.25	1.3	1.35	1.4	1.45	1.5

Flow chart for 40 percent glycol at 20° differential. (*Courtesy of Wirsbo.*)

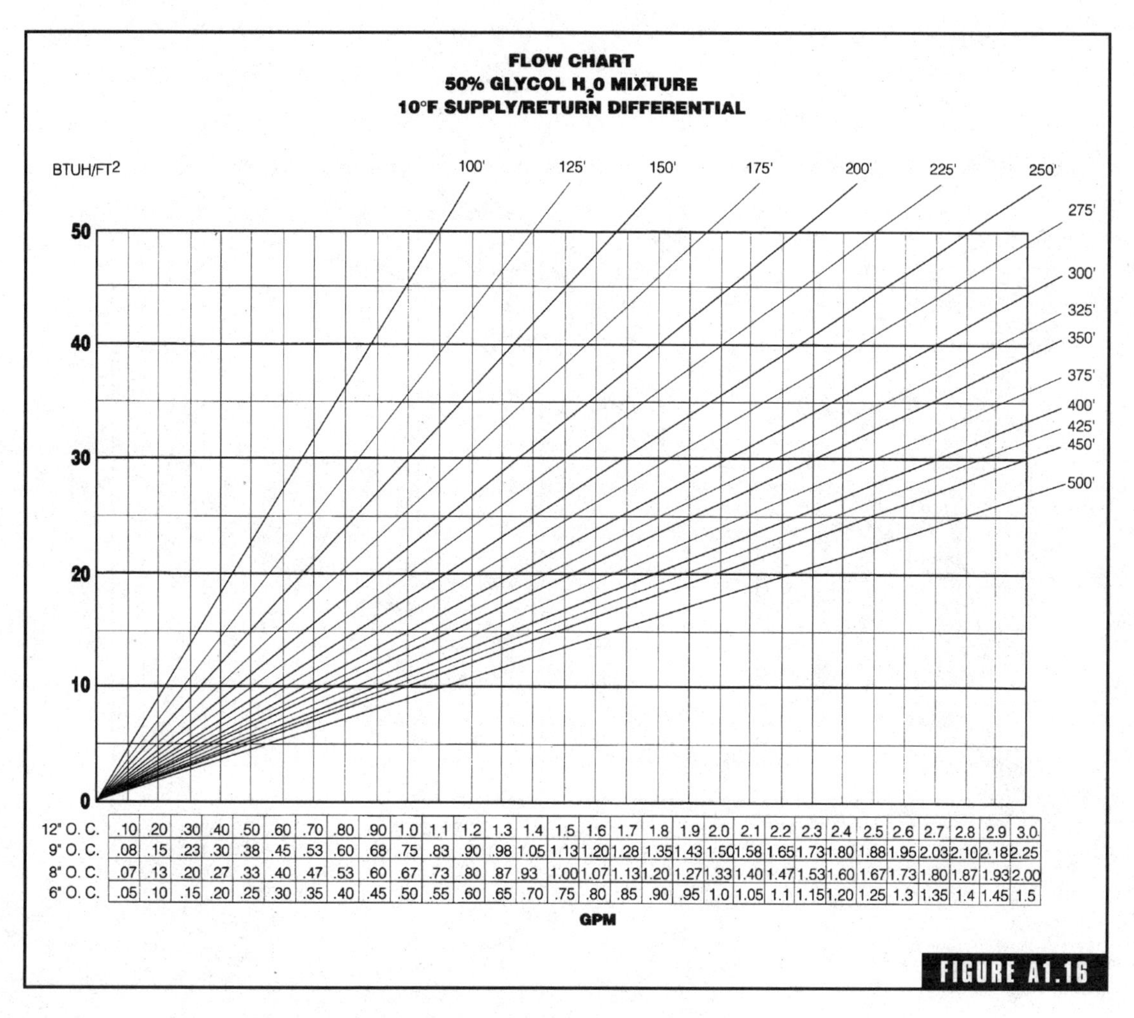

12" O. C.	.10	.20	.30	.40	.50	.60	.70	.80	.90	1.0	1.1	1.2	1.3	1.4	1.5
9" O. C.	.08	.15	.23	.30	.38	.45	.53	.60	.68	.75	.83	.90	.98	1.05	1.13
8" O. C.	.07	.13	.20	.27	.33	.40	.47	.53	.60	.67	.73	.80	.87	.93	
6" O. C.	.05	.10	.15	.20	.25	.30	.35	.40	.45	.50	.55	.60	.65	.70	.75

12" O. C.	1.6	1.7	1.8	1.9	2.0	2.1	2.2	2.3	2.4	2.5	2.6	2.7	2.8	2.9	3.0
9" O. C.	1.20	1.28	1.35	1.43	1.50	1.58	1.65	1.73	1.80	1.88	1.95	2.03	2.10	2.18	2.25
8" O. C.	1.00	1.07	1.13	1.20	1.27	1.33	1.40	1.47	1.53	1.60	1.67	1.73	1.80	1.87	1.93	2.00
6" O. C.	.80	.85	.90	.95	1.0	1.05	1.1	1.15	1.20	1.25	1.3	1.35	1.4	1.45	1.5

Flow chart for 50 percent glycol at 10° differential. (*Courtesy of Wirsbo.*)

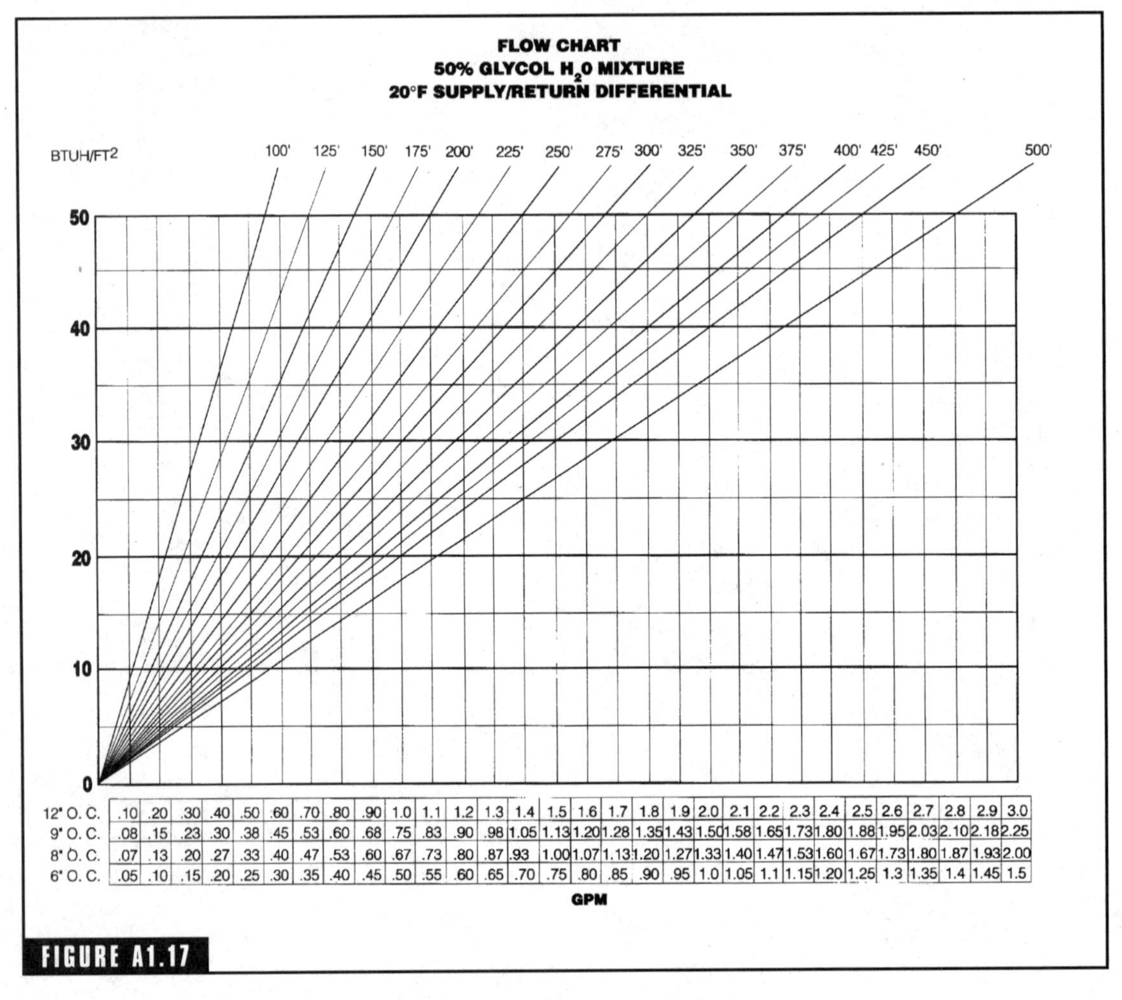

FIGURE A1.17

Flow chart for 50 percent glycol at 20° differential. (*Courtesy of Wirsbo.*)

APPENDIX

B

Temperature Charts

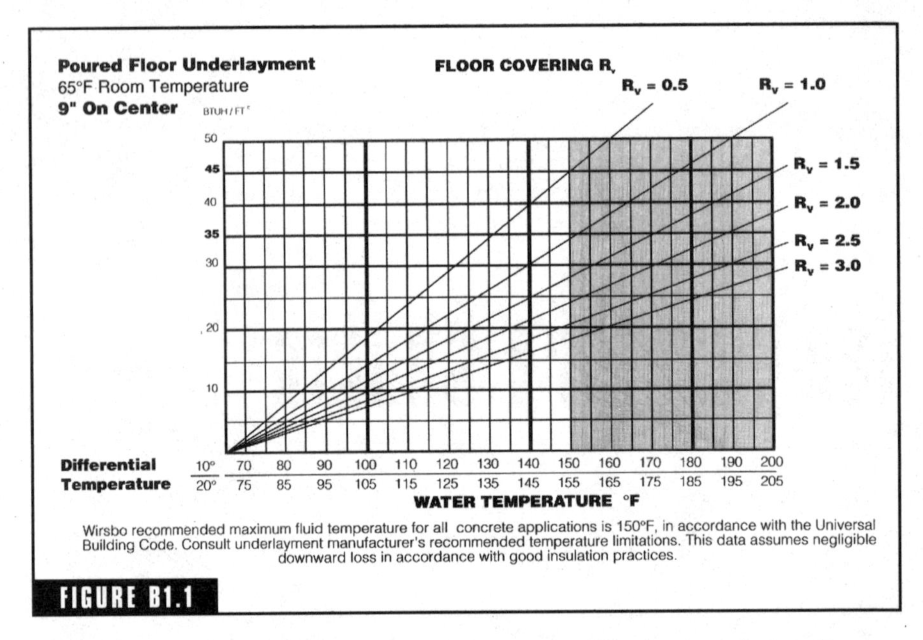

FIGURE B1.1

Poured floor underlayment temperature chart for 9-in. on-center installation. (*Courtesy of Wirsbo.*)

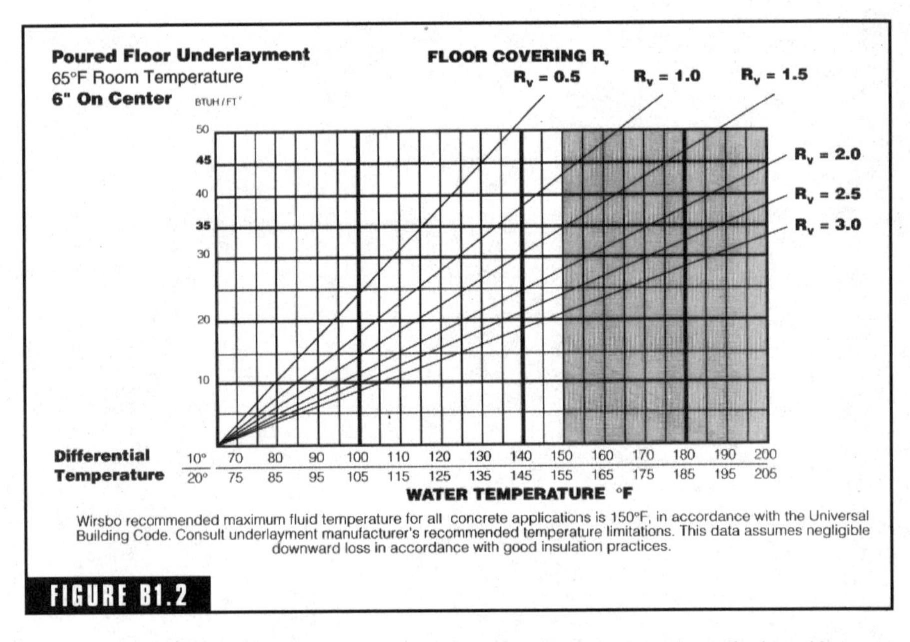

FIGURE B1.2

Poured floor underlayment temperature chart for 6-in. on-center installation. (*Courtesy of Wirsbo.*)

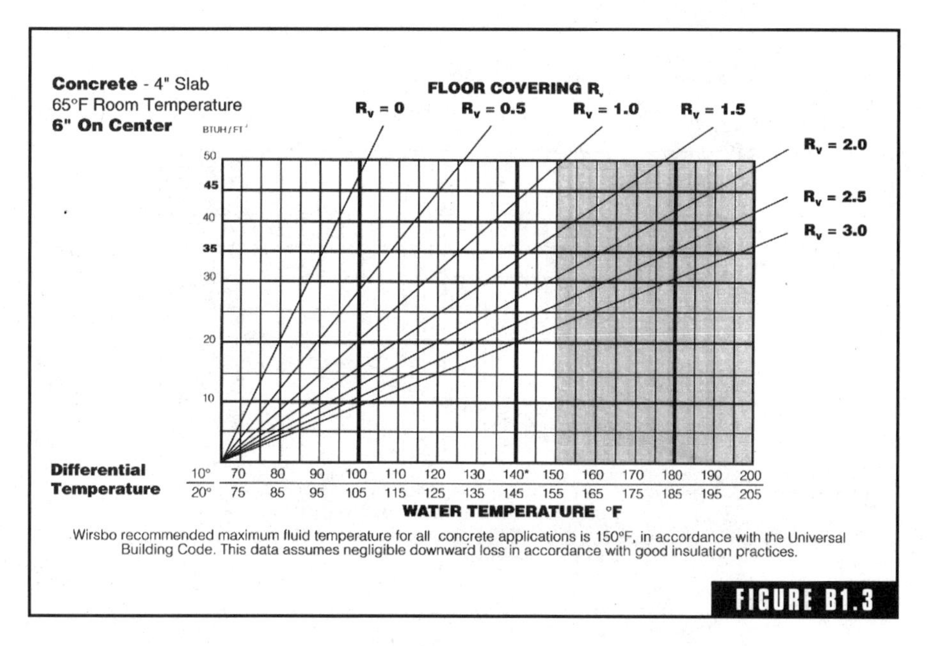

FIGURE B1.3

4-in. concrete slab temperature chart for 6-in. on-center installation. (*Courtesy of Wirsbo.*)

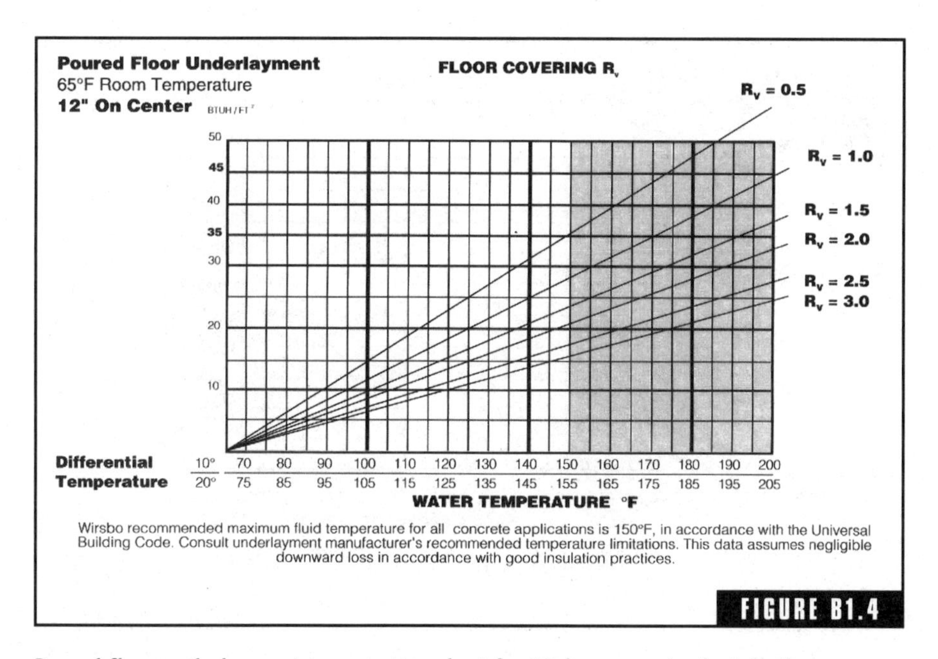

FIGURE B1.4

Poured floor underlayment temperature chart for 12-in. on-center installation. (*Courtesy of Wirsbo.*)

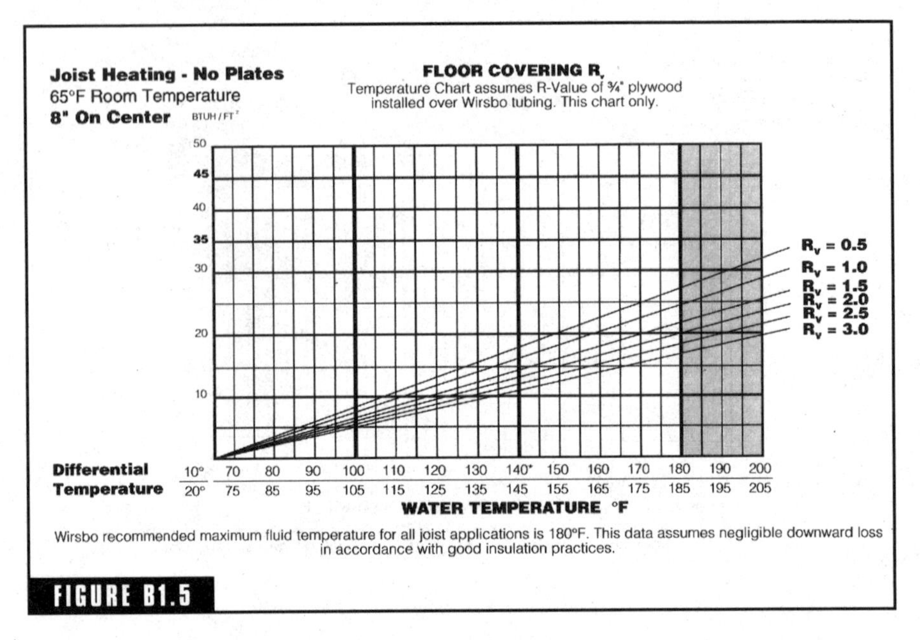

FIGURE B1.5

Joist heating with no temperature plates, temperature chart. (*Courtesy of Wirsbo.*)

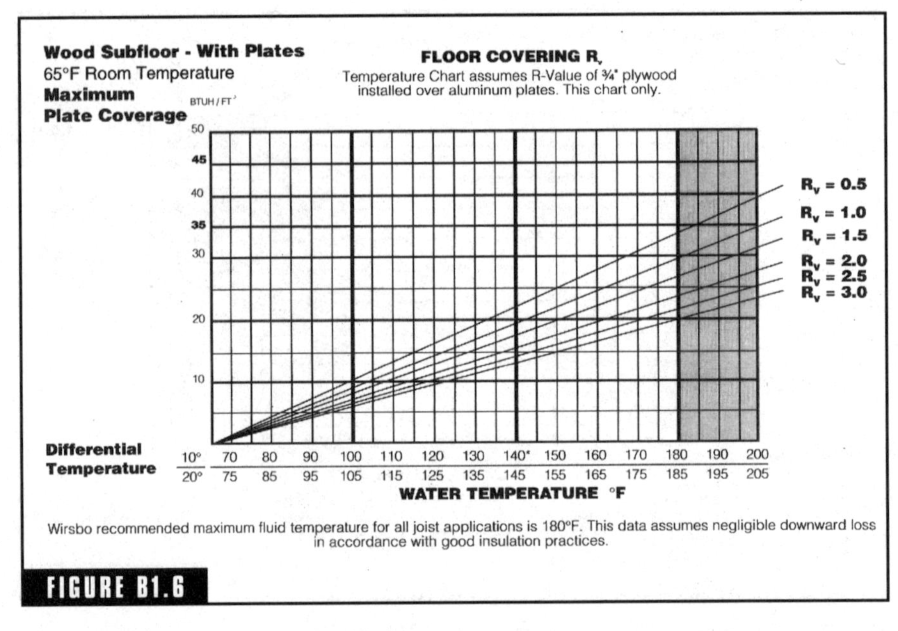

FIGURE B1.6

Wood subfloor, with transfer plate, temperature chart. (*Courtesy of Wirsbo.*)

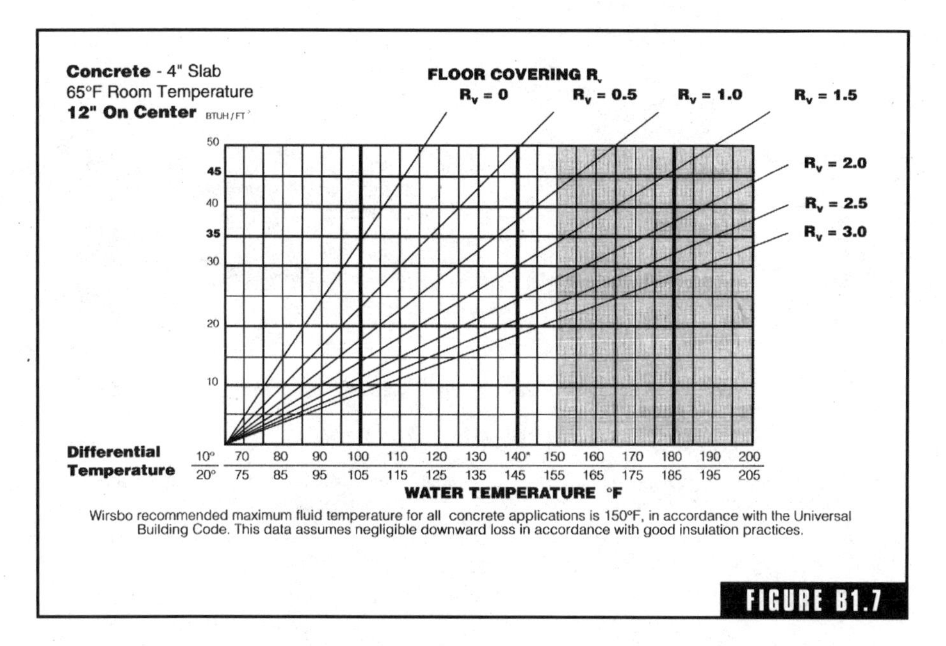

FIGURE B1.7

4-in. concrete slab temperature chart for 12-in. on-center installation. *(Courtesy of Wirsbo.)*

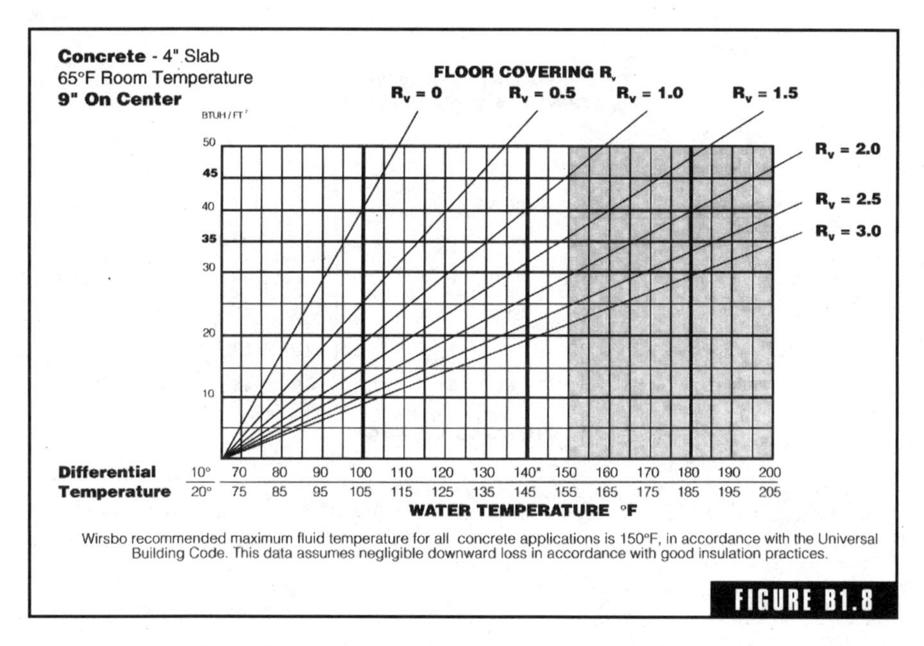

FIGURE B1.8

4-in. concrete slab temperature chart for 9-in. on-center installation. *(Courtesy of Wirsbo.)*

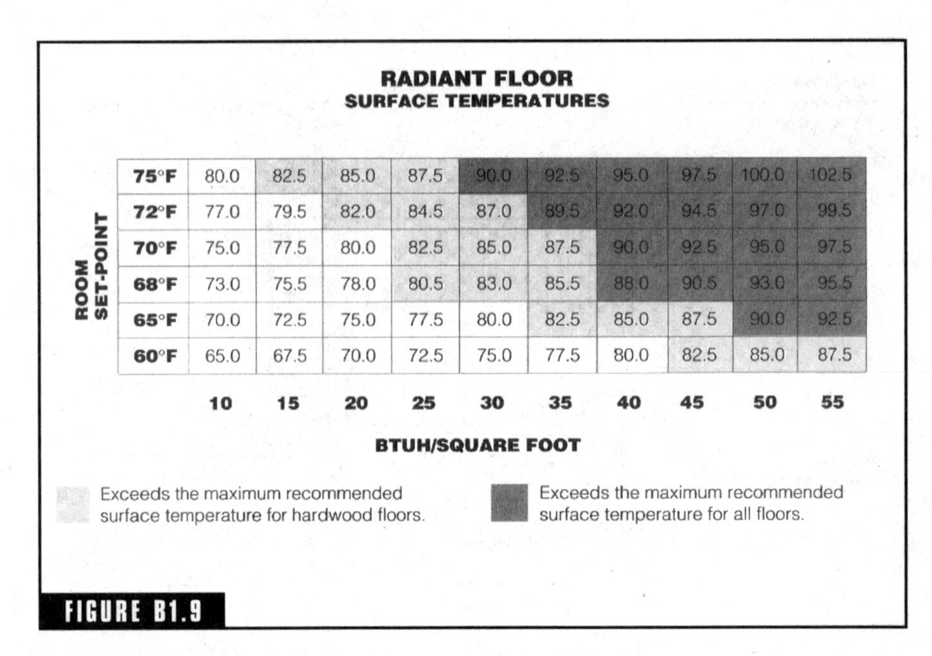

RADIANT FLOOR
SURFACE TEMPERATURES

ROOM SET-POINT	10	15	20	25	30	35	40	45	50	55
75°F	80.0	82.5	85.0	87.5	90.0	92.5	95.0	97.5	100.0	102.5
72°F	77.0	79.5	82.0	84.5	87.0	89.5	92.0	94.5	97.0	99.5
70°F	75.0	77.5	80.0	82.5	85.0	87.5	90.0	92.5	95.0	97.5
68°F	73.0	75.5	78.0	80.5	83.0	85.5	88.0	90.5	93.0	95.5
65°F	70.0	72.5	75.0	77.5	80.0	82.5	85.0	87.5	90.0	92.5
60°F	65.0	67.5	70.0	72.5	75.0	77.5	80.0	82.5	85.0	87.5

BTUH/SQUARE FOOT

Exceeds the maximum recommended surface temperature for hardwood floors.

Exceeds the maximum recommended surface temperature for all floors.

FIGURE B1.9

Radiant floor surface temperatures. (*Courtesy of Wirsbo.*)

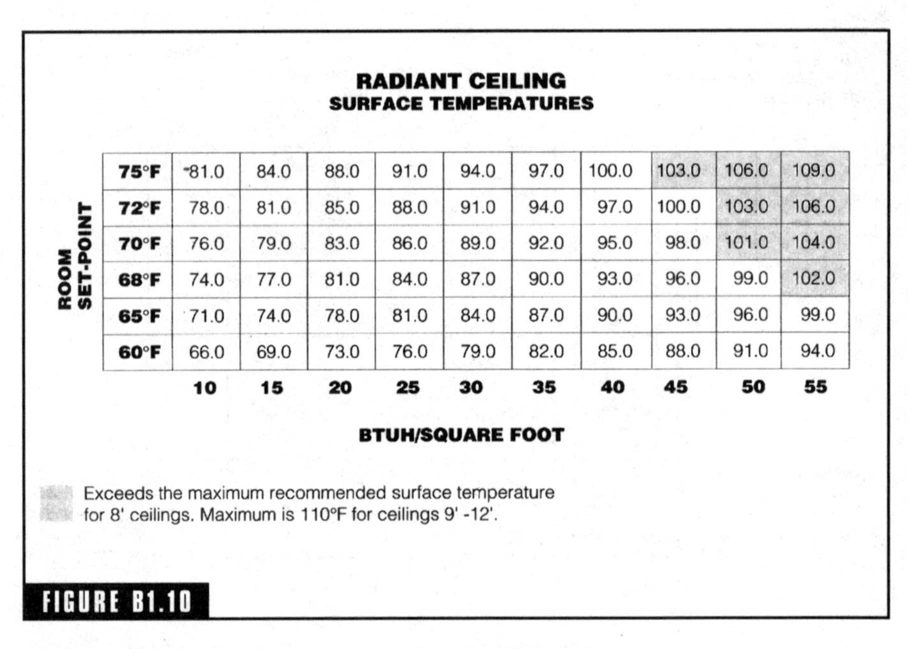

RADIANT CEILING
SURFACE TEMPERATURES

ROOM SET-POINT	10	15	20	25	30	35	40	45	50	55
75°F	81.0	84.0	88.0	91.0	94.0	97.0	100.0	103.0	106.0	109.0
72°F	78.0	81.0	85.0	88.0	91.0	94.0	97.0	100.0	103.0	106.0
70°F	76.0	79.0	83.0	86.0	89.0	92.0	95.0	98.0	101.0	104.0
68°F	74.0	77.0	81.0	84.0	87.0	90.0	93.0	96.0	99.0	102.0
65°F	71.0	74.0	78.0	81.0	84.0	87.0	90.0	93.0	96.0	99.0
60°F	66.0	69.0	73.0	76.0	79.0	82.0	85.0	88.0	91.0	94.0

BTUH/SQUARE FOOT

Exceeds the maximum recommended surface temperature for 8' ceilings. Maximum is 110°F for ceilings 9' -12'.

FIGURE B1.10

Radiant ceiling surface temperatures. (*Courtesy of Wirsbo.*)

FLOOR COVERING R-VALUE	1/8"	1/4"	3/8"	1/2"	5/8"	3/4"	1.00"
Bare Floor	0.0	-	-	-	-	-	-
Plywood (Medium or High Density)	-	-	0.3	0.5	0.7	0.9	1.1
Sheet goods							
Vinyl	0.2	-	-	-	-	-	-
Linoleum (uninsulated)	0.2	-	-	-	-	-	-
Linoleum (insulated)	-	0.4	-	-	-	-	-
Cork	-	0.3	-	0.6	-	0.9	1.2
Tiles and Stones (w/thinset base)							
Ceramic Tile	-	0.3	-	0.5	-	0.8	1.0
Limestone	-	-	-	0.5	-	0.8	1.0
Stone (quarried)	-	-	-	0.4	-	-	0.8
Marble	-	0.2	-	0.4	-	-	0.8
Brick	-	-	-	-	-	-	1.0
CARPET and PADS							
Commercial Glue Down	-	0.6	0.8	-	-	-	-
Level Loop	-	-	1.5	2.0	2.5	**NR**	**NR**
Plush	-	-	0.4	0.8	1.2	1.6	2.4
Shag	-	-	0.8	1.0	1.6	2.4	**NR**
Saxony	-	-	1.3	1.7	2.2	**NR**	**NR**
Wool	-	-	2.5	**NR**	**NR**	**NR**	**NR**
Rubber (solid)	0.3	0.6	0.9	1.2	**NR**	**NR**	**NR**
Rubber (waffled)	0.2	0.4	0.6	0.8	1.0	**NR**	**NR**
Felt	1.0	1.5	**NR**	**NR**	**NR**	**NR**	**NR**
Hair and Jute	-	1.0	1.5	**NR**	**NR**	**NR**	**NR**
Prime Urethane	-	1.0	1.5	2.0	**NR**	**NR**	**NR**
Ultra Prime Urethane	-	-	0.7	-	-	-	-
Bonded Urethane	-	-	1.5	**NR**	**NR**	**NR**	**NR**
Wood Flooring							
Hardwoods							
Ash	-	-	0.4	0.5	0.6	0.8	1.0
Cherry	-	-	0.3	0.4	0.5	0.6	0.8
Elm	-	-	0.4	0.5	0.6	0.8	1.0
Mahogany	-	-	0.3	0.3	0.4	0.5	0.7
Maple	-	-	0.4	0.5	0.6	0.8	1.0
Oak	-	-	0.3	0.4	0.5	0.6	0.8
Walnut	-	-	0.3	0.4	0.5	0.6	0.8
Softwoods							
Fir	-	-	0.4	0.6	0.7	0.9	1.2
Pine	-	-	0.5	0.7	0.8	1.0	1.3
Spruce	-	-	0.5	0.7	0.8	1.0	1.3
Wood Flooring Pad	0.2	0.4	-	-	-	-	-

NR: Not recomended due to the high R-Value of the specific thickness.
All data is from the National Flooring Institute's 1990 study or ASHRAE thermoconductivity values.

FIGURE B1.11

R-values for various floor covering materials. (*Courtesy of Wirsbo.*)

Typical floor coverings

This table shows the most common types of floor coverings, their thickness and resistance to thermal conductivity		Thickness	Thermal conductivity	Resistance to thermal conductivity	Thickness of the overall structure
Designation	**Cross Section**	d	λ	$R_{\lambda}B$	d_{WD}
		mm	W/mk	m^2K/W	mm
Textile floor Covering		10	0.007	max 0.15	10
Parquet adhesive compound		8 2	0.2 0.2	0.04 0.01 0.05	10
Plastic covering, e.g. vinyl		5	0.23	0.022	5
Ceramic floor tiles thin bed of mortar		10 2	1.0 1.4	0.01 0.001 0.011	12
Ceramic floor tiles mortar bed		10 10	1.0 1.4	0.01 0.007 0.017	20
Natural or cast stone tiles here: marble mortar bed		15 10	3.5 1.4	0.004 0.007 0.011	25

FIGURE B1.12

Thickness and resistance to thermal conductivity for various floor coverings. (*Courtesy of Wirsbo.*)

INDEX

Air in radiant systems, 199–200
  air pockets, 201–203
  air vents, 205–208
  dissolved air, 204
  problems, 200–201
  purging process, 208–210
  stationary air pockets, 203–204
Air scoops, 207
Air vents, 205–208
  automatic, 205, 206
  central deaerators, 205, 207–208
  float-type, 206–207
  high-point, 205, 206
  manual, 205–206, 208
Aluminum transfer plates, 180–184
  noise, 181–182, 185
  truss joints, 182
Anthracite coal, 220–221
Antifreeze compounds, 144
Aquastat, 49, 120–121
  triple control, 150
Asphalt in snow-melting system, 194–195
Atmospheric yellow flame, 80–81
Automatic melting system (see ice removal)
Auxiliary loads, 153–154

Backflow prevention, 129–130
Balancing valves, 65
Ball valves, 127
Baseboard systems, 1
  combining infloor with, 43–44
  temperature problem, 45–49
Baseboard tees, 125
Bathrooms, 57, 59
Bedrooms, 56–57
Below-floor systems, 17 (see dry systems)
Bend supports (plastic), 10, 11, 95
Bituminous coal, 221
  semi-, 221–222
Boilers, 24
  auxiliary loads, 153–154
  cast iron, 74–76
  coal-fired, 219–225
  combination, 26, 230–231
  condensing, 46–47, 66
  gas-fired, 26, 80–81, 235–245
  geothermal, 26
  mixing boiler water with potable, 154–155
  non-condensing, 46, 48, 49
  oil-fired, 26, 71–73, 247- 254
  piping, 74
  selection, 81, 84
  setup, 78, 82–83
  wood-fired, 26, 227–232
Boiler blocks, 75
Boiler loop circulator, 49

Bracing supports, 170

Cast iron boilers, 74–76
closed-loop systems, 76
corrosion, 90
packaged units, 76
sectional, 75
Cavitation, 109–111
gaseous, 110–111
Ceilings, 35–36
Centrifugal pumps, 97–98, 132
Check valves, 128–129
Circulating pumps, 33–34, 49, 50, 97–98, 132–133
design, 98–101
end-section pumps, 98–99
high-head, 107
impeller damage, 110
in line, 98
location, 104–105, 133–134
low-head, 107
mounting, 103
parallel installation (lining them up), 109
performance curves, 106–107
position, 101–103
pump headers, 102–103
selection, 111–112
series installation (stacking them up) 108
supports, 102
three-piece circulator, 100–101
used with coil and storage tank, 150–151
wet-rotor circulator, 99–100, 133
Clamp rails, 160
Closed-loop antifreeze solar heating system, 216
Closed-loop systems, 76–77
Coal-fired heating systems, 219–220
maintenance, 223
start up, 222–223
stokers, 222
troubleshooting, 223–225
types of coal, 219–222
Coil design, 195–196
grid, 195
serpentine, 195–196
Combination boilers, 230–231
double-door system, 230
Computer design, 32
software, 37–38
Concrete, 2,3, 4–9
abrasion on tubing, 88
basement floors, 25
benefits for radiant heat, 157
control joints to prevent cracking, 17
copper, bad reaction, 19
heat dispersement in, 25
lightweight, 15, 16–17, 174–175
set point controls to protect, 66
site preparation for, 9, 15
snow-melting system, 194–195
stress, 88–89
thin-slab installations, 14–17
Condensing boiler, 46, 47, 66
Controls, 32, 95, 113–114, 137
aquastats, 120–121
determining type needed, 115–116
electrical functions (switches and relays) 117–118
high limit, 121
injection mixing controls, 95
modulating, 117, 137
on-off, 114–116
outdoor reset, 117, 138
proportional valve controls, 95
relay centers, 121
staged, 116–117, 137
telestats, 95
thermostats, 118–120, 137
three-way tempering valves, 95
zone control modules, 95
zone valves, 95, 121–122
Control joints, 17, 175
Convection, 4
Copper tubing, 123, 124
testing for leaks, 167
Corrosion inhibitors, 91
Cross-linking, 86–87
cold, 87
electronic process, 87
Engel method, 87
hot, 87
Silane method, 87

Deaerators, 205, 207–208
Depth of tubing, 159
Design issues, 22–24, 29–30
  furniture placement, 22–23
  heating zones, 32–33, 35
  manifold systems, 63–67
  physical appearance, 24
  pumps, 98–101
  safety, 23
  slab-on-grade piping systems, 158
Diaphragm-type expansion tank, 134–136, 142–144
  chemical reactions, 144–145
  materials, 144–145
  sizing, 144
Dielectric unions, 126
Direct circulation in a solar heating system, 213–214
Dissolved air, 204
Diverter tees, 125
Double valves, 103
Draft hoods, 81
Draindown solar heating system, 214–215
Dry systems, 14, 179-180
  heat transfer plates, 180–184
  no transfer plates used, 188–189
  obstacles, 187–188
  sleeper systems, 183–184
  tubing installed above subflooring, 183–185
  tubing installed below subflooring, 185–187
Dust, 22

Electrical functions (switches and relays), 117–118
  general purpose, 118
  time-delay, 118
Electronic reset controls, 66–67
End-suction pumps, 99
Engel method of cross linking, 87
  PEX A tubing, 11–12
Entrained air bubbles, 202
Expansion joints, 165
  in lightweight concrete, 175
Expansion tank, 50, 134–136, 139–141
  diaphragm-type, 134–136, 142–144
  matching to system, 144–145
  ratings, 145–146
  selection and installation, 136, 146–148
  standard, 141–142
  sizing, 144

Fan-assisted heating units, 1
Ferrous components, 30–31
  PEX, used with, 93
  preventing corrosion, 90, 91
Fittings, 94, 124–126
  baseboard tees, 125
  compression, 94
  crimp, 94
  dielectric unions, 126
  diverter tees, 125
  thermocycle tests, 94
Flanges, 103–104, 134
  brass, 103
  bronze, 103
  cast iron, 103
  isolation, 103
Flat-plate solar collectors, 212–213
Flooring
  allowing radiant heat to rise, 17
  carpet, 17
  tile, 17
Floor loads, 170–171
  bracing supports, 170
  joists, 171
Floor joists, 36, 172
  drilled holes, 186–187
Flow-check valves, 130, 131
Foam insulation, 159, 161–162
Forced air, 1
Fuel oil, 77, 78
  fill pipes, 79
  storage tanks, 78–79
  vent pipes, 79

Garages, 44, 59
Gas-fired boilers, 235
  atmospheric yellow flame, 80
  draft hoods, 81
  power burner, 80

safety measures, 235–237
troubleshooting, 237–245
Gate valves, 127
Globe valves, 127
Gravity drainback solar heating system, 215–216
Grid coils, 195
Gypsum, 175–177
bonding agents, 175
installation in thin-slab systems, 175–177
sealants, 15, 16
underlayments, 175–176

Half unions, 103
Head factors, 105–106
Headroom, 173–174
Heat exchangers, 50, 91
independent tanks, 151–152
Heat sources, 26
boilers, 26, 66
solar collection, 66
water heaters, 37, 66
Heat transfer plates
above-floor systems, 17–18
between floor joists, 36
dry systems, 17
Heating curve, 2, 3
Heating systems
fin-tube baseboard, 1, 2
thin-slab radiant, 3
Heating zones, 32–33, 35
bathrooms, 57, 59
bedrooms, 56–57
circulators, 33
garages, 59
general living space, 60–61
kitchens, 60
primary living space, 55–56
specialty rooms, 59
telestats, 33
zone valves, 33
High density polyethylene (HDPE), 87
Hydronic boilers, 149–150
Hydronic heat pumps, 66, 97
Hydronic heating systems, 97
boilers, 149–150
components, 123
heat pumps, 66, 97

Ice removal, 191–193
Impeller damage, 110
Independent tanks, 151–153
Injector mixing controls, 95
Inline pump, 98
Installation of multiple pumps, 108–109
parallel (lining them up), 109
series (stacking them up), 108
Installation of radiant heat systems, 157
basement floors, 25
concrete, 157–158
design, 158
direct piping, 66–67
ferrous components, 30–31
manifold systems, 63–67
other materials, 95
problems, 158
slab-on-grade, 158–162
Insulation
batt, 14
foam, 8, 9, 10
glass fiber, 14

Kitchens, 60

Leaks, 94, 165
Life span of system, 26
Lightweight concrete, 174–175

Maintenance, 24
Manifold devices, 35, 63–64
accessories, 65
balancing valves, 65
connections, 68–69
location, 39, 63, 67–69
multiple, 55, 56, 65
preassembled, 68, 93
single, 56
slab-on-grade systems, 163–164
supply and return, 63, 64
Manifold risers, 9–11
bend supports, 10
block guides 10

feed and return pipes, 10
mock walls, 10
thin-slab installations, 14-15
Manifold station, 64–65
Microbubbles, 203
MIX port, 49
Mixing tanks, 49–50
Mixing valves, 130–131
motorized, 51–52
Mock walls, 10
Modulating controls, 117, 137
Motorized mixing valves, 51–52
Mounting circulators, 103
double valves, 103
flanges, 103
half unions, 103
isolation flanges, 103

Non-condensing boiler, 46, 48, 51
Non-ferrous components, 50

Oil burners, 79–80
Oil-fired boilers, 71–73, 247–248
cast iron, 74–76
leaks, 253
noise, 248–249, 251
oil supply, 252–253
operation, 79–80, 251
pressure, 253–254
smoky joe, 249–250
steel, 76–77
unwanted cutoffs, 254
On-off controls, 114–116
Outdoor heating, 61
Outdoor reset controls, 117, 138
Oxygen diffusion, 89–92
corrosion inhibitors, 91
oxygen diffusion barrier, 91

Parallel installation (lining them up), 109
PEX tubing, 1, 4–5, 20, 25, 85–87, 123–124
cable ties, 95
clips, 95
cold cross-linking, 87
Engel method, 87
electronic process, 87
fittings, 94
hot cross-linking, 87
ice removal systems, 195
installed between floor joists, 36
PEX-A, 87
PEX-C, 87
preventing chemical reactions, 89–92
silane method, 87
temperature and pressure ratings, 92
tubing size and capacity, 92–93
PEX-A tubing, 87
PEX-C tubing, 87
Pilot lights on gas-fired boilers, 240–242
Piping
boiler, 74
copper, 45, 123–124
materials, 123–124
PEX, 123
return and feed, 10
spacing, 7
thin-slab systems, 14–15
Plastic clips, 160
Polybutylene tubing (PB), 12, 85–86, 124
cross-linked(PEX), 1, 4–5, 20, 25, 85–87, 123–124
fittings, 94
problems with, 19
Polyethylene vapor barrier, 163
Power burner, 80
Preassembled manifolds, 68
Pressure-reducing valves, 128, 129
Pressure-relief valves, 129
Primary living space, 55–56
Priority zoning, 58
Problems in installation, 158
Problems with materials
copper tubing, 19
polybutylene (PB) tubing, 19
Pump design, 98–101
Pump headers, 102–103
Pump placement, 101–103, 133–134
Purging air from systems, 208–210
boiler drain, 208–210

forced-water method, 208
gravity method, 208
purge valve, 208–210

Radiant heat
comfort, 20–21
controls, 32, 95
design, 22–23
dust, 22
efficiency, 21–22
heat sources, 26
ice removal, 191–193
leaks, 94
life span, 26
maintenance, 24
materials, 85–86
stress, effect on materials, 87–89
Type 1 Systems, 46
Type 2 Systems, 46
Type 3 Systems, 46
Radiant ice melting systems, 191–193
coil design, 195–196
precautions, 194–195
tubing selection, 196
typical installation, 193
Radiant panel loop circulator, 49
Relay centers, 118
Remodeling with radiant heat, 41, 187–188
Romans' use of radiant heat, 1, 19

Sealants, 15–16
Sectional boilers, 75
Securing the tubing, 159–160
clamp rails, 160
clips, 160
ties, 159–160
Serpentine coils, 195–196
Setpoint controls, 66
Sidearms, 49–50
Site preparation, 9, 11
Slab-on-grade systems, 3, 157–159
depth of tubing, 161
expansion joints, 165
foam insulation, 159, 161–162
sample installation, 163–165
spacing, 159–161
testing for leaks, 165–168
tubing, 164–165
Sleeper sections, 17–18, 183–184
Sleeves for tubing, 12, 95
Solar heating systems, 211–212
cost, 217–218
flat-plate solar collectors, 212–213
installation, 213–216
planning, 216–217
storing solar energy, 213
Solar heating system installation, 213–216
closed-loop antifreeze, 216
direct circulation, 213–214
draindown, 214–215
gravity drainback, 215–216
Spacing between loops of radiant piping, 6–7, 159–161
Specialty rooms, 59
Staged controls, 116–117, 137
Stationary air pockets, 203–204
Steel boilers, 76–77
closed-loop system, 77
firetubes, 76–77
turbulators, 76
Storage tanks, 150–151
boosting system efficiency, 153–155
construction, 153
Storing solar energy, 213
Stratification, 2
Structural support, 172–173
Surges in heat output, 115–116
System performance, 105
head factors, 105–106

Tankless coils, 149–150
combining with tank, 150–151
Telestats, 33, 55, 56, 65, 95
Tempering valve, 46–47, 49, 93
Testing, 12, 14
leaks, 14
Thermocycling tests, 94
Thermostatic mixing valve, 153
Thermostats, 118–120, 137
differential, 119
location, 119–120
set-point temperature, 119

Thin-slab systems, 5, 169–170
  above-floor systems, 17–18
  adjustments to building procedures, 171–172
  below-floor systems, 18
  bracing supports, 170
  concrete, 15–17
  construction, 15
  door openings, 173
  dry system, 17
  floor loads, 170–171
  gypsum, 15–16, 171–177
  headroom, 173–174
  installation, 14–18
  joists, 170, 171
  lightweight concrete, 16–17, 174–175
  plumbing fixtures, 173
  structural support, 172–173
Three-piece circulators, 100–101, 133
Ties, 159–160
Transitional threshold, 173
Troubleshooting gas boilers, 235–245
  gas leak at regulator vent, 244
  long flames, 244
  motor runs but doesn't heat, 243
  motor won't run, 242–243
  no pilot light, 240
  pilot light goes off during standby period, 240–241
  pilot light goes off when motor starts, 241–242
  short flames, 243–244
  valve won't close, 244
Troubleshooting oil-fired boilers, 247–248
  bad oil pressure, 253
  leaks, 253
  noise, 248–249, 251
  no oil, 252–253
  odors, 250–251
  pressure pulsation, 253–254
  smoky joe, 249–250
  unwanted cutoffs, 254
Tubing
  air pressure, 14
  above-floor systems, 17–18
  below-floor systems, 18
  concrete, installation in, 157–159, 163–165
  corrosion, 50
  depth, 7–8
  layout in an ice melting system, 195–196
  manifold, 64
  oxygen diffusion barrier, 31, 89, 91
  PEX, 20, 85–87, 123–124
  routes, 36–39
  selection for radiant melting system, 196
  site preparation, 9
  size, 31–32
  sleeves, 12
  snow-melting system, 194–195
  testing, 12, 14
Type 1 system, 46
Type 2 system, 46–47, 49–50
Type 3 system, 46, 52

Valves
  backflow prevention, 129–130
  balancing, 65
  ball, 127
  bypass, 132
  check, 128
  flow checks, 130
  gate, 126
  globe, 127
  heating, 126–132
  lockshield-balancing, 132
  metered balancing, 132
  mixing, 130–131
  pressure-reducing, 128, 129
  pressure-relief, 128
  tempering, 46–47, 49
  thermostatic mixing, 153
  zone, 33, 55- 56, 121–122, 131–132
Venturi spreads, 81
Volute, 98, 103

Water heating, 37
  domestic, 149–150, 152, 153–155
Water temperature
  controls, 66–67

problems combining baseboard
with radiant, 45–47, 49
Wet-rotor circulator, 99–100, 133
Wiring for zone valves, 54, 58, 60, 61
Wood-fired boilers, 227–231
combination, 230–231
solid-fuel, 228–230
Wood-fired heating systems, 227–228
boilers, 228–231
efficiency, 231
installation with an oil-fired
boiler, 232–233
transfer storage tanks, 231–232

Zone control modules, 95
Zone valves, 33, 55, 56, 95, 131–132
wiring, 58, 60, 121–122